Reviews in Modern Astronomy

16

Further Titles in Astronomy

Holliday, K.
Introductory Astronomy
1998, 314 pages, Softcover.
ISBN 0-471-98332-2

Liddle, A.
An Indroduction to Modern Cosmology
2003, 192 pages, Soft- and Hardcover.
ISBN 0-470-84835-9 SC
ISBN 0-470-84834-0 HC

Coles, P. / Lucchin, F.
Cosmology, 2nd edition
2002, 552 pages, Hardcover.
ISBN 0-471-48909-3

Mark, H.
Enyclopedia of Space Science and Technology
2003, 1258 pages, Hardcover.
ISBN 0-471-32408-6

Shore, S.
The Tapestry of Modern Astrophysics
2002, 888 pages, Hardcover.
ISBN 0-471-16816-5

Foukal, P.
Solar Astrophysics, 2nd edition
October 2003, approx. 475 pages
ISBN 3-527-40374-4

Weigert, A. / Wendker, H. J. / Wisotzki, L.
Astronomie und Astrophysik.
Ein Grundkurs
4th edition, Textbook
January 2004, approx. 380 pages, Softcover.
ISBN 3-527-358-2

Diver, D.
A Plasma Formulary for Physics, Technology and Astrophysics
2001, 204 pages, Hardcover.
ISBN 3-527-40294-2

Plait, PC
**Bad Astronomy – Misconceptions & Misuses Revealed
from Astrology to the Moon Landing "Hoax"**
2002, 288 pages, Softcover.
ISBN 0-471-40796-6

Maran
Astronomy For Dummies
1999, 360 pages, Softcover.
ISBN 0-7645-5155-8

Moché, D. L.
A Self-Teaching Guide
5th edition
2000, XII, 352 pages, Softcover.
ISBN 0-471-38353-8

Scientific American / Carlson, S. (eds.)
The Amateur Astronomer
2001, XIV, 272 pages, Softcover.
ISBN 0-471-38282-5

Schielicke, R. E. (Ed.)
**Reviews in Modern Astronomy 15
JENAM 2001: Astronomy with Large Telescopes from Ground and Space**
2002, X, 294 pages, Hardcover.
ISBN 3-527-40404-X

**Astronomische Nachrichten /
Astronomical Notes**
ISSN 0004-6337

Reinhard E. Schielicke (Ed.)

Reviews in Modern Astronomy 16

The Cosmic Circuit of Matter

WILEY-VCH GmbH & Co.

Edited on behalf of the *Astronomische Gesellschaft* by
Dr. *Reinhard E. Schielicke*
Universitäts-Sternwarte Jena
Schillergäßchen 2, D-07745 Jena
Germany

This book was carefully produced. Nevertheless, author and publisher do not warrant the information contained therein to be free of errors. Readers are advised to keep in mind that statements, data, illustrations, procedural details or other items may inadvertently be inaccurate.

Cover picture:
Giant galactic nebula NGC 3603
This picture illustrates the entire stellar life cycle of stars, starting with the Bok globules and giant gaseous pillars (evidence of embryonic stars), followed by circumstellar disks around young stars, and progressing to aging, massive stars in a young starburst cluster. The blue super-giant with its ring and bipolar outflow (upper left of center) marks the end of the life cycle.
Credit: Wolfgang Brandner (JPL/IPAC), Eva K. Grebel (Univ. Washington), You-Hua Chu (Univ. Illinois Urbana-Champaign), and NASA

Library of Congress Card No. applied for.

British Library Cataloguing-in-Publication Data:
A catalogue record for this book is available from the British Library.

Die Deutsche Bibliothek – CIP Cataloguing-in-Publication Data:
A catalogue record for this publication is available from Die Deutsche Bibliothek.

ISBN 3-527-40451-1

© 2003 WILEY-VCH GmbH & Co. KGaA, Weinheim
Printed on acid-free paper.

All rights reserved (including those of translation into other languages). No part of this book may be reproduced in any form – by photoprinting, microfilm, or any other means – nor transmitted or translated into a machine language without written permission from the publishers. Registered names, trademarks, etc. used in this book, even when not specifically marked as such, are not considered unprotected by law.

Printing: Strauss Offsetdruck GmbH, Mörlenbach
Bookbinding: Großbuchbinderei Schäffer GmbH & Co. KG, Grünstadt

Printed in the Federal Republic of Germany.

Preface

The annual series *Reviews in Modern Astronomy* of the ASTRONOMISCHE GESELLSCHAFT was established in 1988 in order to bring the scientific events of the meetings of the Society to the attention of the worldwide astronomical community. *Reviews in Modern Astronomy* is devoted exclusively to the Karl Schwarzschild Lectures, the Ludwig Bierman Award Lectures, the invited reviews, and to the Highlight Contributions from leading scientists reporting on recent progress and scientific achievements at their respective research institutes.

The Karl Schwarzschild Lectures constitute a special series of invited reviews delivered by outstanding scientists who have been awarded the Karl Schwarzschild Medal of the Astronomische Gesellschaft, whereas excellent young astronomers are honoured by the Ludwig Biermann Prize.

Volume 16 continues the series with thirteen invited reviews and Highlight Contributions which were presented during the International Scientific Conference of the Society on "The Cosmic Circuit of Matter", held at Berlin, September 24 to 29, 2002.

The Karl Schwarzschild medal 2002 was awarded to Professor Charles H. Townes, Berkeley, USA. His lecture with the title "The Behavior of Stars Observed by Infrared Interferometry" opened the meeting.

The talk presented by the Ludwig Biermann Prize winner 2002, Dr Ralf S. Klessen, Potsdam, Germany, dealt with the topic "Star Formation in Turbulent Interstellar Gas".

Other contributions to the meeting published in this volume discuss, among other subjects, the Solar atmosphere, formation of stars, substellar objects, galaxies, and clusters of galaxies.

The editor would like to thank the lecturers for stimulating presentations. Thanks also to the local organizing committee from the "Zentrum für Astronomie und Astrophysik" of the Technische Universität Berlin, Germany, chaired by Erwin Sedlmayr.

Jena, April 2003 *Reinhard E. Schielicke*

The Astronomische Gesellschaft awards the **Karl Schwarzschild Medal**. Awarding of the medal is accompanied by the Karl Schwarzschild lecture held at the scientific annual meeting and the publication.

Recipients of the Karl Schwarzschild Medal are

1959 Martin Schwarzschild:
Die Theorien des inneren Aufbaus der Sterne.
Mitteilungen der AG 12, 15

1963 Charles Fehrenbach:
Die Bestimmung der Radialgeschwindigkeiten mit dem Objektivprisma.
Mitteilungen der AG 17, 59

1968 Maarten Schmidt:
Quasi-stellar sources.
Mitteilungen der AG 25, 13

1969 Bengt Strömgren:
Quantitative Spektralklassifikation und ihre Anwendung auf Probleme der Entwicklung der Sterne und der Milchstraße.
Mitteilungen der AG 27, 15

1971 Antony Hewish:
Tree years with pulsars.
Mitteilungen der AG 31, 15

1972 Jan H. Oort:
On the problem of the origin of spiral structure.
Mitteilungen der AG 32, 15

1974 Cornelis de Jager:
Dynamik von Sternatmosphären.
Mitteilungen der AG 36, 15

1975 Lyman Spitzer, jr.:
Interstellar matter research with the Copernicus satellite.
Mitteilungen der AG 38, 27

1977 Wilhelm Becker:
Die galaktische Struktur aus optischen Beobachtungen.
Mitteilungen der AG 43, 21

1978 George B. Field:
Intergalactic matter and the evolution of galaxies.
Mitteilungen der AG 47, 7

1980 Ludwig Biermann:
Dreißig Jahre Kometenforschung.
Mitteilungen der AG 51, 37

1981 Bohdan Paczynski:
Thick accretion disks around black holes.
Mitteilungen der AG 57, 27

1982   Jean Delhaye:
        Die Bewegungen der Sterne
        und ihre Bedeutung in der galaktischen Astronomie.
        Mitteilungen der AG 57, 123

1983   Donald Lynden-Bell:
        Mysterious mass in local group galaxies.
        Mitteilungen der AG 60, 23

1984   Daniel M. Popper:
        Some problems in the determination
        of fundamental stellar parameters from binary stars.
        Mitteilungen der AG 62, 19

1985   Edwin E. Salpeter:
        Galactic fountains, planetary nebulae, and warm H I.
        Mitteilungen der AG 63, 11

1986   Subrahmanyan Chandrasekhar:
        The aesthetic base of the general theory of relativity.
        Mitteilungen der AG 67, 19

1987   Lodewijk Woltjer:
        The future of European astronomy.
        Mitteilungen der AG 70, 21

1989   Sir Martin J. Rees:
        Is there a massive black hole in every galaxy.
        Reviews in Modern Astronomy 2, 1

1990   Eugene N. Parker:
        Convection, spontaneous discontinuities,
        and stellar winds and X-ray emission.
        Reviews in Modern Astronomy 4, 1

1992   Sir Fred Hoyle:
        The synthesis of the light elements.
        Reviews in Modern Astronomy 6, 1

1993   Raymond Wilson:
        Karl Schwarzschild and telescope optics.
        Reviews in Modern Astronomy 7, 1

1994   Joachim Trümper:
        X-rays from Neutron stars.
        Reviews in Modern Astronomy 8, 1

1995   Henk van de Hulst:
        Scaling laws in multiple light scattering under very small angles.
        Reviews in Modern Astronomy 9, 1

1996   Kip Thorne:
        Gravitational Radiation – A New Window Onto the Universe.
        Reviews in Modern Astronomy 10, 1

1997   Joseph H. Taylor:
       Binary Pulsars and Relativistic Gravity.
       not published

1998   Peter A. Strittmatter:
       Steps to the LBT – and Beyond.
       Reviews in Modern Astronomy 12, 1

1999   Jeremiah P. Ostriker:
       Historical Reflections
       on the Role of Numerical Modeling in Astrophysics.
       Reviews in Modern Astronomy 13, 1

2000   Sir Roger Penrose:
       The Schwarzschild Singularity:
       One Clue to Resolving the Quantum Measurement Paradox.
       Reviews in Modern Astronomy 14, 1

2001   Keiichi Kodaira:
       Macro- and Microscopic Views of Nearby Galaxies.
       Reviews in Modern Astronomy 15, 1

2002   Charles H. Townes:
       The Behavior of Stars Observed by Infrared Interferometry.
       Reviews in Modern Astronomy 16, 1

The **Ludwig Biermann Award** was established in 1988 by the Astronomische Gesellschaft to be awarded in recognition of an outstanding young astronomer. The award consists of financing a scientific stay at an institution of the recipient's choice.

Recipients of the Ludwig Biermann Award are

  1989   Dr. Norbert Langer (Göttingen),
  1990   Dr. Reinhard W. Hanuschik (Bochum),
  1992   Dr. Joachim Puls (München),
  1993   Dr. Andreas Burkert (Garching),
  1994   Dr. Christoph W. Keller (Tucson, Arizona, USA),
  1995   Dr. Karl Mannheim (Göttingen),
  1996   Dr. Eva K. Grebel (Würzburg) and
         Dr. Matthias L. Bartelmann (Garching),
  1997   Dr. Ralf Napiwotzki (Bamberg),
  1998   Dr. Ralph Neuhäuser (Garching),
  1999   Dr. Markus Kissler-Patig (Garching),
  2000   Dr. Heino Falcke (Bonn),
  2001   Dr. Stefanie Komossa (Garching),
  2002   Dr. Ralf S. Klessen (Potsdam).

Contents

Karl Schwarzschild Lecture:
The Behavior of Stars Observed by Infrared Interferometry
By Charles H. Townes (With 17 Figures) 1

Ludwig Biermann Award Lecture:
Star Formation in Turbulent Interstellar Gas
By Ralf S. Klessen (With 11 Figures) 23

Dynamics of Small Scale Motions in the Solar Photosphere
By Arnold Hanslmeier (With 4 Figures) 55

The Interstellar Medium and Star Formation: The Impact of Massive Stars
By José Franco, Stan Kurtz, and Guillermo García-Segura (With 12 Figures) .. 85

Circuit of Dust in Substellar Objects
By Christiane Helling (With 10 Figures) 115

Hot Stars: Old-Fashioned or Trendy?
By A. W. A. Pauldrach (With 24 Figures) 133

Gas and Dust Mass Loss of O-rich AGB-stars
By Franz Kerschbaum, Hans Olofsson, Thomas Posch,
David González Delgado, Per Bergman, Harald Mutschke,
Cornelia Jäger, Johann Dorschner, and Fredrik Schöier (With 9 Figures) 171

Finding the Most Metal-poor Stars of the Galactic Halo
with the Hamburg/ESO Objecrive-prism Survey
By Norbert Christlieb (With 6 Figures) 191

A Tale of Bars and Starbursts:
Dense Gas in the Central Regions of Galaxies
By Susanne Hüttemeister (With 8 Figures) 207

Tip-AGB Mass-Loss on the Galactic Scale
By Klaus-Peter Schröder (With 6 Figures) 227

The Dusty Sight of Galaxies:
ISOPHOT Surveys of Normal Galaxies, ULIRGS, and Quasars
By Ulrich Klaas (With 13 Figures) 243

Abundance Evolution with Cosmic Time
By James W. Truran (With 7 Figures) 261

Matter and Energy in Clusters of Galaxies
as Probes for Galaxy and Large-Scale Structure Formation in the Universe
By Hans Böhringer (With 24 Figures) 275

Index of Contributors 303

General Table of Contents 305

General Index of Contributors 317

Karl Schwarzschild Lecture

The Behavior of Stars Observed by Infrared Interferometry

Charles H. Townes

University of California
Department of Physics, Berkeley, CA 94720-7300, USA
cht@ssl.berkeley.edu

History

Successful stellar interferometry was initiated by Michelson and Pease who were the first to successfully measure the size of a star, α Orionis, in 1921 (Michelson and Pease, 1921). Since that time, a number of additional somewhat similar measurements have been made. In addition, intensity interferometry was initiated and successfully used by Hanbury Brown et al. (Hanbury Brown et al., 1967). But it is only during the last two decades that stellar interferometry has been growing rapidly, undertaken or planned by a number of different scientific groups. This growth and optimism about its use is largely because of newly developed technology which can make such measurements more precise and sensitive. This includes the use of lasers for accurate distance measurement and control, new sensitive detectors, particularly at infrared wavelengths, adaptive optics, and computers which can automatically control complex systems. There are now many different interferometers built or under construction, operating at a variety of wavelengths. A system using heterodyne detection in the mid-infrared ($\sim 10~\mu$m) region will be described here with results it has obtained in studying stars and their behavior.

Michelson and Pease's pioneering measurement gave a diameter for α Orionis of 48.5 mas $\pm 10\%$. Michelson also commented that this size could be too small because of limb darkening, the effect of gas immediately surrounding the star shading the edges of the star more than the center. However, he had no immediate estimate of the magnitude of this effect at the optical wavelengths where he and Pease made the measurement. Other astronomers used interferometry from time to time to measure stellar diameters, but the technique was difficult enough that there was no steady stream of data. Measurements of α Ori over the years from 1921 to about 1980 by a number of different individuals is summarized in Figure 1. The diameters are plotted as a function of the phase of what was thought to be a roughly 5-year cyclic variation in size, a variation which is not very convincing from the data shown. Figure 1 shows that subsequent measurements of α Ori were not substantially more accurate

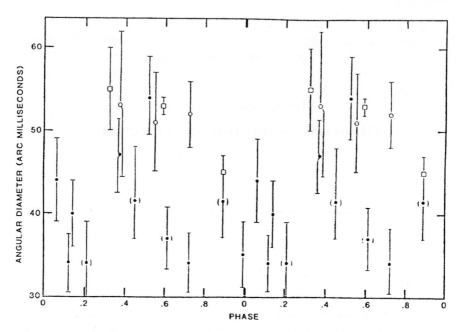

Figure 1: Measured angular diameters of α Ori made between 1921 and 1980 plotted against the phase of an apparent 5 year cycle in luminosity (from White et al., 1980). Points surrounded by parentheses were measured under what was reported as poor seeing conditions. Evidence for a cydic change in luminosity is not strong, nor is evidence for systematic changes in size, which this diagram, made by White, was designed to examine.

than Michelson's first one, and that his value of 48.5 mas is close to the average (∼ 45 mas) of all subsequent measurements at optical wavelengths. Today new technology allows a substantial improvement in the precision of such measurements.

An Interferometer for the Mid-IR by Heterodyne Detection

There are good reasons for present use of a variety of interferometers using a variety of wavelengths. The system to be described here operates at mid-infrared wavelengths and uses heterodyne detection, which provides measurements at a variety of wavelengths, each over the relatively narrow frequency range of $\sim 6 \times 10^9$ Hz or ~ 0.2 cm^{-1}. Reasons why this type of interferometry is useful for measuring stars include the following:

1. Theoretical expectations are that limb darkening of the older stars is much less at longer wavelengths than in the visible region. Estimates give the apparent size of α Ori about 15 % smaller than reality when measured at visible wavelengths, but less than 1 % smaller at wavelengths near 10 μm.

2. It is now well recognized that older stars can have rather large spots of varying intensity. In particular, such a spot has been found on α Ori with about 15 % of the total optical luminosity. Such bright spots gives an interferometric measurement of the star a misleading diameter unless the star is extensively mapped at very high angular resolution. A surface temperature variation of 10 % can change the optical luminosity by about a factor of two, but would change the 10 μm intensity by only about 10 %, due to the nature of the Planck function. Hence the star can be expected to be more uniform over its surface at mid-IR wavelengths than in visible light and the apparent size closer to reality.

3. Many old stars emit gas, from which dust is formed so that the stars are surrounded by dust. Mid-IR wavelengths not only penetrate the dust more easily than do shorter waves and thus see the star more clearly, they can also detect and measure objects which are cooler than the star, down to temperatures of a few hundred Kelvin, and thus can study the dust itself. The emission of dust, its distribution about the star, and its motions are important parts of the study of stellar behavior, and mid-IR wavelengths are well adapted to such study.

4. Although dust is generally not closer than a few stellar radii from the star since the temperature must be below about 1300 K for it to condense, there are frequently atoms and molecules surrounding the star whose radiation or radiation absorption can substantially distort a measurement of stellar size. Hence it is often important to do interferometry over a region of limited bandwidth which can avoid coincidence with spectral lines. The narrow bandwidth of heterodyne detection readily allows this.

5. To study stellar behavior, one also wants to study the gas emitted by and surrounding a star. Hence a narrow bandwidth which coincides with a spectral line is useful in examining the distribution of radiation emitted or absorbed by gas.

It must be emphasized again that there are good reasons for many types of interferometers, such as an interferometer at shorter wavelengths and using a broad band of wavelengths in order to study the less intense stars, or to examine the change in intensity over a star's surface. However, the above lists some of the reasons why the interferometer to be discussed here is useful. Further description and discussion of our heterodyne system have been published (Hale et al., 2000).

Technical Design

A schematic of the design used for an individual telescope of our initial two-telescope interferometer is shown in Figure 2. A flat 2.03 m alt mirror on an alt-azimuth mount sends starlight into a f 3.16 parabolic mirror of 1.65 m diameter which focuses it through a hole in the flat mirror onto an optics table. There a Schwarzschild mirror combination produces a small diameter f 80 beam on an optics table. This beam is mixed with local oscillator power from a CO_2 laser and sent to a detector in a liquid

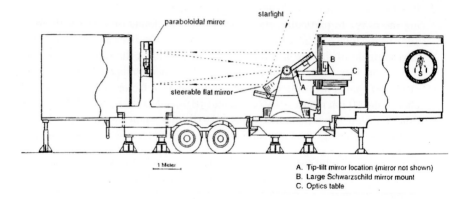

A. Tip-tilt mirror location (mirror not shown)
B. Large Schwarzschild mirror mount
C. Optics table

Figure 2: A schematic of one of the 1.65 meter mobile telescopes used in the infrared Spatial Interferometer (ISI).

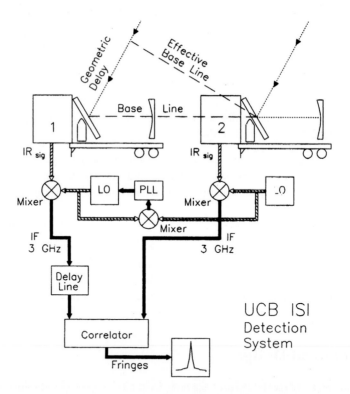

Figure 3: A schematic of the ISI system using two telescopes and heterodyne detection. The two local oscillators (LO) are CO_2 lasers. These are locked in phase by mixing their signals, with the phase lock circuit (PLL) making one LO follow the other in frequency and phase. The local oscillators, mixers, and correlator which produces an interference fringe (shown as a peaked signal) are actually in the trailers which enclose the telescopes, but are shown here outside the trailers for clarity of their diagram.

N$_2$ Dewar. This mirror arrangement is chosen so that the mirrors can be on rather rigid mounts close to the ground and avoid vibration. The trailer allows the telescope to be moved so that baselines of various lengths and orientations may be used.

A schematic of two telescopes operating as an interferometer is shown in Figure 3. The mixers are actually located on the optics table with each telescope, but shown separated here for clarity. They respond fast enough to produce frequencies up to about 3×10^9 Hz, so that the total bandwidth, including two sidebands, is approximately 6×10^9 Hz. The signals from the two telescopes are then sent by cable to the correlator which allows them to beat together, adding if they are in phase and subtracting if they are out of phase, thus producing a fringe. The entire system is completely analogous to microwave or radio interferometry, which of course also use heterodyne detection. The CO_2 laser local oscillators in the two telescopes must be kept in phase. This is done by sending a beam of CO_2 radiation from one telescope to the other, beating this together with a signal from the other CO_2 laser, and locking the two together in phase. Since the path length between the two telescopes is long and may vary, it too must be controlled. This is done by reflecting part of the beam coming from one telescope to the other directly back on exactly the same path. On arrival back at its origin, this beam is interferred with the initial wave, and the path length automatically adjusted so that the total round trip is exactly an integral number of wavelengths. Overall, the system keeps the two lasers in phase to a precision of about 5°. Relative phases of the two oscillators are actually changed steadily to maintain a constant fringe frequency, or rate of change in phase between the two telescope signals, so that the correlation yields a convenient constant beat frequency of 100 Hz.

The delay line shown is a device which automatically switches in or out small lengths of cable in order to maintain approximate equality of the total path lengths from the star through the two telescopes and circuitry to the correlator. For our narrow bandwidth, these path lengths need to be equal only to an accuracy of about 5 millimeters.

Figure 4 is a photo of the three telescopes now used. They are shown on minimum length baselines with separations of 4, 8, and 12 meters. A 4 meter baseline for an interferometer provides approximately the same resolution as a 10 m circular telescope. This distance was hence chosen to overlap the resolution of images taken by a 10 m Keck telescope. Other positions for the telescopes provide baselines as long as about 75 m, and resolve objects as small as 15 mas.

Michelson's "Visibility"

Michelson used and defined a quantity called "visibility" of an interferometer. Figure 5 illustrates its meaning and significance. The circular objects represent a star, and the lobes represent the intensity pattern received by an interferometer. This intensity varies sinusoidally from maximum when the two signals received by the two telescopes are in phase to minimum when they are out of phase. In the upper figure a small star is illustrated passing through rather widely spared lobes of the interferometer's response. In this case two signals will add and subtract completely, giving a

Figure 4: A picture of three ISI (Infrared Spatial Interferometer) telescopes in position on Mt. Wilson with their shortest baselines. The telescopes here are separated by distances of 4 m (numbers 1 and 2), 8 m (number 2 and 3), and 12 m (numbers 1 and 3).

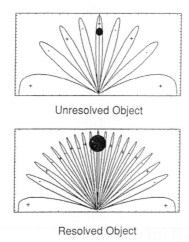

Figure 5: A diagram of the lobes and maximum and minimum interference intensity on the sky, illustrating the visibility, or fractional change in total signal as an object such as a star moves across the lobes. In the upper figure the circular object (star) is as small as the width of a lobe, and hence its signal in the interferometer goes from essentially zero to essentially maximum. This is defined as a visibility of unity. In the lower picture, a larger object (star) extends over both maxima and minima of lobes, hence its interferometric intensity does not vary so much, i.e., its visibility is less than unity. A real stellar interferometer has many millions of lobes across the sky rather than the rather widely spaced lobes illustrated here.

maximum fringe signal, which Michelson defined as a visibility of unity. The lower figure illustrates a larger star and more closely spaced lobes of responsivity. In this case, some parts of the star will give maximum interference signal and some minimum, so that the total variation in the signal is less than maximum as the star moves through the lobes. For sufficiently closely spaced lobes, there would be no variation in the signal at all, giving a "visibility" of zero. This visibility actually represents the amplitude of components of a Fourier spectrum of intensity on the sky. For two telescopes, one can obtain the amplitude of the Fourier spectrum but not the phases, since it is impossible to be sure of the relative distance from a star to two separated telescopes with an accuracy of a fraction of a wavelength. If the object being observed has reflection symmetry, then the phase is zero, e.g., one can assume all cosine functions for the separate Fourier components. A circular star has such symmetry and hence phase information is not needed. However, to image more complex objects phase is needed. It can be obtained if three or more telescopes are used. For three telescopes, the sum of measured phases of fringe variation between telescopes 1 and 2, 2 and 3, and 3 and 1, known as phase closure, is zero if the object is symmetric but otherwise is not zero and gives phase information for the Fourier spectrum. Since it has become very clear that many dust shells around stars are not symmetric, we have now built a third telescope to obtain this phase closure and hence complete an accurate mapping of asymmetric infrared intensity distributions.

Phenomena at Astronomical Objects

Seeing Conditions

Actual fringe power obtained from observation of an astronomical object (α Ori) is shown in Figure 6. The signal in this figure at 100 Hz frequency and of very narrow bandwidth represents the visibility of α Ori measured under exceptionally good seeing conditions. The other fringe record is from α Ori when the seeing was exceptionally poor. Its bandwidth is about 20 Hz wide, indicating a major change in relative phase of the signal arriving at the two telescopes due to atmospheric fluctuations in a time as short as about 1/20 s. This illustrates the importance of good seeing for accurate interferometry, or measurement of visibility. Nevertheless, the α Ori signal is strong enough that even in the very poor seeing case one can integrate the total curve and obtain a rather accurate measurement of the fringe power. For very weak signals, a narrow type of fringe obtained under good seeing helps to provide accurate measurements.

Sars Surrounded by Dust

A simple model of a star surrounded by dust in illustrated in Figure 7. As gas emitted by the star expands, dust forms at a distance where the temperature is sufficiently low, perhaps 1000 to 1300 K. A temperature of 1300 K corresponds roughly to a distance of 3 stellar radii from a stellar surface temperature of 2600 K. Hence there is a distinct separation between star and dust shell which allows interferometry to separate rather

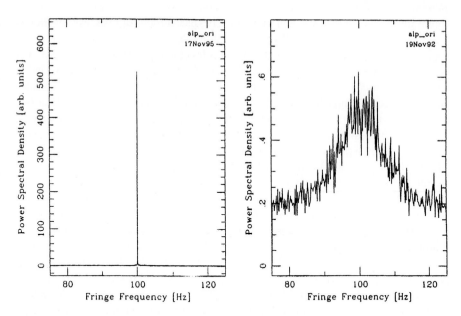

Figure 6: Plots of actual fringe power obtained by the ISI from a star (α Ori) as a function frequency. Light path delays are arranged so that the fringe signal oscillates at 100 Hz as the star moves across the sky if the atmospheric paths do not fluctuate due to "seeing". The very narrow bandwidth fringe signal above was obtained during a period of excellent seeing. The broad fringe was obtained during very poor seeing and its width indicates that relative pathlengths from the star through the atmosphere to the two telescopes was fluctuating as much as a wavelength on time scales as short as about 1/20 s.

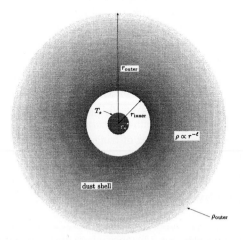

Figure 7: An idealized model of a star (the central circle) surrounded by a spherical dust shell beginning a few stellar radii from the star. The dust density ϱ decreases with distance r from the star approximately as $1/r^2$.

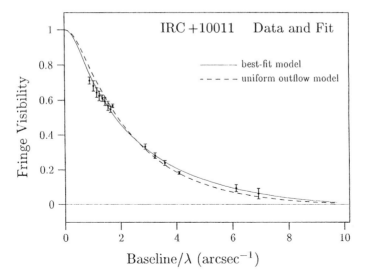

Figure 8: A plot of visibility measurements of the star IR C+10011 as a function of lobe spacing (from Lipman et al., 2000). The lobe spacing is in units of arc sec^{-1} and the dust shell is resolved, but not the star, which provides only a small fraction of the total infrared energy. The dashed line represents a visibility curve for an idealized model with dust density decreasing as $1/r^2$ with distance r from the star. The solid line represents the best fit of measured points and indicates only a slightly different rate of dust density change with distance.

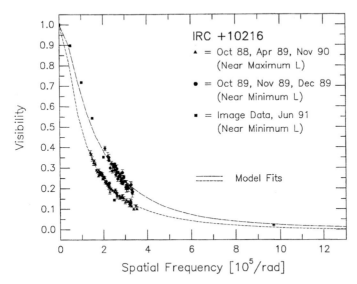

Figure 9: Visibility measurements of the variable star IR C+10216 at 11 μm wavelength (from Danchi et al., 1994). The scale for lobe spacing is 10^5 lobes per radian. The lower curve was taken near the time of maximum stellar luminosity, the upper one near minimum luminosity. This shows the effect of varying luminosity on infrared intensity distribution, and also indicates the formation of some new dust during the time of low luminosity.

completely the Fourier components due to each. In this standard model, the dust is assumed to move our radially at a constant velocity so that the density decreases as $1/r^2$, where r is the radial distance, and the temperature decreases approximately as $1/r^{1/2}$. Details of temperature depend, of course, on dust and gas densities and absorption spectra.

Some stars fit the simple model of Figure 7 rather well. Figure 8 shows actual visibility measurements for the star CIT3 (or IR C10011) as a function of lobe spacing. The dashed curve represents expected results for the idealized model of Figure 7. This solid curve is an optimum fit of the visibility data, and indicates that the dust density decreases with radius somewhat more slowly than $1/r^2$.

Stellar Luminosity Changes

There are many additional phenomena and complications which are manifested by actual measurements. For example, stellar luminosities change with time, particularly for variable stars, and this changes visibility measurements. Figure 9 represents visibility curves for the intense variable star IR C+10216. The upper curve gives the visibility at minimum luminosity and the lower at maximum luminosity. High luminosity warms the dust shell around the star out to larger radii, and thus reduces the visibility at a given resolution. The change in visibility seen in Figure 9 is largely explained by this simple heating or cooling of surrounding dust. However, this does not completely explain the difference between the two stellar phases. It is necessary to assume condensation of additional dust at the inner radius of the dust shell when the luminosity is low. This formation of additional dust is of course not unexpected, but with these interferometric measurements it can now be proven and evaluated.

Gas and Dust Emission

A substantial fraction of old stars have also been found to emit gas and dust episodically rather than continuously, as assumed in the model discussed above. Figure 10 represents the visibility of NML Tau (or TK Tau). The fact that the visibility curve, or the Fourier transform, has wiggles is an obvious implication that there is structure in the radial distribution of infrared intensity surrounding the star. Furthermore, the visibility curve has changed during the course of only one year, as shown in the figure. Infrared intensities as a function of distance from the star obtained by fitting the visibility curve are plotted in the figure for the years 1992 and 1993. This shows that there are three shells of dust surrounding the star, and that the outer one has moved a significant amount over the course of one year. The expansion velocity of gas around the star has been measured by the Doppler shift. With this and the angular change in the size of the outer shell of dust, one can evaluate the distance of this star as approximately 265 pc. In addition, the observed motion of this outer shell indicates the time between emissions of the three successive shells, which is 12 years. Thus the star has had episodic emissions repeated on a 12 year time scale even though its luminosity variation has a period of 1.29 years. A number of other stars have been shown to have discrete dust shells. This episodic behavior is a little surprising, but

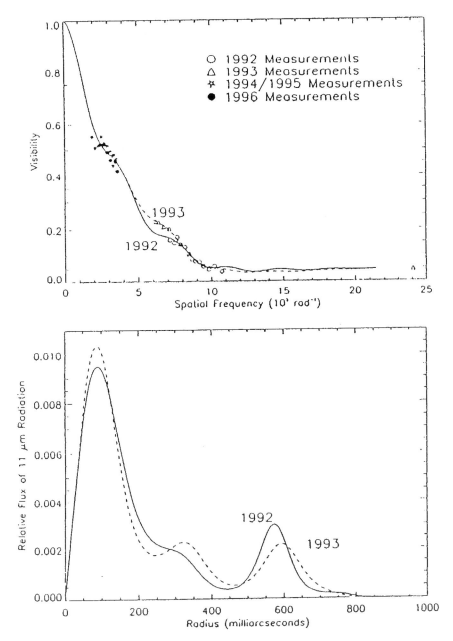

Figure 10: Visibility measurements of the star NML Tau made at two different years, which shows a shell structure of dust around the star and motion outwards of the dust (from Hale et al., 1997). The upper figure shows measured visibility points for two successive years and curves fitting these points. The lower figure shows the theoretical infrared intensity as a function of distance from the star deduced from the visibility measurements. There is evidence for three discrete shells of dust and motion of at least the outer shell over one year's time. This indicates that emissions of the dust shells were separated by approximately 12 years.

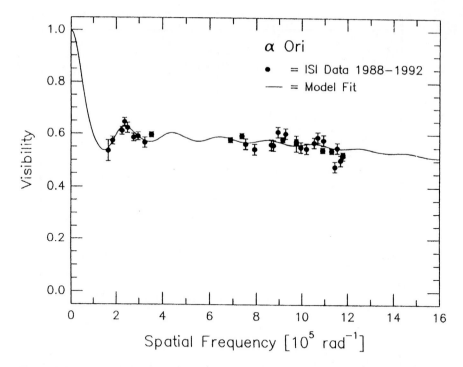

Figure 11: Visibility curve for α Ori at relatively low resolution where the dust shells are resolved but not the star itself (from Danchi et al., 1994). The oscillation in visibility shown by the curve is the result of addition and then subtraction of signals from the star and from a thin dust shell surrounding the star at a distance of 1 arc sec.

it turns out to be common. Thirty to forty percent of the old stars we have observed have emitted discrete shells. α Ori has been known for some time to have emitted two dust shells with a time separation of about 100 years. In general, the time scale between observed episodic emissions is in the range of a decade to a century. Why this happens and why this time scale remains to be understood.

Amount of Material Emission and its Asymmetries

The amount of material emitted by stars is another important parameter. Among the stars we have measured, the normal range is 10^{-5} to 10^{-7} solar masses per year.

It is also clear from our present measurements that dust is not always emitted spherically, which simple models assume. Sometimes the emission is quite asymmetrical. This is one reason for the importance of our new third telescope. Its use will not only give us visibility measurements on three different baselines simultaneously rather than only one; it will provide phase closure so that a complete mapping of intensity distribution can be made, including asymmetries.

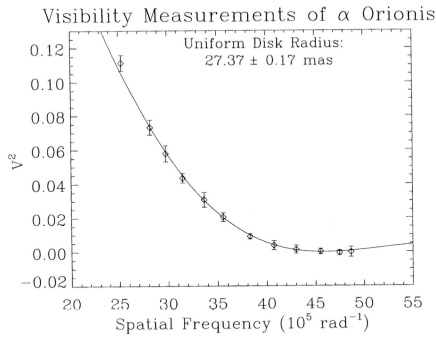

Figure 12: The square of the visibility of α Ori at relatively high resolution, including the resolution where the visibility becomes zero (from Weiner et al., 2000). The theoretical fit to measured points gives a diameter of 54.74 ± 0.34 mas for the star.

Figure 11 shows the visibility curve for α Ori at relatively low resolutions, where the dust shells surrounding it have been resolved, but the star is unresolved. In fact, the dust shells are at such large radii (approximately 1 and 2 arc seconds) that they are resolved at the lowest resolution we have used and the visibility due to radiation from the star alone is primarily measured. However, there is a striking lump at a spatial lobe frequency of approximately 3×10^5 per radian. This is due to the fact that when the star is on a maximum of one of the interferometer lobes the shell around it is near the maximum of the next lobe, so their intensities add. At approximately twice the lobe spacing, the two intensities are out of phase and subtract, giving a minimum. This provides a good measurement of the position of the rather thin shell around α Ori, at a radius of 1 arc second from the star. To resolve the star itself and measure its diameter, higher resolution is needed than what is shown on this figure, and this will now be discussed.

Higher Resolution Measurements

Figure 12 shows the square of the visibility of α Ori at higher lobe spacings, where the star is resolved. The square of the visibility is plotted here because that is what is actually measured in typical interferometry, and the probably errors are hence somewhat more uniform and easily interpreted than for visibility itself. At the resolutions

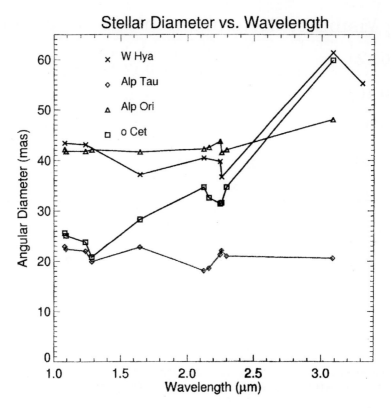

Figure 13: Measurements of apparent stellar sizes as a function of wavelength using broadband interferometry (from Tuthill et al., 1999). Stars surrounded by gas and dust produce apparent sizes which vary substantially with wavelength, e. g. o Ceti, while those having little gas and dust have apparent sizes which are much less dependent on wavelength, e. g., α Tau.

shown in this figure, the dust shells are completely resolved and have essentially no influence on the shape of the curve, though they do affect the scale of visibility. As indicated in Figure 12, the diameter of α Ori is determined to be 54.74 ± 0.34 mas. This is 13 % larger than Michelson's measurement, or 22 % larger than the average measurements shown in Figure 1, rather close to the difference due to limb darkening which had been estimated from theoretical calculations. This is also some confirmation that, in accordance with present calculations, the measurement made at mid-IR wavelengths is not appreciably limb-darkened and represents the real stellar diameter.

It is important to recognize the substantial errors that may be made in measurement of stellar sizes due not only to limb darkening by normal atmospheres around stars, but also due to surrounding molecules or dust which may not be well known or evaluated. The problem is illustrated by the plot of interferometrically measured stellar diameters as a function of wavelength shown in Figure 13. It is particularly severe for o Ceti, a star surrounding by much gas and dust, and for which near infrared measurements show an apparent diameter variation with wavelength of as

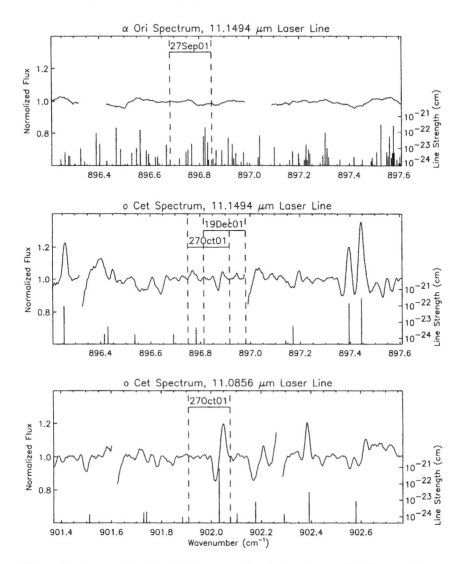

Figure 14: Spectra of α Ori and o Cet surrounding the bandpass at 11.149 μm used to measure their diameters, and at 10.086 μm where there is a strong H_2O line. The spectra were measured by J. H. Lacy and M. J. Richter.

much as a factor of 3. These measurements are made by members of our Berkeley group, using masking on the Keck telescope to simulate interferometry with a member of telescopes. They are made over bandwidths which are moderately broad and are known to include a number of molecular lines. On the other hand α Tau, which is surrounded by relatively little gas and dust, has a measured diameter which does not vary much with wavelength as can be seen in the same Figure 13. Normal limb darkening is still presumably present, so these measured values are all probably somewhat smaller than the real diameter of α Tau.

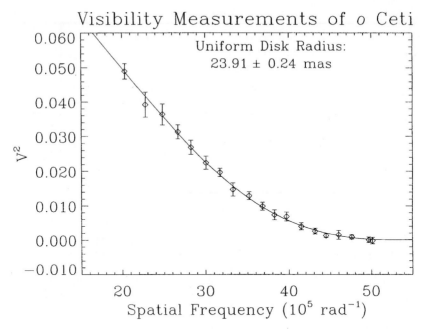

Figure 15: The square of the visibility curve of o Ceti at high resolution, resolving the star. The theoretical curve fitting the data gives a diameter for o Ceti of 47.82 ± 0.48 mas (from Weiner et al., to be published).

Measurements over a Narrow Frequency Range

Figure 13 illustrates the importance of making measurements over a narrow frequency range which does not include the spectral frequencies of material immediately surrounding the star. Spectra have been measured for stars of interest by John Lacy et al. and his associates at the University of Texas. Figure 14 shows spectra of α Ori and o Ceti over a range of frequencies which include the frequencies near 11.15 μm wavelengths at which we measured diameters of these two stars. Most of the lines seen in this figure are due to water vapor. Depending on the time of year, the earth's motion shifts the stellar frequencies somewhat, as shown in this figure. In the case of α Ori, there are no apparent spectral lines within the bandwidth of our measurement. In the case of o Ceti, during part of the year there is a very weak absorption line. The more intense lines seen in the spectrum are strong enough to produce appreciable errors in the interferometric measurement of stellar diameter. However, the weak line within the range of measurement should not, in accordance with calculations, produce an error more than a small fraction of one percent. One can see in Figure 14 that there is both emission from H_2O and, at a somewhat shifted frequency due to Doppler effects, some absorption. The absorbing H_2O must be further away from the star than the emitting H_2O, and hence absorbing lines should have less effect on modifying the apparent stellar diameter.

Table 1: ISI Measurements from 1999 to 2001 of the Diameter of o Ceti. Wavelengths 10.884, 11.086, and 11.171 μm include strong H_2O lines within their bandwidths. The wavelength 11.159 μm does not.

Dates	Phase	λ (μm)	A[a]	Diameter (mas)
22, 26 Oct 99	0.99	11.149	0.276	46.56 ± 1.43
10–19 Nov 99	0.06	11.149	0.327	49.25 ± 0.55
30 Sep 00, 03–06 Oct 00	0.03	11.149	0.310	47.63 ± 0.80
17–20 Oct 00	0.08	11.149	0.326	48.84 ± 0.91
01 Nov 00	0.12	11.149	0.345	48.25 ± 0.94
28 Nov 00	0.20	11.149	0.433	47.35 ± 1.24
17, 21 Dec 00	0.27	11.149	0.465	46.48 ± 0.84
18–31 Jul 01	0.92	11.149	0.456	50.67 ± 0.63
01–08 Aug 01	0.95	10.884	0.392	57.90 ± 1.52
22, 24 Aug 01	0.01	11.149	0.384	53.18 ± 0.67
25–27 Sep 01	0.11	11.149	0.411	53.99 ± 0.53
04, 05 Oct 01	0.13	11.149	0.348	51.44 ± 0.67
11 Oct 01	0.15	11.149	0.346	54.27 ± 1.67
23, 24 Oct 01	0.19	11.149	0.366	55.06 ± 0.58
25, 26 Oct 01, 02 Nov 01	0.20	11.086	0.384	62.24 ± 0.84
02 Nov 01	0.22	11.149	0.324	52.11 ± 0.95
06–09 Nov 01	0.24	11.149	0.369	52.23 ± 0.50
10–16 Nov 01	0.26	11.171	0.355	55.88 ± 0.74
13, 14 Dec 01	0.35	11.149	0.243	49.31 ± 1.04
19 Dec 01	0.36	11.149	0.293	49.92 ± 0.79

[a] The parameter A is the fraction of the total flux which is emitted by the stellar disk as defined in equation 1. It has an uncertainty for all measurements of roughly 0.04.

The square of visibility of o Ceti at high angular resolution as measured in a narrow spectral band around a wavelength of 11.149 microns is plotted in Figure 15. This provides a diameter for the star of 47.82 ± 0.48 mas, considerably larger than the various values which have been previously estimated. To show the effects of spectral lines, Table 1 lists diameters of o Ceti as measured in regions with spectral lines, at wavelengths 10.884, 11.086, and 11.171 μm, as well as measurements made at essentially the same time at a wavelength of 11.149 μm. Figure 16 shows a plot of a single night of visibility measurements at 11.086 and 11.149 μm respectively, showing the very marked difference between the two, with the measurement at 11.086 μm much distorted by the H_2O line. This illustrates the importance of either very careful interpretation and extensive measurements if broadband interferometry is used on stars with substantial surrounding material, or of interferometry only over selected narrow-band frequency ranges where there are no spectral lines.

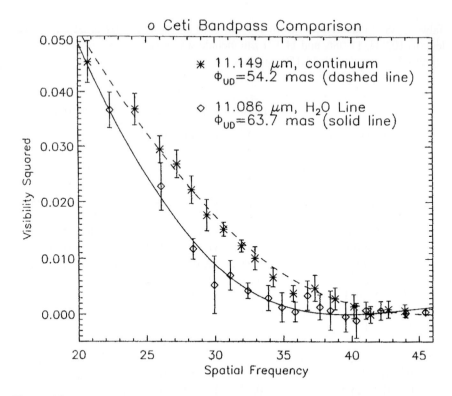

Figure 16: o Cet visibility curves with band passes centered at 11.149 μm, where there are no strong spectral lines, and at 11.086 μm, which contains a strong H_2O line. The two measurements were made within a few days of each other and demonstrate the large effect of spectral lines on apparent stellar diameter.

Measurements of Spectral Lines

Spectral lines are in themselves not uninteresting to interferometry. Narrow band interferometry on the exact frequency of a line can determine position and distribution of the molecule producing the line. Such work has been done by Monnier in our group (Monnier et al., 2000). With radio-frequency filters, he narrowed the spectral range of radiation detection substantially below 6×10^9 Hz to span only the width of a line, and did interferometry on lines of NH_3 and SiH_4 gas surrounding the stars IRC+10216 and VX Sgr. His interferometric analysis showed that these molecules occur at distances further than about 60 stellar radii from the star. The formation of NH_3 around stars or in interstellar clouds has in fact been somewhat of a puzzle. This work indicates that NH_3 and SiH_4 are formed at temperatures below about 500 K, and hence where many atoms and molecules would be absorbed on dust grains. We can hence conclude that these molecules are probably formed by chemical interactions on the surfaces of dust grains.

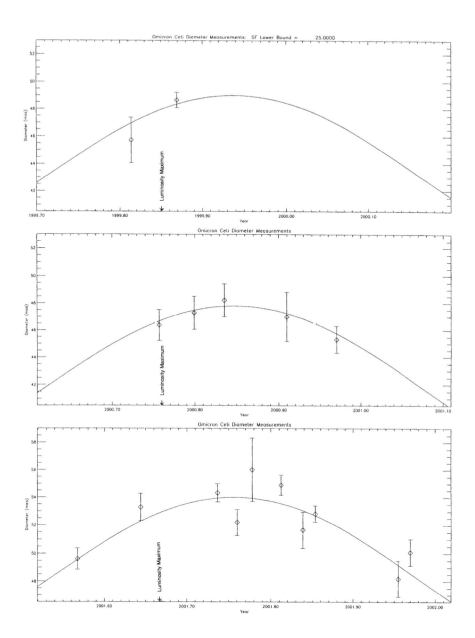

Figure 17: Variation in apparent diameter of o Ceti as a function of its luminosity phase, which cycles with a period of approximately 330 days (from Weiner et al., to be published). It is primarily the lowest curve, containing data taken in 2001, which provides a convincing case of variation in size with phase. The data is fitted to a sinusoidal curve with an amplitude of 15 % of the diameter. Note also that the apparent diameter is substantially larger in the year 2001 than in 2000. These dates involved orientations of the ISI interferometer differing by $35°$, and hence may indicate an elongation of the star.

Variable Stars with Cycles in Size

It has long been recognized that variable stars such as Miras change size as they cycle in luminosity over time periods of one to two years. Theoretical estimates have been made of the likely changes in size, assuming the lowest mode of oscillation, and some experimental results have indicated detection of size change. But none of these measurements have been clear or direct measurements of the stellar diameters. We have now, however, observed o Ceti over some period of time and believe we have direct and reliable measurements as shown in Figure 17. Reliable indication of size change with phase of the star is given only by the measurements during the year 2001, shown in the central curve. Observations were made over about one-half of the stellar period of 330 days, and indicate an approximately sinusoidal change during that time of about 15 %. If the curve continues sinusoidally, this indicates a change in size of ± 15 %, approximately what theoretical calculations have indicated.

The phase of size variation compared with phase of luminosity variation is also of interest. Figure 17 indicates the time of maximum visual luminosity (defined as phase = 0 or 1), and shows that the maximum size occurs approximately 1/8 of a cycle after the maximum luminosity. This again is in reasonable agreement with what at least some theorists have predicted.

Measurements and Theoretical Expectations

Although interferometric measurements show some agreement with theoretical expectations, they also show profound and puzzling disagreements. One is simply the star's diameter. If the stellar oscillation mode is the lowest radial one, as expected, and the stars internal structure agrees with present theory, its period could apparently not be as short as it is if the diameter is as large as measured. Perhaps the star oscillates in the next higher radial mode, or perhaps the internal density structure of the star is not as expected. Some indications of the latter have been given by second order dynamical calculations of Ya'ari and Tuchman (1999).

Another startling observation can be seen immediately in Figure 17. There the maximum diameter in 2001 is shown as approximately 54 mas, but in 2000 it was approximately 48 mas. The star does not repeat its characteristics exactly from cycle to cycle, but no such major changes have been indicated by other observations during 2000–2001. We believe this change is due to non-sphericity of the star, because the orientation of the interferometer baseline was changed between the years 2000 and 2001 by about 35°. If the star is elongated this could account for the observed change in size. The size of α Ori was measured at approximately the same time on the same baselines, and it was found not to change – it is indeed approximately circular. Perhaps o Ceti even oscillates in an elongating mode!

o Ceti as Examinee

As the result of these two unexpected results, the large size of o Ceti and its apparent non-sphericity, we have examined possible errors carefully (cf. Weiner et al., 2000).

One question is how the stellar diameter is really defined, and what part of the star are we really measuring. The size of a star is normally defined as the Rosseland radius, which is the radius from which the major radiation emitted by the star encounters an optical depth of unity. We accept that definition. And in accordance with present theories of any possible ionization layers or other absorbing material surrounding o Ceti, calculations show that the interferometry measurements at 11.149 μm wavelength should be seeing the Rosseland radius, within an accuracy of at least 1%.

One suggestion concerning the apparent non-sphericity of o Ceti is that there is a nearby clump of dust which moved into view when we rotated out baseline. We have made other interferometric studies of the dust distribution around o Ceti. This distribution is indeed somewhat complex, but it seems quite unreasonable and unlikely that dust could be close enough to the star and of such a character that it could distort the star's apparent shape as we have observed.

It is quite important to continue careful and precise observations of o Ceti, particularly to confirm its apparent elongation. Measurements on a few other Mira stars have been made, not with the accuracy we have obtained an o Ceti, but they too seem larger than expected. To check o Ceti's elongation, we look forward to use of three telescopes simultaneously, both for phase closure and measurements of asymmetric objects, and also to examine o Ceti carefully from two directions simultaneously. This should rather completely determine if it is elongated, as well as the mode of oscillation.

Outlook

It is clear that modern interferometry offers the possibility of measuring stellar characteristics and time-varying behavior with new precision and in such a way that it will test, challenge, and stimulate theoretical considerations. Many more types of stars and many wavelengths different from those of our own measurements will be involved in measurements by the many different interferometers which are becoming available or are foreseen.

References

Danchi, W.C., Bester, M., Degiacomi, C.G., Greenhill, L.J., Townes, C.H. 1994, ApJ 107, 1469

Hale, D.D.S., Bester, M., Danchi, W.C., Fitelson, W., Hoss, S., Townes, C.H. 2000, ApJ 537, 998

Hale, D.D.S., Bester, M., Danchi, W.C., Hoss, S., Lipman, E., Monnier, J.D., Tuthill, P.G., Townes, C.H., Johnson, M., Lopez, B., Geballe, T.R. 1997, ApJ 490, 407

Hanbury-Brown, R., Davis, J., Allen, L.R., Rome, J.M. 1967, MNRAS 137, 393

Lipman, E.A., Hale, D.D.S., Monnier, J.D., Townes, C.H. 2000, ApJ 532, 467

Michelson, A.A., Pease, F.G. 1921, ApJ 53, 249

Monnier, J.D., Danchi, W.C., Hale, D.S., Tuthill, P.G., Townes, C.H. 2000, ApJ 543, 868

Tuthill, P.G., Monnier, J.D., Danchi, W.C. 1999, in: Working on the Fringe: Optical and IR Interferometry from Ground and Space, eds. S. Unwin, R. Stachnik, PASPC 194, 188

Weiner, J., Danchi, W.C., Hale, D.D.S., McMahon, J., Townes, C.H., Monnier, J.D., Tuthill, P.G. 2000, ApJ 544, 1097

Weiner, J., Hale, D.D.S., Townes, C.H. ApJ, to be published

White, N.M. 1980, ApJ 242, 646

Ya'ari, A., Tuchman, Y. 1999, ApJ 514, L135

Ludwig Biermann Award Lecture

Star Formation in Turbulent Interstellar Gas

Ralf S. Klessen

Astrophysikalisches Institut Potsdam,
An der Sternwarte 16, D-14482 Potsdam, Germany
rklessen@aip.de

Abstract

Understanding the star formation process is central to much of modern astrophysics. For several decades it has been thought that stellar birth is primarily controlled by the interplay between gravity and magnetostatic support, modulated by ambipolar diffusion. Recently, however, both observational and numerical work has begun to suggest that supersonic interstellar turbulence rather than magnetic fields controls star formation. Supersonic turbulence can provide support against gravitational collapse on global scales, while at the same time it produces localized density enhancements that allow for collapse on small scales. The efficiency and timescale of stellar birth in Galactic molecular clouds strongly depend on the properties of the interstellar turbulent velocity field, with slow, inefficient, isolated star formation being a hallmark of turbulent support, and fast, efficient, clustered star formation occurring in its absence.

1 Introduction

Stars are important. They are the primary source of radiation (with competition from the 3 K black body radiation of the cosmic microwave background and from accretion processes onto black holes in active galactic nuclei, which themselves are likely to have formed from stars), and of all chemical elements heavier than the H and He that made up the primordial gas. The Earth itself consists primarily of these heavier elements, called metals in astronomical terminology. Metals are produced by nuclear fusion in the interior of stars, with the heaviest elements produced during the passage of the final supernova shockwave through the most massive stars. To reach the chemical abundances observed today in our solar system, the material had to go through many cycles of stellar birth and death. In a literal sense, we are star dust.

Stars are also our primary source of astronomical information and, hence, are essential for our understanding of the universe and the physical processes that govern its evolution. At optical wavelengths almost all natural light we observe in the sky originates from stars. In daytime this is more than obvious, but it is also true at night. The Moon, the second brightest object in the sky, reflects light from our Sun, as do the planets, while virtually every other extraterrestrial source of visible light is a star or a collection of stars. Throughout the millenia, these objects have been the observational targets of traditional astronomy, and define the celestial landscape, the constellations. When we look at a dark night sky, we can also note dark patches of obscuration along the band of the Milky Way. These are clouds of dust and gas that block the light from stars further away.

Since about half a century ago we know that these clouds are associated with the birth of stars. The advent of new observational instruments and techniques gave access to astronomical information at wavelengths far shorter and longer that visible light. It is now possible to observe astronomical objects at wavelengths ranging high-energy γ-rays down to radio frequencies. Especially useful for studying these dark clouds are radio and sub-mm wavelengths, at which they are transparent. Observations now show that *all* star formation occurring in the Milky Way is associated with dark clouds.

These clouds are dense enough, and well enough protected from dissociating UV radiation by self-shielding and dust scattering in their surface layers for hydrogen to be mostly in molecular form in their interior. The density and velocity structure of molecular clouds is extremely complex and follows hierarchical scaling relations that appear to be determined by supersonic turbulent motions (e. g. Williams, Blitz, & McKee 2000). Molecular clouds are large, and their masses exceed the threshold for gravitational collapse by far when taking only thermal pressure into account. Naively speaking, they should be contracting rapidly and form stars at very high rate. This is generally not observed. The star formation efficiency of molecular clouds in the solar neighborhood is estimated to be of order of a few percent (e. g. Elmegreen 1991, McKee 1999).

For many years it was thought that support by magnetic pressure against gravitational collapse offered the best explanation for the low rate of star formation. In this so called "standard theory of star formation", developed by Shu (1977; and see Shu, Adams, & Lizano 1987), Mouschovias & Spitzer (1976), Nakano (1976), and others, interstellar magnetic fields prevent the collapse of gas clumps with insufficient mass to flux ratio, leaving dense cores in magnetohydrostatic equilibrium. The magnetic field couples only to electrically charged ions in the gas, though, so neutral atoms can only be supported by the field if they collide frequently with ions. The diffuse interstellar medium (ISM) with number densities n of order unity remains ionized highly enough so that neutral-ion collisional coupling is very efficient (see Mouschovias 1991a, b). In dense cores, where $n > 10^5$ cm^{-3}, ionization fractions drop below parts per million. Neutral-ion collisions no longer couple the neutrals tightly to the magnetic field, so the neutrals can diffuse through the field in a process known in astrophysics as ambipolar diffusion. This allows gravitational collapse to proceed in the face of magnetostatic support, but on a timescale as much as an order of magnitude longer than the free-fall time, drawing out the star formation process.

Recently, however, both observational and theoretical results have begun to cast doubt on the "standard theory" (for a recent compilation see Mac Low & Klessen 2003). While theoretical considerations point against singular isothermal spheres as starting conditions of protostellar collapse as postulated by the theory (see Whitworth et al. 1996, Nakano 1998, Desch & Mouschovias 2001), there is a series of observational findings that put other fundamental assumptions of the "standard theory" into question as well. For example, the observed magnetic field strengths in molecular cloud cores appear too weak to provide support against gravitational collapse (Crutcher 1999, Bourke et al. 2001). At the same time, the infall motions measured around star forming cores extend too broadly (e. g. Tafalla et al. 1998 or Williams et al. 1999 for L1544), while the central density profiles of cores are flatter than expected for isothermal spheres (e. g. Bacmann et al. 2000). Furthermore, the chemically derived ages of cloud cores are comparable to the free-fall time instead of the much longer ambipolar diffusion timescale (Bergin & Langer 1997). Observations of young stellar objects also appear discordant. Accretion rates appear to decrease rather than remain constant, far more embedded objects have been detected in cloud cores than predicted, and the spread of stellar ages in young clusters does not approach the ambipolar diffusion time (as discussed in the review by André et al. 2000).

These inconsistencies suggest to look beyond the standard theory, and we do so by seeking inspiration from the classical dynamical picture of star formation which we reconsider in the light of the recent progress in describing and understanding molecular cloud turbulence. Rather than relying on quasistatic evolution of magnetostatically supported objects, a new dynamical theory of star formation invokes supersonic interstellar turbulence to control the star formation process. We argue that this is both sufficient to explain star formation in Galactic molecular clouds, and more consistent with observations.

Our line of reasoning leads us first to a general introduction of the concept of turbulence (Section 2), which is then followed by an analysis of its decay properties (Section 3). As our arguments rely to a large degree on results from numerical models we give a brief introduction into numerical simulations of supersonic turbulence (Section 4). We then discuss how local collapse can occur in globally stable interstellar gas clouds (Section 5) and investigate the physical processes that may prevent or promote this collapse (Section 6) leading to either more clustered or more isolated modes of star formation (Section 7). We deal with the timescales of star formation (Section 8) and discuss how the inclusion of magnetic fields may influence molecular cloud fragmentation (Section 9). We also discuss specific predictions of the new theory of turbulent star formation for protostellar mass accretion rates (Section 10) and for the resulting stellar mass spectra (Section 11). We then speculate about physical scales of interstellar turbulence in our Galaxy (Section 12), and ask what sets the overall efficiency of star formation (Section 13) and what terminates the process on scales of individual star forming regions (Section 14). At the end of this review (Section 15), we summarize our results and conclude that indeed the hypothesis that stellar birth is controlled by the complex interplay between supersonic turbulence and self-gravity offers an attractive pathway towards a consistent and comprehensive theory of star formation.

2 Turbulence

At this point, we should briefly discuss the concept of turbulence, and the differences between supersonic, compressible (and magnetized) turbulence, and the more commonly studied incompressible turbulence. We mean by turbulence, in the end, nothing more than the gas flow resulting from random motions at many scales. We furthermore will use in the discussion below only the very general properties and scaling relations of turbulent flows, focusing mainly on effects of compressibility. For a more detailed discussion of the complex statistical characteristics of turbulence, we refer the reader to the book by Lesieur (1997).

Most studies of turbulence treat incompressible turbulence, characteristic of most terrestrial applications. Root-mean-square (rms) velocities are subsonic, and density remains almost constant. Dissipation of energy occurs entirely in the centers of small vortices, where the dynamical scale ℓ is shorter than the length on which viscosity acts ℓ_{visc}. Kolmogorov (1941) described a heuristic theory based on dimensional analysis that captures the basic behavior of incompressible turbulence surprisingly well, although subsequent work has refined the details substantially. He assumed turbulence driven on a large scale L, forming eddies at that scale. These eddies interact to from slightly smaller eddies, transferring some of their energy to the smaller scale. The smaller eddies in turn form even smaller ones, until energy has cascaded all the way down to the dissipation scale ℓ_{visc}.

In order to maintain a steady state, equal amounts of energy must be transferred from each scale in the cascade to the next, and eventually dissipated, at a rate

$$\dot{E} = \eta v^3 / L, \tag{1}$$

where η is a constant determined empirically. This leads to a power-law distribution of kinetic energy $E \propto v^2 \propto k^{-10/3}$, where $k = 2\pi/\ell$ is the wavenumber, and density does not enter because of the assumption of incompressibility. Most of the energy remains near the driving scale, while energy drops off steeply below ℓ_{visc}. Because of the local nature of the cascade in wavenumber space, the viscosity only determines the behavior of the energy distribution at the bottom of the cascade below ℓ_{visc}, while the driving only determines the behavior near the top of the cascade at and above L. The region in between is known as the inertial range, in which energy transfers from one scale to the next without influence from driving or viscosity. The behavior of the flow in the inertial range can be studied regardless of the actual scale at which L and ℓ_{visc} lie, so long as they are well separated. The behavior of higher order structure functions $S_p(\vec{r}) = \langle \{v(\vec{x}) - v(\vec{x} + \vec{r})\}^p \rangle$ in incompressible turbulence has been successfully modeled by She & Leveque (1994) by assuming that dissipation occurs in the filamentary centers of vortex tubes.

Gas flows in the ISM vary from this idealized picture in a number of important ways. Most significantly, they are highly compressible, with Mach numbers $\mathcal{M}$ ranging from order unity in the warm, diffuse ISM, up to as high as 50 in cold, dense molecular clouds. Furthermore, the equation of state of the gas is very soft due to radiative cooling, so that pressure $P \propto \rho^\gamma$ with the polytropic index falling in the range $0.4 < \gamma < 1.2$ (e. g. Scalo et al. 1998, Ballesteros-Paredes, Vázquez-Semadeni, &

Scalo 1999b, Spaans & Silk 2000). Supersonic flows in highly compressible gas create strong density perturbations. Early attempts to understand turbulence in the ISM (von Weizsäcker 1943, 1951, Chandrasekhar 1949) were based on insights drawn from incompressible turbulence. Although the importance of compressibility was already understood, how to incorporate it into the theory remained unclear. Furthermore, compressible turbulence is only one physical process that may cause the strong density inhomogeneities observed in the ISM. Others are thermal phase transitions (Field, Goldsmith, & Habing 1969, McKee & Ostriker 1977, Wolfire et al. 1995) or gravitational collapse (e. g. Wada & Norman 1999).

In supersonic turbulence, shock waves offer additional possibilities for dissipation. Shock waves can transfer energy between widely separated scales, removing the local nature of the turbulent cascade typical of incompressible turbulence. The spectrum may shift only slightly, however, as the Fourier transform of a step function representative of a perfect shock wave is k^{-2}, so the associated energy spectrum should be close to $\rho v^2 \propto k^{-4}$, as was indeed found by Porter, Pouquet, & Woodward (1994). However, even in hypersonic turbulence, the shock waves do not dissipate all the energy, as rotational motions continue to contain a substantial fraction of the kinetic energy, which is then dissipated in small vortices. However, Boldyrev (2002) has proposed a theory of structure function scaling based on the work of She & Leveque (1994) using the assumption that dissipation in supersonic turbulence primarily occurs in sheet-like shocks, rather than linear filaments. First comparisons to numerical models show good agreement with this model (Boldyrev, Nordlund, & Padoan 2002a), and it has been extended to the density structure functions by Boldyrev, Nordlund, & Padoan (2002b).

The driving of interstellar turbulence is neither uniform nor homogeneous. Controversy still reigns over the most important energy sources at different scales, but it appears likely that isolated and correlated supernovae dominate (Mac Low & Klessen 2003). However, it is not yet understood at what scales expanding, interacting blast waves contribute to turbulence. Analytic estimates have been made based on the radii of the blast waves at late times (Norman & Ferrara 1996), but never confirmed with numerical models (much less experiment).

Finally, interstellar gas is magnetized. Although magnetic field strengths are difficult to measure, with Zeeman line splitting being the best quantitative method, it appears that fields within an order of magnitude of equipartition with thermal pressure and turbulent motions are pervasive in the diffuse ISM, most likely maintained by a dynamo driven by the motions of the interstellar gas. A model for the distribution of energy and the scaling behavior of strongly magnetized, incompressible turbulence based on the interaction of shear Alfvén waves is given by Goldreich & Sridhar (1995, 1997) and Ng & Bhattacharjee (1996). The scaling properties of the structure functions of such turbulence was derived from the work of She & Leveque (1994) by Müller & Biskamp (2000; also see Biskamp & Müller 2000) by assuming that dissipation occurs in current sheets. A theory of very weakly compressible turbulence has been derived by using the Mach number $\mathcal{M} \ll 1$ as a perturbation parameter (Lithwick & Goldreich 2001), but no further progress has been made towards analytic models of strongly compressible magnetohydrodynamic (MHD) turbulence with $\mathcal{M} \gg 1$.

With the above in mind, we propose that stellar birth is regulated by interstellar turbulence and its interplay with gravity. Turbulence, even if strong enough to counterbalance gravity on global scales, will usually provoke local collapse on small scales. Supersonic turbulence establishes a complex network of interacting shocks, where converging flows generate regions of high density. This density enhancement can be sufficient for gravitational instability. Collapse sets in. However, the random flow that creates local density enhancements also may disperse them again. For local collapse to actually result in the formation of stars, collapse must be sufficiently fast for the region to 'decouple' from the flow, i. e. it must be shorter than the typical time interval between two successive shock passages. The shorter this interval, the less likely a contracting region is to survive. Hence, the efficiency of star formation depends strongly on the properties of the underlying turbulent velocity field, on its lengthscale and strength relative to gravitational attraction. This principle holds for star formation throughout all scales considered, ranging from small local star forming regions in the solar neighborhood up to galaxies as a whole (see Mac Low & Klessen 2003).

3   Decay and Maintenance of Supersonic Motions

We first consider the question of how to maintain the observed supersonic motions in molecular clouds. As described above, magnetohydrodynamic waves were generally thought to provide the means to prevent the dissipation of interstellar turbulence. However, numerical models have now shown that they probably do not. One-dimensional simulations of decaying, compressible, isothermal, magnetized turbulence by Gammie & Ostriker (1996) showed quick decay of kinetic energy K in the absence of driving, but found that the quantitative decay rate depended strongly on initial and boundary conditions because of the low dimensionality. Mac Low et al. (1998), Stone, Ostriker & Gammie (1998), and Padoan & Nordlund (1999) measured the decay rate in direct numerical simulations in three dimensions, using a number of different numerical methods. They uniformly found rather faster decay, with Mac Low et al. (1998) characterizing it as $K \propto t^{-\eta}$, with $0.85 < \eta < 1.1$. A resolution and algorithm study is shown in Figure 1. Magnetic fields with strengths ranging up to equipartition with the turbulent motions (ratio of thermal to magnetic pressures as low as $\beta = 0.025$) do indeed reduce η to the lower end of this range, but not below that, while unmagnetized supersonic turbulence shows values of $\eta \approx 1 - 1.1$.

Stone et al. (1998) and Mac Low (1999) showed that supersonic turbulence decays in less than a free-fall time under molecular cloud conditions, regardless of whether it is magnetized or unmagnetized. The hydrodynamical result agrees with the high-resolution, transsonic, decaying models of Porter et al. (1994). Mac Low (1999) showed that the formal dissipation time $\tau_d = K/\dot{K}$ scaled in units of the free fall time t_{ff} is

$$\tau_d/\tau_{\text{ff}} = \frac{1}{4\pi\xi}\left(\frac{32}{3}\right)^{1/2}\frac{\kappa}{\mathcal{M}_{\text{rms}}} \simeq 3.9\frac{\kappa}{\mathcal{M}_{\text{rms}}}, \qquad (2)$$

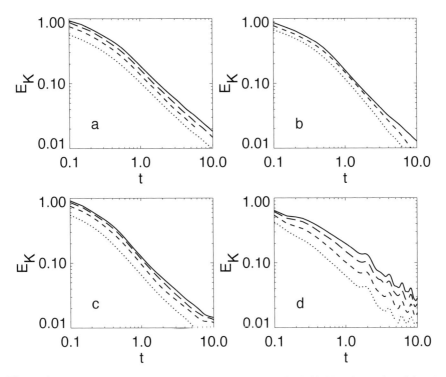

Figure 1: Decay of 3-dimensional supersonic turbulence for initial Mach number $\mathcal{M} = 5$ and isothermal equation of state. ZEUS models have 32^3 (*dotted*), 64^3 (*short dashed*), 128^3 (*long dashed*), or 256^3 (*solid*) zones. SPH models have 7000 (*dotted*), 50,000 (*short dashed*), or 350,000 (*solid*) particles. The panels show *a)* hydro runs with ZEUS, *b)* hydro runs with SPH, *c)* $A = 5$ MHD runs with ZEUS, and *d)* $A = 1$ MHD runs with ZEUS. $A = v_{\rm rms}/v_{\rm A} = v_{\rm rms}/(B^2/4\pi\rho)^{-1/2}$ is the Alfvén number (From Mac Low et al. (1998)).

where $\xi = 0.21/\pi$ is the energy-dissipation coefficient, $\mathcal{M}_{\rm rms} = v_{\rm rms}/c_{\rm s}$ is the rms Mach number of the turbulence, and κ is the ratio of the driving wavelength to the Jeans wavelength $\lambda_{\rm J}$, which is the critical scale for gravitational collapse to set in (Jeans 1902). In molecular clouds, $\mathcal{M}_{\rm rms}$ is typically observed to be of order 10 or higher. If the ratio $\kappa < 1$, as is probably required to maintain gravitational support (Léorat et al. 1990), then even strongly magnetized turbulence will decay long before the cloud collapses and not markedly retard the collapse.

Either observed supersonic motions must be continually driven, or molecular clouds must be less than a single free-fall time old. Observational evidence does suggest that clouds are a few free-fall times old, on average, though perhaps not more than two or three, so there is likely some continuing energy input into the clouds (Ballesteros-Paredes, Hartmann, & Vázquez-Semadeni 1999a, Fukui et al. 1999, Elmegreen 2000).

4 Modeling Turbulence in Self-Gravitating Gas

This leads to the question of what effects supersonic turbulence will have on self-gravitating clouds. Can turbulence alone delay gravitational collapse beyond a free-fall time? Most analytical approaches to that question are based on the assumption that the turbulent flow is close to incompressible, and are therefore not applicable to interstellar turbulence. However, some more recent models have made certain progress in recovering the velocity structure of compressible turbulence as well (Boldyrev 2002, Boldyrev, Nordlund, & Padoan 2002a,b).

Numerical models of highly compressible, self-gravitating turbulence have shown the importance of density fluctuations generated by the turbulence to understanding support against gravity. Early models were done by Bonazzola et al. (1987), who used low resolution (32×32 collocation points) calculations with a two-dimensional spectral code to support their analytical results. The hydrodynamical studies by Passot et al. (1988), Léorat et al. (1990), Vázquez-Semadeni, Passot, & Pouquet (1995) and Ballesteros-Paredes et al. (1999b), were also restricted to two dimensions, and were focused on the interstellar medium at kiloparsec scales rather than molecular clouds, although they were performed with far higher resolution (up to 800×800 points). Magnetic fields were introduced in these models by Passot, Vázquez-Semadeni, & Pouquet (1995), and extended to three dimensions with self-gravity (though at only 64^3 resolution) by Vázquez-Semadeni, Passot, & Pouquet (1996). One-dimensional computations focused on molecular clouds, including both MHD and self-gravity, were presented by Gammie & Ostriker (1996) and Balsara, Crutcher & Pouquet (2001). Ostriker, Gammie, & Stone (1999) extended their work to 2.5 dimensions more recently.

These models at low resolution, low dimension, or both, suggested several important conclusions. First, gravitational collapse, even in the presence of magnetic fields, does not generate sufficient turbulence to markedly slow continuing collapse. Second, turbulent support against gravitational collapse may act at some scales, but not others.

More recently, three-dimensional high-resolution computations by Klessen (2000), Klessen, Heitsch, & Mac Low (2000) and Heitsch, Mac Low, & Klessen (2001a) have confirmed both of these results. These authors used two different numerical methods: ZEUS-3D (Stone & Norman 1992a, b), an Eulerian MHD code; and an implementation of smoothed particle hydrodynamics (SPH; Benz 1990, Monaghan 1992), a Lagrangian hydrodynamics method using particles as an unstructured grid. Both codes were used to examine the gravitational stability of three-dimensional hydrodynamical turbulence at high resolution. The use of both Lagrangian and Eulerian methods to solve the equations of self-gravitating hydrodynamics in three dimensions (3D) allowed them to attempt to bracket reality by taking advantage of the strengths of each approach. This gives some protection against interpreting numerical artifacts as physical effects (for a detailed discussion see Klessen et al. 2000).

The computations discussed here were done on periodic cubes, with an isothermal equation of state, using up to 256^3 zones (with one model at 512^3 zones) or 80^3 SPH particles. To generate turbulent flows Gaussian velocity fluctuations are introduced with power only in a narrow interval $k-1 \leq |\vec{k}| \leq k$, where $k = L/\lambda_\mathrm{d}$ counts the number of driving wavelengths λ_d in the box (Mac Low et al. 1998). This offers a simple approximation to driving by mechanisms acting on that scale. To drive the turbulence, this fixed pattern is normalized to maintain constant kinetic energy input rate $\dot{E}_\mathrm{in} = \Delta E/\Delta t$ (Mac Low 1999). Self-gravity is turned on only after a state of dynamical equilibrium has been reached.

5 Local versus Global Collapse

First we examine the question of whether gravitational collapse can generate enough turbulence to prevent further collapse. Hydrodynamical SPH models initialized at rest with Gaussian density perturbations show fast collapse, with the first collapsed objects forming in a single free-fall time (Klessen, Burkert, & Bate 1998; Klessen & Burkert 2000, 2001). Models set up with a freely decaying turbulent velocity field behaved similarly (Klessen 2000). Further accretion of gas onto collapsed objects then occurs over the next free-fall time, defining the predicted spread of stellar ages in a freely-collapsing system. The turbulence generated by the collapse (or virialization) does not prevent further collapse contrary to what sometimes has been suggested (e. g. by Elmegreen 1993). The presence of magnetic fields does not change that conclusion (Balsara et al. 2001) as accretion down filaments aligned with magnetic field lines onto cores can occur readily. This allows high mass-to-flux ratios to be maintained even at small scales, which is necessary for supercritical collapse to continue after fragmentation occurs.

Second, we examine whether continuously driven turbulence can provide support against gravitational collapse. The models of driven, self-gravitating turbulence by Klessen et al. (2000) and Heitsch et al. (2001a) show that *local* collapse occurs even when the turbulent velocity field carries enough energy to counterbalance gravitational contraction on global scales. An example of local collapse in a globally supported cloud is given in Figure 2. A hallmark of global turbulent support is isolated, inefficient, local collapse.

Thus, highly compressible turbulence does both, it promotes as well as prevents collapse. Its net effect is to inhibit collapse globally, while at the same time promoting it locally. The resolution to this apparent paradox lies in the requirement that any substantial turbulent support must come from supersonic flows, as otherwise pressure support would be at least equally important. Supersonic flows compress the gas in shocks. In isothermal gas with density ρ the postshock gas has density $\rho' = \mathcal{M}^2\rho$, where $\mathcal{M}$ is the Mach number of the shock. The turbulent Jeans length $\lambda_\mathrm{J} \propto \rho'^{-1/2}$ in these density enhancements, so it drops by a factor of $\mathcal{M}$ in isothermal shocks making shock compressed gas clumps more susceptible to gravitational collapse. On the other hand, if we consider the system on scales exceeding the lengthscale of turbulence (i.e. in the limit of microturbulence), we can follow the classical picture that treats turbulence as an additional pressure and define an effective sound speed $c_\mathrm{s,eff}^2 = c_\mathrm{s}^2 + v_\mathrm{rms}^2/3$ (Chandrasekhar 1949). The critical mass for gravitational collapse, the Jeans mass $M_\mathrm{J} \propto \rho^{-1/2}c_\mathrm{s}^3$, then strongly increases with the turbulent rms velocity dispersion v_rms, so that for $v_\mathrm{rms} \gg c_\mathrm{s}$ turbulence ultimately does inhibit

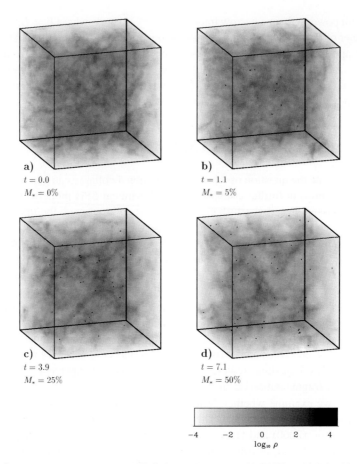

Figure 2: Density cubes for model $B2h$ from Klessen et al. (2000), which is driven at intermediate wavelengths, shown (a) at the time when gravity is turned on, (b) when the first collapsed cores are formed and have accreted $M_* = 5\%$ of the mass, (c) when the mass in dense cores is $M_* = 25\%$, and (d) when $M_* = 50\%$. Time is measured in units of the global system free-fall timescale τ_{ff}, dark dots indicate the location of the collapsed cores.

collapse on global scales. Between these two scales, there is a broad intermediate region, especially for long wavelength driving, where local collapse can occur despite global support.

Klessen et al. (2000) demonstrated that turbulent support can completely prevent collapse only when it can support not just the average density, but also these high-density shocked regions, a point that was appreciated already by Elmegreen (1993) and Vázquez-Semadeni et al. (1995). Two criteria must be fulfilled: the rms velocity must be sufficiently high for the turbulent Jeans criterion to be met in these regions, and the driving wavelength $\lambda_{\mathrm{d}} < \lambda_{\mathrm{J}}(\rho')$. If these two criteria are not fulfilled, the high-density regions collapse, although the surrounding flow remains turbulently supported. The efficiency of collapse depends on the properties of the supporting turbulence. Sufficiently strong driving on short enough scales can prevent local collapse

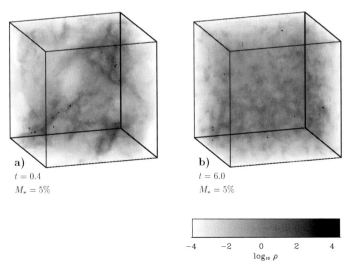

Figure 3: Density cubes for (a) a model of large-scale driven turbulence ($\mathcal{B}1h$) and (b) a model of small-scale driven turbulence ($\mathcal{B}3$) at dynamical stages where the core mass fraction is $M_* = 5\,\%$. Compare with Figure 2b. Together they show the influence of different driving wavelengths for otherwise identical physical parameters. Larger-scale driving results in collapsed cores in more organized structure, while smaller-scale driving results in more randomly distributed cores. Note the different times at which $M_* = 5\,\%$ is reached. (From Klessen et al. 2000).

for arbitrarily long periods of time, but such strong driving may be rather difficult to arrange in a real molecular cloud. Furthermore, if we assume that stellar driving sources have an effective wavelength close to their separation, then the condition that driving acts on scales smaller then the Jeans wavelength in 'typical' shock generated gas clumps requires the presence of an extraordinarily large number of stars evenly distributed throughout the cloud, with typical separation 0.1 pc in Taurus, or only 350 AU in Orion. This is not observed. Very small driving scales seem also to be at odds with the observed large-scale velocity fields at least in some molecular clouds (e. g. Ossenkopf & Mac Low 2002).

6 Promotion and Prevention of Local Collapse

The origin of local collapse can also be understood in terms of a timescale argument. Roughly speaking, the lifetime of a clump is determined by the interval between two successive passing shocks: the first creates it, while if the second is strong enough, it disrupts the clump again if it has not already collapsed (Klein, McKee & Colella 1994, Mac Low et al. 1994). If its lifetime is long enough, a Jeans unstable clump can contract to sufficiently high densities to effectively decouple from the ambient gas flow. It then becomes able to survive the encounter with further shock fronts (e. g. Krebs & Hillebrandt 1983), and continues to accrete from the surrounding gas,

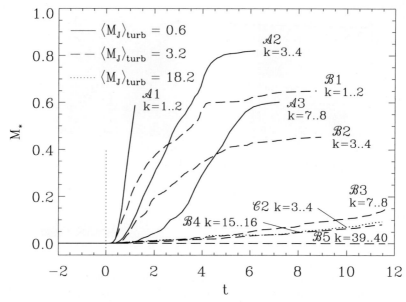

Figure 4: Fraction M_* of mass accreted in dense cores as function of time for different models of self-gravitating supersonic turbulence. The models differ by driving strength and driving wavenumber, as indicated in the figure. The mass in the box is initially unity, so the solid curves are formally unsupported, while the others are formally supported. The figure shows how the efficiency of local collapse depends on the scale and strength of turbulent driving. Time is measured in units of the global system free-fall timescale τ_ff. Only a model driven strongly at scales smaller than the Jeans wavelength λ_J in shock-compressed regions shows no collapse at all. (From Klessen et al. 2000).

forming a dense core. The weaker the passing shocks, and the greater the separation between them, the more likely that collapse will occur. Equivalently, weak driving and long typical driving wavelengths enhance collapse. The influence of the driving wavelength is more pronounced, however, because individual shocks sweep up more mass when the typical wavelength is longer, so density enhancements resulting from the interaction of shocked layers will have larger masses, and so are more likely to exceed their local Jeans limit. Turbulent driving mechanisms that act on large scales will produce large coherent structures (filaments of compressed gas with embedded dense cores) on relatively short timescales compared to small-scale driving even if the total kinetic energy in the system is the same. Examples of the density structure of long and small-wavelength driving, respectively, are given in Figure 3, which can be directly compared to Figure 2b.

Further insight of how local collapse proceeds comes from examining the mass growth rates in each model. Figure 4 shows the accretion history for three sets of models from Klessen et al. (2000). The driving strength increases from $\mathcal{A}$ over $\mathcal{B}$ to $\mathcal{C}$, but is held constant for each set of models with the effective driving wavelength λ_d being varied. All models show local collapse, except at the extreme end, when $\lambda_\mathrm{d} < \lambda_J(\rho')$ (model $\mathcal{B}5$).

The cessation of strong accretion onto cores occurs long before all gas has been accreted. This is because the time that dense cores spend in shock-compressed, high-density regions decreases with increasing driving wavenumber and increasing driving strength. In the case of long wavelength driving, cores form coherently in high-density regions associated with one or two large shock fronts that can accumulate a considerable fraction of the total mass of the system. The overall accretion rate is high and cores spend sufficient time in this environment to accrete a large fraction of the total mass in the region. Any further mass growth has to occur from chance encounters with other dense regions. In the case of short wavelength driving, the network of shocks is tightly knit. Cores form in shock generated clumps of small masses because individual shocks are not able to sweep up much matter. Furthermore, in this rapidly changing environment the time interval between the formation of clumps and their destruction is short. The period during which individual cores are located in high-density regions where they are able to accrete at high rate is short as well. So altogether, the global accretion rates are small and saturate at lower values of M_* as the driving wavelength is decreased.

7 Clustered versus Isolated Star Formation

Different star formation regions present different distributions of protostars and pre-main sequence stars. In some regions, such as the Taurus molecular cloud, stars form isolated from other stars, scattered throughout the cloud (Mizuno et al. 1995). In other regions, they form in clusters, as in L1630 in Orion (Lada 1992), or even more coherently in starburst regions such as 30 Doradus (Walborn et al. 1999; for a review see Zinnecker et al. 1993).

Numerical simulations of self-gravitating turbulent clouds demonstrate that the *length scale* and *strength* at which energy is inserted into the system determine the structure of the turbulent flow and therefore the locations at which stars are most likely to form. Large-scale driving leads to large coherent shock structures (see e. g. Figure 3a). Local collapse occurs predominantly in filaments and layers of shocked gas and is very efficient in converting gas into stars. This leads to what we can identify as 'clustered' mode of star formation: stars form in coherent aggregates and clusters. Even more so, this applies to regions of molecular gas that have become decoupled from energy input. As turbulence decays, these regions begin to contract and form dense clusters of stars with very high efficiency on about a free-fall time scale (Klessen et al. 1998, Klessen & Burkert 2000). The same holds for insufficient support, i. e. for regions where energy input is not strong enough to completely balance gravity. They too will contract to form dense stellar clusters.

The 'isolated' mode of star formation occurs in molecular cloud regions that are supported by driving sources that act on *small* scales and in an incoherent or stochastic manner. In this case, individual shock induced density fluctuations form at random locations and evolve more or less independently of each other. The resulting stellar population is widely dispersed throughout the cloud and, as collapsing clumps are exposed to frequent shock interaction, the overall star formation rate is low.

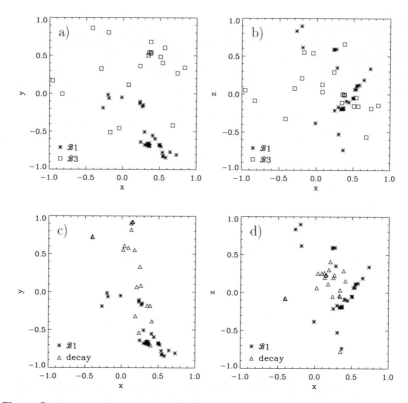

Figure 5: Comparison of collapsed core locations between two globally stable models with different driving wavelength projected into (a) the xy-plane and into (b) the xz-plane. $\mathcal{B}1$ with $k = 1 - 2$ is driven at large scales, and $\mathcal{B}3$ with $k = 7 - 8$ is driven at small ones. Plots (c) and (d) show the core locations for model $\mathcal{B}1$ now contrasted with a simulation of decaying turbulence from Klessen (2000). The snapshots are selected such that the mass accumulated in dense cores is $M_* \lesssim 20\,\%$. Note the different times needed for the different models to reach this point. For model $\mathcal{B}1$ data are taken at $t = 1.1$, for $\mathcal{B}3$ at $t = 12.3$. The simulation of freely decaying turbulence is shown at $t = 1.1$. All times are normalized to the global free-fall timescale of the system. (From Klessen et al. 2000).

These points are illustrated in Figure 5, which shows the distribution of collapsed cores in several models with strong enough turbulence to formally support against collapse. Coherent, efficient local collapse occurs in model $\mathcal{B}1$, where the turbulence is driven strongly at long wavelengths (compare with Figure 3). Incoherent, inefficient collapse occurs in model $\mathcal{B}3$, on the other hand, where turbulence is driven at small scales. Individual cores form independently of each other at random locations and random times, are widely distributed throughout the entire volume, and exhibit considerable age spread. In the decaying turbulence model, once the kinetic energy level has decreased sufficiently, all spatial modes of the system contract gravitationally, including the global ones (Klessen 2000). As in the case of large-scale shock compression, stars form more or less coevally in a limited volume with high efficiency.

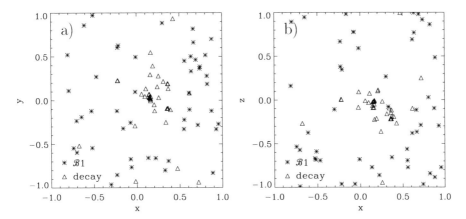

Figure 6: Core positions for model $B1$ ($k = 1 - 2$) and the decay model when the core mass fraction is $M_* \approx 65\%$, projected into (a) the xy-plane and (b) the xz-plane (compare with Figure 5c and d). For $B1$ the time is $t = 8.7$ and for decay model $t = 2.1$. Whereas the cluster in $B1$ is completely dissolved and the stars are widely dispersed throughout the computational volume, the cluster in the decay simulation remains bound. (From Klessen et al. 2000).

Despite the fact that both turbulence driven on large scales and freely decaying turbulence lead to star formation in aggregates and clusters, Figure 6 suggests a possible way to distinguish between them. Decaying turbulence typically leads to the formation of a bound stellar cluster, while aggregates associated with large-scale, coherent, shock fronts often have higher velocity dispersions that result in their complete dispersal. Note, however, that at the late stages of dynamical evolution shown in Figure 6, the model becomes less appropriate, as feedback from newly formed stars is not included. Ionization and outflows from the stars formed first will likely retard or even prevent the accretion of the remaining gas onto protostars, possibly preventing a bound cluster from forming even in the case of freely decaying turbulence.

The control of star formation by supersonic turbulence gives rise to a continuous but articulated picture. There may not be physically distinct modes of star formation, but qualitatively different behaviors do appear over the range of possible turbulent flows. The apparent dichotomy between a clustered mode of star formation and an isolated one, as discussed by Lada (1992) for L1630 and Strom, Strom, & Merrill (1993) for L1641, disappears, if a different balance between turbulent strength and gravity holds at the relevant length scales in these different clouds.

Turbulent flows tend to have hierarchical structure (Lesieur 1997) which may explain the hierarchical distribution of stars in star forming regions shown by statistical studies of the distribution of neighboring stars in young stellar clusters (e. g. in Taurus, see Larson 1995). Hierarchical clustering seems to be a common feature of all star forming regions (e. g. Efremov & Elmegreen 1998). It is a natural outcome of turbulent fragmentation.

8 Timescales of Star Formation

Turbulent control of star formation predicts that stellar clusters form predominantly in regions that are insufficiently supported by turbulence or where only large-scale driving is active. In the absence of driving, molecular cloud turbulence decays more quickly than the free-fall timescale τ_{ff} (Eq. 2), so dense stellar clusters will form on the free-fall timescale. Even in the presence of support from large-scale driving, substantial collapse still occurs within a few free-fall timescales, see Figure 7a. If the dense cores followed in these models continue to collapse on a short timescale to build up stellar objects in their centers, then this directly implies the star formation timescale. Therefore the age distribution will be roughly τ_{ff} for stellar clusters that form coherently with high star formation efficiency. When scaled to low densities, say $n(\mathrm{H}_2) \approx 10^2 \, \mathrm{cm}^{-3}$ and $T \approx 10$ K, the global free-fall timescale in the models is $\tau_{\mathrm{ff}} = 3.3 \times 10^6$ years. If star forming clouds such as Taurus indeed have ages of order τ_{ff}, as suggested by Ballesteros-Paredes et al. (1999), then the long star formation time computed here is quite consistent with the very low star formation efficiencies seen in Taurus (e. g. Leisawitz et al. 1989, Palla & Stahler 2000, Hartmann 2001), as the cloud simply has not had time to form many stars. In the case of high-density regions, $n(\mathrm{H}_2) \approx 10^5 \, \mathrm{cm}^{-3}$ and $T \approx 10$ K, the dynamical evolution proceeds much faster and the corresponding free-fall times drops to $\tau_{\mathrm{ff}} = 10^5$ years. These values are indeed supported by observational data such as the formation timescale of the Orion Trapezium cluster. It is inferred to stem from gas of density $n(\mathrm{H}_2) \lesssim 10^5 \, \mathrm{cm}^{-3}$, and is estimated to be less than 10^6 years old (Hillenbrand & Hartmann 1998). The age spread in the models increases with increasing driving wavenumber k and increasing $\langle M_\mathrm{J}\rangle_{\mathrm{turb}}$, as shown in Figure 7. Long periods of core formation for globally supported clouds appear consistent with the low efficiencies of star-formation in regions of isolated star formation, such as Taurus, even if they are rather young objects with ages of order τ_{ff}.

9 Effects of Magnetic Fields

So far, we concentrated on the effects of purely hydrodynamic turbulence. How does the picture discussed here change, if we consider the presence of magnetic fields? Magnetic fields may alter the dynamical state of a molecular cloud sufficiently to prevent gravitationally unstable regions from collapsing (McKee 1999). They have been hypothesized to support molecular clouds either magnetostatically or dynamically through MHD waves.

Mouschovias & Spitzer (1976) derived an expression for the critical mass-to-flux ratio in the center of a cloud for magnetostatic support. Assuming ideal MHD, a self-gravitating cloud of mass M permeated by a uniform flux Φ is stable if the mass-to-flux ratio

$$\frac{M}{\Phi} < \left(\frac{M}{\Phi}\right)_{\mathrm{cr}} \tag{3}$$

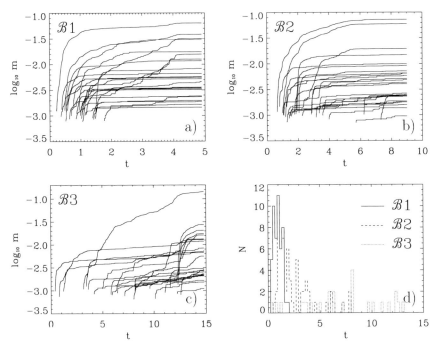

Figure 7: Masses of individual protostars as function of time in SPH models (a) $\mathcal{B}1$ driven at large scales with $k = 1-2$ driving, (b) $\mathcal{B}2$ with $k = 3-4$ driving, i. e. at intermediate scales, and (c) $\mathcal{B}3$ with $k = 7-8$ small-scale driving. The curves represent the formation and accretion histories of individual protostars. For the sake of clarity, only every other core is shown in (a) and (b), whereas in (c) the evolution of every single core is plotted. Time is given in units of the global free-fall time $\tau_{\rm ff}$. Note the different timescale in each plot. In the depicted time interval models $\mathcal{B}1$ and $\mathcal{B}2$ reach a core mass fraction $M_* = 70\%$, and both form roughly 50 cores. Model $\mathcal{B}3$ reaches $M_* = 35\%$ and forms only 25 cores. Figure (d) compares the distributions of formation times. The age spread increases with decreasing driving scale showing that clustered core formation should lead to a coeval stellar population, whereas a distributed stellar population should exhibit considerable age spread. (From Klessen et al. 2000).

with $(M/\Phi)_{\rm cr} = c_\Phi G^{-1/2}$. The exact value depends on the geometry and the field and density distribution of the cloud. A cloud is termed *subcritical* if it is magnetostatically stable and *supercritical* if it is not. Mouschovias & Spitzer (1976) determined that $c_\Phi = 0.13$ for spherical clouds. Assuming a constant mass-to-flux ratio in a cylindrical region results in $c_\Phi = 1/(2\pi) \approx 0.16$ (Nakano & Nakamura 1978). Without any other mechanism of support such as turbulence acting along the field lines, a magnetostatically supported cloud will collapse to a sheet which then will be supported against further collapse. Fiege & Pudritz (2000) discussed a sophisticated version of this magnetostatic support mechanism, in which poloidal and toroidal fields aligned in the right configuration could prevent a cloud filament from fragmenting and collapsing.

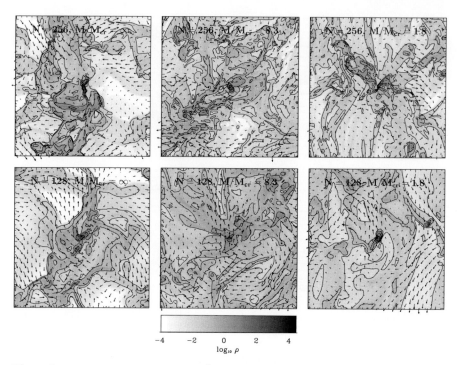

Figure 8: Two-dimensional slices of 256^3 models from Heitsch et al. (2001a) driven at large scales with wavenumbers $k = 1 - 2$ hard enough that the mass in the box represents only 1/15 $\langle M_{\rm J}\rangle_{\rm turb}$, and with initially vertical magnetic fields strong enough to give critical mass fractions as shown. The slices are taken at the location of the zone with the highest density at the time when 10 % of the total mass has been accreted onto dense cores. The plot is centered on this zone. Arrows denote velocities in the plane. The length of the largest arrows corresponds to a velocity of $v \sim 20c_s$. The density greyscale is given in the colorbar. As fields become stronger, they influence the flow more, producing anisotropic structure. (From Heitsch et al. 2001a).

Investigation of the second alternative, support by MHD waves, concentrates mostly on the effect of Alfvén waves, as they (1) are not as subject to damping as magnetosonic waves and (2) can exert a force along the mean field, as shown by Dewar (1970) and Shu et al. (1987). This is because Alfvén waves are transverse waves, so they cause perturbations $\delta\vec{B}$ perpendicular to the mean magnetic field $\vec{B}$. McKee & Zweibel (1995) argue that Alfvén waves can even lead to an isotropic pressure, assuming that the waves are neither damped nor driven. However, in order to support a region against self-gravity, the waves would have to propagate outwardly, rather than inwardly, which would only further compress the cloud. Thus, as Shu et al. (1987) comment, this mechanism requires a negative radial gradient in wave sources in the cloud.

It can be demonstrated (e. g. Heitsch et al. 2001a) that supersonic turbulence does not cause a magnetostatically supported region to collapse, and vice versa, that in the absence of magnetostatic support, MHD waves cannot completely prevent collapse,

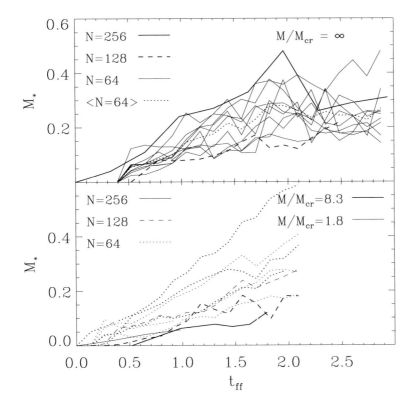

Figure 9: *Upper panel:* Core-mass accretion rates for 10 different low-resolution models ($N = 64^3$ cells) of purely hydrodynamic turbulence with equal parameter set but different realizations of the turbulent velocity field. The thick line shows a "mean accretion rate", calculated from averaging over the sample. For comparison, higher-resolution runs with identical parameters but $N = 128^3$ and $N = 256^3$ are shown as well. The latter one can be regarded as an envelope for the low resolution models. *Lower panel:* Mass accretion rates for various models with different magnetic field strength and resolution. Common to all models is the occurrence of local collapse and star formation regardless of the detailed choice of parameters, as long as the system is magnetostatically supercritically (Heitsch et al. 2001a).

although they can retard it to some degree. In the case of a subcritical region with $M < M_{cr}$ sheets of high density gas form perpendicular to the field lines. Turbulence can shift the sheets along the field lines without changing the mass-to-flux ratio, but collapse does not occur, because the shock waves cannot sweep gas across field lines and the entire region is initially supported magnetostatically.

A supercritical cloud with $M > M_{cr}$ could only be stabilized by MHD wave pressure. This is insufficient to completely prevent gravitational collapse, as shown in Figure 8. The effect of the magnetic field on the morphology of the cloud is week, and collapse occurs in all models of unmagnetized and magnetized turbulence regardless of the numerical resolution and magnetic field strength as long as the system is magnetically supercritical. This is shown more quantitatively in Figure 9.

Effects of numerical resolution make themselves felt in different ways in hydrodynamical and MHD models. In the hydrodynamical case, higher resolution results in thinner shocks and thus higher peak densities. These higher density peaks form cores with deeper potential wells that accrete more mass and are more stable against disruption. Higher resolution in the MHD models, on the other hand, better resolves short-wavelength MHD waves, which apparently can delay collapse, but not prevent it. This result extends to models with 512^3 zones (Heitsch et al. 2001b).

10 Mass Growth of Protostellar Cores

Supersonic turbulence is able to produce star forming regions that vary enormously in scale. The most likely outcome of turbulent molecular cloud fragmentation in the Milky Way are stellar aggregates or clusters (Adams & Myers 2001). The number density of protostars and protostellar cores in the extreme cases the can be high enough for mutual dynamical interaction to become important. This has important consequences for the mass growth history of individual stars and the subsequent dynamical evolution of the nascent stellar cluster, because this introduces a further degree of stochasticity to the star formation process in addition to the statistical chaos associated with turbulence and turbulent fragmentation in the first place.

Klessen (2001a) considers the formation of a nascent star cluster for the case where turbulence is decayed and has left behind random Gaussian fluctuations in the density structure. As the system contracts gravitationally, a dense cluster of protostellar cores builds up on a timescale of about two to three free-fall times. The protostellar accretion rates in this environment are strongly time variable, as illustrated in Figure 10, which is a direct result of the mutual dynamical interaction and competition between protostellar cores. While gas clumps collapse to build up protostars, they may merge as they follow the flow pattern towards the cluster potential minimum. The timescales for both processes are comparable. The density and velocity structure of merged gas clumps generally differs significantly from their progenitor clumps, and the predictions for isolated cores are no longer valid. More importantly, these new larger clumps contain multiple protostars, which subsequently compete with each other for the accretion from a common gas reservoir. The most massive protostar in a clump is hereby able to accrete more matter than its competitors (also Bonnell et al. 1997, Klessen & Burkert 2000, Bonnell et al. 2001). Its accretion rate is enhanced through the clump merger, whereas the accretion rate of low-mass cores typically decreases. Temporary accretion peaks in the wake of clump mergers are visible in abundance in Figure 10. Furthermore, the small aggregates of cores that build up are dynamically unstable and low-mass cores may be ejected. As they leave the high-density environment, accretion terminates and their final mass is reached.

The typical density profiles of gas clumps that give birth to protostars exhibit a flat inner core, followed by a density fall-off $\rho \propto r^{-2}$, and are truncated at some finite radius (see Figure 13 in Klessen & Burkert 2000), which in the dense centers of clusters often is due to tidal interaction with neighboring cores. As result, a short-lived initial phase of strong accretion occurs when the flat inner part of the pre-stellar clump collapses. This corresponds to the class 0 phase of protostellar evolution

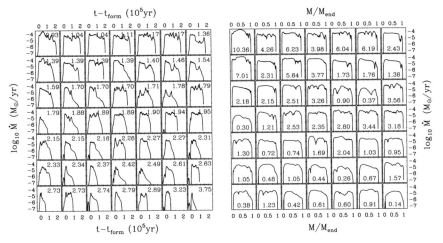

Figure 10: Time-varying protostellar mass accretion rates in a dense cluster environment. The left panel shows accretion rate $\dot{M}$ versus time after formation $t - t_{\mathrm{form}}$ for 49 randomly selected protostellar cores in a numerical model of molecular cloud fragmentation from Klessen & Burkert (2000). Formation time t_{form} is defined by the first occurance of a hydrostatic object in the interior of a collapsing gas clump. To link individual accretion histories to the overall cluster evolution, t_{form} is indicated in the upper right corner of each plot and measures the elapsed time since the start of the simulation. The free-fall timescale of the considered molecular region is $\tau_{\mathrm{ff}} \approx 10^5$ years. The right panel plots for the same cores $\dot{M}$ as function of the accreted mass M with respect to the final mass M_{end}, which is indicated in the center of each plot. Note that the mass range spans two orders of magnitude. (From Klessen 2001a).

(André et al. 2000). If these cores were to remain isolated and unperturbed, the mass growth rate would gradually decline in time as the outer envelope accretes onto the center. This is the class I phase. Once the truncation radius is reached, accretion fades and the object enters the class II phase. This behavior is expected from analytical models (e. g. Henriksen et al. 1997) and agrees with other numerical studies (e. g. Foster & Chevalier 1993). However, collapse does not start from rest for the density fluctuations considered here, and the accretion rates exceed the theoretically predicted values even for the most isolated objects in the simulation.

The most massive protostars begin to form first and continue to accrete at high rate throughout the entire cluster evolution. As the most massive gas clumps tend to have the largest density contrast, they are the first to collapse and constitute the center of the nascent cluster. These protostars are fed at high rate and gain mass very quickly. As their parental clumps merge with others, more gas is fed into their 'sphere of influence'. They are able to maintain or even increase the accretion rate when competing with lower-mass objects (e. g. core 1 and 8 in Figure 10). Low-mass stars, on average, tend to form somewhat later in the dynamical evolution of the system (as indicated by the absolute formation times in Figure 10), and typically have only short periods of high accretion.

As high-mass stars are associated with large core masses, while low-mass stars come from low-mass cores, the stellar population in clusters is predicted to be mass segregated right from the beginning. High-mass stars form in the center, lower-mass stars tend to form towards the cluster outskirts. This is in agreement with recent observational findings for the cluster NGC 330 in the Small Magellanic Cloud (Sirianni et al. 2002). Dynamical effects during the embedded phase of star cluster evolution will enhance this initial segregation even further.

11 Mass Spectra from Turbulent Fragmentation

As discussed before, a full understanding of turbulent molecular cloud fragmentation should in principle allow for a prediction of the distribution of stellar masses (e. g. Larson 1981, Fleck 1982, Elmegreen 1993, Padoan 1995, Padoan & Nordlund). However, a complete theory of compressible interstellar turbulence is still out of reach, and we have to resort to numerical modeling instead to make some progress. To illustrate this point we examine the mass spectra of gas clumps and collapsed cores from models of self-gravitating, isothermal, supersonic turbulence driven with different wavelengths (Klessen 2001b). In the absence of magnetic fields and more accurate equations of state, these models can only be illustrative, not definitive, but nevertheless they offer insight into the processes acting to form the initial stellar mass function (IMF; for a review see Kroupa 2002). Figure 11 plots for four different models the mass distribution of gas clumps, of the subset of gravitationally unstable clumps, and of collapsed cores, at four different evolutionary phases. In the initial phase, before local collapse begins to occur, the clump mass spectrum is not well described by a single power law. During subsequent evolution, as clumps merge and grow bigger, the mass spectrum extends towards larger masses, approaching a power law with slope $\alpha \approx -1.5$. Local collapse sets in and results in the formation of dense cores most quickly in the freely collapsing model. The influence of gravity on the clump mass distribution weakens when turbulence dominates over gravitational contraction on the global scale, as in the other three models. The more the turbulent energy dominates over gravity, the more the spectrum resembles the initial case of pure hydrodynamic turbulence. This suggests that the clump mass spectrum in molecular clouds will be shallower in regions where gravity dominates over turbulent energy. This may explain the observed range of slopes for the clump mass spectrum in different molecular cloud regions (e. g. Kramer et al. 1998).

Like the distribution of Jeans-unstable clumps, the mass spectrum of dense protostellar cores resembles a log-normal in the model without turbulent support and in the one with long-wavelength turbulent driving, with a peak at roughly the average thermal Jeans mass $\langle m_{\rm J} \rangle$ of the system. However, models supported at shorter wavelength have mass spectra much flatter than observed, suggesting that clump merging and competitive accretion are important factors leading to a log-normal mass spectrum. The protostellar clusters discussed here only contain between 50 and 100 cores. This allows for comparison with the IMF only around the characteristic mass scale, typically about 1 $M_\odot$, since the numbers are too small to study the very low- and high-mass end of the distribution. Focusing on low-mass star formation,

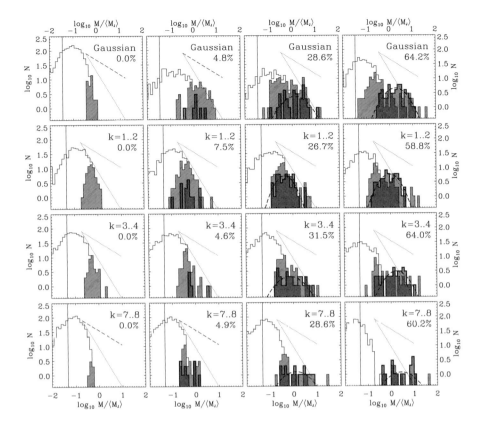

Figure 11: Mass spectra of dense collapsed cores (hatched thick-lined histograms), of gas clumps (thin lines), and of the subset of Jeans unstable clumps (thin lines, hatched distribution) for four turbulence models. Gaussian density perturbations without turbulence leads to global collapse, while three models with turbulence are nominally supported on driven on long, intermediate or short scales, respectively, as indicated by the driving wavenumbers k. Masses are binned logarithmically and normalized to the average Jeans mass $\langle m_J \rangle$. The left column gives the initial state of the system, the second column shows the mass spectra when $m_* \approx 5\%$ of the mass is accreted onto dense cores, the third column shows $m_* \approx 30\%$, and the last one $m_* \approx 60\%$. For comparison with power-law spectra ($dN/dm \propto m^\nu$), a slope $\alpha = -1.5$ typical for the observed clump mass distribution, and the Salpeter (1955) slope $\alpha = -2.33$ for the IMF, are indicated by the dotted lines. The vertical line shows the resolution limit of the numerical model. In columns 3 and 4, the long dashed curve shows the best log-normal fit. (From Klessen 2001b).

however, Bate, Bonnell, & Bromm (2002) demonstrate that brown dwarfs are a natural and frequent outcome of turbulent fragmentation. In this model, brown dwarfs form when dense molecular gas fragments into unstable multiple systems that eject their smallest members from the dense gas before they have been able to accrete to stellar masses. Numerical models with sufficient dynamic range to treat the full range of stellar masses remain yet to be done.

12 Scales of Interstellar Turbulence

Turbulence has self-similar properties only within a certain range of scales. The upper end typically is associated with the global extent of the system or with the scale at which energy is inserted. The lower scale is the energy dissipation scale of the system where turbulent kinetic energy is converted into random motion, into heat. In purely hydrodynamic systems this is the scale where molecular viscosity becomes important.

In interstellar clouds this situation may be different. It was first shown by Zweibel & Josafatsson (1983) that ambipolar diffusion would be the most important dissipation mechanism in typical molecular clouds with very low ionization fractions $x = \rho_i/\rho_n$, where ρ_i is the density of ions, ρ_n is the density of neutrals, and $\rho = \rho_i + \rho_n$. The ambipolar diffusion strength is defined as

$$\lambda_{\mathrm{AD}} = v_{\mathrm{A}}^2/\nu_{ni}, \tag{4}$$

where $v_{\mathrm{A}}^2 = B^2/4\pi\rho_n$ approximates the effective Alfvén speed for the coupled neutrals and ions if $\rho_n \gg \rho_i$, and $\nu_{ni} = \gamma\rho_i$ is the rate at which each neutral is hit by ions. The coupling constant depends on the cross-section for ion-neutral interaction, and for typical molecular cloud conditions has a value of $\gamma \approx 9.2 \times 10^{13}$ cm^3 s^{-1}g^{-1} (e.g. Smith & Mac Low 1997). Zweibel & Brandenburg (1997) define an ambipolar diffusion Reynolds number as

$$R_{\mathrm{AD}} = \tilde{L}\tilde{V}/\lambda_{\mathrm{AD}} = \mathcal{M}_{\mathrm{A}}\tilde{L}\nu_{ni}/v_{\mathrm{A}}, \tag{5}$$

which must fall below unity for ambipolar diffusion to be important, where $\tilde{L}$ and $\tilde{V}$ are the characteristic length and velocity scales, and $\mathcal{M}_{\mathrm{A}} = \tilde{V}/v_{\mathrm{A}}$ is the characteristic Alfvén Mach number. In our situation we again can take the rms velocity as typical value for $\tilde{V}$. By setting $R_{\mathrm{AD}} = 1$, we can derive a critical lengthscale below which ambipolar diffusion is important

$$\tilde{L}_{cr} = \frac{v_{\mathrm{A}}}{\mathcal{M}_{\mathrm{A}}\nu_{ni}} \approx (0.041 \text{ pc}) \left(\frac{B}{10\,\mu\mathrm{G}}\right) \mathcal{M}_{\mathrm{A}}^{-1} \left(\frac{x}{10^{-6}}\right)^{-1} \left(\frac{n_n}{10^3 \text{ cm}^{-3}}\right)^{-3/2}, \tag{6}$$

with the magnetic field strength B, the ionization fraction x, the neutral number density n_n, and where we have taken $\rho_n = \mu n_n$, with $\mu = 2.36\,m_{\mathrm{H}}$. This is consistent with typical sizes of protostellar cores (e.g. Bacmann et al. 2000), if we assume that ionization and magnetic field both depend on the density of the region and follow the empirical laws $n_i = 3 \times 10^{-3}$ cm^{-3} $(n_n/10^5$ cm$^{-3})^{1/2}$ (e.g. Mouschovias 1991b) and $B \approx 30\,\mu\mathrm{G}\,(n_n/10^3$ cm$^{-3})^{1/2}$ (e.g. Crutcher 1999).

On large scales, an maximum upper limit to the turbulent cascade in the Milky Way is given by the extent and thickness of the Galactic disk. This is indeed the true upper scale, if molecular clouds are created by converging large-scale flows generated by the collective influence of recurring supernovae explosions. For individual molecular clouds this means, that turbulent energy is fed in at scales beyond the size of the considered cloud. The bulk of its turbulent energy content is then generated in the event of cloud assembly, which then dissipates rapidly resulting in short cloud life times. The same compressional motions that created the cloud in the first place, however, may also act as continuing source of kinetic energy during some initial period (e. g. Walder & Folini 2000, Hartmann, Ballesteros-Paredes, & Bergin 2001), thus extending the overall cloud lifetime to a few crossing times. This energy cascades down to supply turbulence on smaller scales within the cloud. This picture of molecular cloud turbulence being driven by large-scale, external sources is strongly support by analysis of velocity structure which is always is dominated by the large-scale modes in all clouds observed (Ossenkopf & Mac Low 2002).

13 Efficiency of Star Formation

The *global* star formation efficiency in normal molecular clouds is usually estimated to be of the order of a few per cent. Their life times may be on the order of a few crossing times, i.e. a few 10^6 years (Ballesteros-Paredes et al. 1999a, Fukui et al. 1999, Elmegreen 2000). In this case nearly all models of interstellar turbulence discussed above are consistent with the observed overall efficiencies. If molecular clouds survive for several tens of their free-fall time τ_{ff} (i.e. a few 10^7 years as proposed by Blitz & Shu 1980), turbulence models are more strongly constrained. However, even in this case models with parameters reasonable for Galactic molecular clouds can maintain global efficiencies below $M_* = 5\%$ for $10\,\tau_{ff}$ (Klessen et al. 2000). Furthermore, it needs to be noted that the *local* star formation efficiency in molecular clouds can reach very high values. For example, the Trapezium star cluster in Orion is likely to have formed with an efficiency of about 50 % (Hillenbrand & Hartmann 1998), compared to a value of 5 % proposed for Taurus-Aurigae.

14 Termination of Local Star Formation

It remains quite unclear what terminates stellar birth on scales of individual star forming regions, and even whether these processes are the primary factor determining the overall efficiency of star formation in a molecular cloud. Three main possibilities exist. First, feedback from the stars themselves in the form of ionizing radiation and stellar outflows may heat and stir surrounding gas up sufficiently to prevent further collapse and accretion. Second, accretion might peter out either when all the high density, gravitationally unstable gas in the region has been accreted in individual stars, or after a more dynamical period of competitive accretion, leaving any remaining gas to be dispersed by the background turbulent flow. Third, background flows may sweep through, destroying the cloud, perhaps in the same way that it was created. Most likely the astrophysical truth lies in some combination of all three possibilities.

If a stellar cluster formed in a molecular cloud contains OB stars, then the radiation field and stellar wind from these high-mass stars strongly influence the surrounding cloud material. The UV flux ionizes gas out beyond the local star forming region. Ionization heats the gas, raising its Jeans mass, and possibly preventing further protostellar mass growth or new star formation. The termination of accretion by stellar feedback has been suggested at least since the calculations of ionization by Oort & Spitzer (1955). Whitworth (1979) and Yorke et al. (1989) computed the destructive effects of individual blister H II regions on molecular clouds, while in series of papers, Franco et al. (1994), Rodriguez-Gaspar et al. (1995), and Diaz-Miller et al. (1998) concluded that indeed the ionization from massive stars may limit the overall star forming capacity of molecular clouds to about 5 %. Matzner (2002) analytically modeled the effects of ionization on molecular clouds, concluding as well that turbulence driven by HII regions could support and eventually destroy molecular clouds. The key question facing these models is whether HII region expansion couples efficiently to clumpy, inhomogeneous molecular clouds, a question probably best addressed with numerical simulations.

Bipolar outflows are a different manifestation of protostellar feedback, and may also strongly modify the properties of star forming regions (Norman & Silk 1980, Adams & Fatuzzo 1996). Recently Matzner & McKee (2000) modeled the ability of bipolar outflows to terminate low-mass star formation, finding that they can limit star formation efficiencies to 30–50 %, although they are ineffective in more massive regions. How important these processes are compared to simple exhaustion of available reservoirs of dense gas (Klessen et al. 2000) remains an important question.

The models relying on exhaustion of the reservoir of dense gas argue that only dense gas will actually collapse, and that only a small fraction of the total available gas reaches sufficiently high densities, due to cooling (Schaye 2002), gravitational collapse and turbulent triggering (Elmegreen 2002), or both (Wada, Meurer, & Norman 2002). This of course pushes the question of local star formation efficiency up to larger scales, which may indeed be the correct place to ask it.

Other models focus on competitive accretion in local star formation, showing that the distribution of masses in a single group or cluster can be well explained by assuming that star formation is fairly efficient in the dense core, but that stars that randomly start out slightly heavier tend to fall towards the center of the core and accrete disproportionately more gas (Bonnell et al. 1997, 2001). These models have recently been called into question by the observation that the stars in lower density young groups in Serpens simply have not had the time to engage in competitive accretion, but still have a normal IMF (Olmi & Testi 2002).

Finally, star formation in dense clouds created by turbulent flows may be terminated by the same flows that created them. Ballesteros-Paredes et al. (1999a) suggested that the coordination of star formation over large molecular clouds, and the lack of post-T Tauri stars with ages greater than about 10 Myr tightly associated with those clouds, could be explained by their formation in a larger-scale turbulent flow. Hartmann et al. (2001) make the detailed argument that these flows may disrupt the clouds after a relatively short time, limiting their star formation efficiency that way. It can be argued that field supernovae are the most likely driver for this background turbulence in spiral galaxies like the Milky Way (Mac Low & Klessen 2003).

15 Summary: The Control of Star Formation by Supersonic Turbulence

In this review we have proposed that star formation is regulated by interstellar turbulence and its interplay with gravity. We have discussed that this new approach can explain the same observations successfully described by the so called "standard theory", while also addressing (and resolving!) its inconsistencies with other observed properties of Galactic star forming regions.

The key point to this new understanding of star formation in Galactic molecular clouds lies in the properties of interstellar turbulence. Turbulence is observed in the interstellar medium almost ubiquitously and is typically supersonic as well as super-Alfvénic. It is energetic enough to counterbalance gravity on global scales, but at the same time it may provoke local collapse on small scales. This apparent paradox can be resolved when considering that supersonic turbulence establishes a complex network of interacting shocks, where converging flows generate regions of high density. This density enhancement can be sufficiently large for gravitational instability to set in. The same random flow that creates density enhancements, however, may disperse them again. For local collapse to result in stellar birth, it must progress sufficiently fast for the region to 'decouple' from the flow. Typical collapse timescales are hereby of the same order as the lifetimes of shock-generated density fluctuations in the turbulent gas. This makes the outcome highly unpredictable. As stars are born through a sequence of stochastic events, any theory of star formation is in essence a statistical one with quantitative predictions only possible for an ensemble of stars.

In the new picture, the efficiency of protostellar core formation, the growth rates and final masses of the protostars, essentially all properties of nascent star clusters depend on the intricate interplay between gravity on the one hand side and the turbulent velocity field in the cloud on the other. The star formation rate is regulated not just at the scale of individual star-forming cores through ambipolar diffusion balancing magnetostatic support, but rather at all scales (Elmegreen 2002), via the dynamical processes that determine whether regions of gas become unstable to prompt gravitational collapse. The presence of magnetic fields does not alter that picture significantly, as long as they are too weak for magnetostatic support, which is indicated by observations (Crutcher 1999, Bourke et al. 2001). In particular, magnetic fields cannot prevent the decay of interstellar turbulence, which in turn needs to be continuously driven or else stars form quickly and with high efficiency

Inefficient, isolated star formation will occur in regions which are supported by turbulence carrying most of its energy on very small scales. This typically requires an unrealistically large number of driving sources and appears at odds with the measured velocity structure in molecular clouds which in almost all cases is dominated by large-scale modes. The dominant pathway to star formation therefore seems to involve cloud regions large enough to give birth to aggregates or clusters of stars. This is backed up by careful stellar population analysis indicating that most stars in the Milky Way formed in open clusters with a few hundred member stars (Adams & Myers 2001).

Clusters of stars build up in molecular cloud regions where self-gravity overwhelms turbulence, either because such regions are compressed by a large-scale shock, or because interstellar turbulence is not replenished and decays on short timescales. Then, many gas clumps become gravitationally unstable synchronously and start to collapse. If the number density is high, collapsing gas clumps may merge to produce new clumps which now contain multiple protostars. Mutual dynamical interactions become common, with close encounters drastically altering the protostellar trajectories, thus changing the mass accretion rates. This has important consequences for the IMF. Already in their infancy, i.e. already in the deeply embedded phase, very dense stellar clusters are expected to be strongly influenced by collisional dynamics.

Acknowledgments

This review would not have been possible without long-term collaboration and exchange of ideas with M.-M. Mac Low. Special thanks also to P. Bodenheimer and D. Lin for many vivid scientific discussions and for their warm hospitality at UC Santa Cruz; and thanks to J. Ballesteros-Paredes, F. Heitsch, P. Kroupa, E. Vázquez-Semadeni, and H. Zinnecker.

I want to express my gratitudes to the members of the Astronomische Gesellschaft for awarding of the Ludwig Biermann Preis to me, in particular, I want to thank G. Hensler and A. Burkert. I furthermore acknowledge support by the Emmy Noether Program of the Deutsche Forschungsgemeinschaft (DFG: KL1358/1).

References

Adams, F. C., and M. Fatuzzo 1996, ApJ 464, 256

Adams, F. C., and P. C. Myers 2001, ApJ 553, 744

André, P., D. Ward-Thompson, and M. Barsony 2000, in *Protostars and Planets IV*, edited by V. Mannings, A. P. Boss, and S. S. Russell (University of Arizona Press, Tucson), p. 59

Bacmann, A., P. André, J.-.L Puget, A. Abergel, S. Bontemps, and D. Ward-Thompson, 2000, A&A 361, 555

Ballesteros-Paredes, J., L. Hartmann, and E. Vázquez-Semadeni, 1999a, ApJ 527, 285

Ballesteros-Paredes, J., E. Vázquez-Semadeni, and J. Scalo, 1999b, ApJ 515, 286

Balsara, D. S., R. M. Crutcher, and A. Pouquet, 2001, ApJ 557, 451

Bate, M. R., I. A. Bonnell, and V. Bromm, 2002, MNRAS 332, L65

Benz, W., 1990, in *The Numerical Modelling of Nonlinear Stellar Pulsations*, edited by J. R. Buchler (Kluwer, Dordrecht), 269

Bergin, E. A., and W. D. Langer, 1997, ApJ 486, 316

Biskamp, D., and W.-C. Müller, 2000, Phys. Plasmas 7, 4889

Blitz, L., and F. H. Shu, 1980, ApJ 238, 148

Boldyrev, S., 2002, ApJ 569, 841

Boldyrev, S., Å. Nordlund, and P. Padoan, 2002a, ApJ 573, 678

Boldyrev, S., Å. Nordlund, and P. Padoan, 2002b, Phys. Rev. Lett., submitted (astro-ph/0203452)

Bonazzola, S., E. Falgarone, J. Heyvaerts, M. Perault, and J. L. Puget, 1987, A&A 172, 293

Bonnell, I. A., M. R. Bate, C. J. Clarke, and J. E. Pringle, 1997, MNRAS 285, 201

Bonnell, I. A., M. R. Bate, C. J. Clarke, and J. E. Pringle, 2001, MNRAS 323, 785

Bourke, T. L., P. C. Myers, G. Robinson, and A. R. Hyland, 2001, ApJ 554, 916

Chandrasekhar, S., 1949, ApJ 110, 329

Crutcher, R. M., 1999, ApJ 520, 706

Desch, S. J. and T. C. Mouschovias, 2001, ApJ 550, 314

Dewar, R. L., 1970, Phys. Fluids 13, 2710

Diaz-Miller, R. I., J. Franco, and S. N. Shore, 1998, ApJ 501, 192.

Efremov, Y. N., and B. G. Elmegreen, 1998, MNRAS 299, 588

Elmegreen, B. G., 1991, in *NATO ASIC Proc. 342: The Physics of Star Formation and Early Stellar Evolution*, edited by C. J. Lada and N. D. Kylafis (Kluwer Academic Publishers), p. 35

Elmegreen, B. G., 1993, ApJ 419, L29

Elmegreen, B. G., 2000, MNRAS 311, L5

Elmegreen, B. G., 2002, ApJ 577, 206

Fiege, J. D., and R. E. Pudritz, 2000a, MNRAS 311, 85

Field, G. B., D. W. Goldsmith, and H. J. Habing, 1969, ApJ 155, L49

Fleck, R. C., 1982, MNRAS 201, 551

Foster, P. N., and R. A. Chevalier, 1993, ApJ 416, 303

Franco, J., S. N. Shore, and G. Tenorio-Tagle, 1994, ApJ 436, 795

Fukui, Y. et al., 1999, PASJ 51, 745

Gammie, C. F., and E. C. Ostriker, 1996, ApJ 466, 814

Goldreich, P. and S. Sridhar, 1995, ApJ 438, 763

Goldreich, P. and S. Sridhar, 1997, ApJ 485, 680

Hartmann, L., 2001, AJ 121, 1030

Hartmann, L., J. Ballesteros-Paredes, and E. A. Bergin, 2001, ApJ 562, 852

Heitsch, F., M. Mac Low, and R. S. Klessen, 2001a, ApJ 547, 280

Heitsch, F., E. G. Zweibel, M.-M. Mac Low, P. Li, and M. L. Norman, 2001b, ApJ 561, 800

Hillenbrand, L. A., and L. W. Hartmann, 1998, ApJ 492, 540

Hendriksen, R. N., P. André, and S. Bontemps, 1997, A&A 323, 549

Jeans, J. H., 1902, Phil. Trans. A. 199, 1

Klein, R. I., C. F. McKee, and P. Colella, 1994, ApJ 420, 213

Klessen, R. S., 2000, ApJ 535, 869

Klessen, R. S., and A. Burkert, 2000, ApJS 128, 287

Klessen, R. S., and A. Burkert, 2001, ApJ 549, 386

Klessen, R. S., A. Burkert, and M. R. Bate, 1998, ApJ 501, L205

Klessen, R. S., F. Heitsch, and M.-M. Mac Low, 2000, ApJ 535, 887

Kolmogorov, A. N., 1941, Dokl. Akad. Nauk SSSR 30, 301 (reprinted in *Proc. R. Soc. Lond. A* 434, 9–13 [1991])

Kramer, C., J. Stutzki, R. Rohrig, U. Corneliussen, 1998, A&A 329, 249

Krebs, J., and W. Hillebrandt, 1983, A&A 128, 411

Kroupa, P., 2002, Science 295, 82

Lada, E. A., 1992, ApJ 393, L25

Larson, R. B., 1981, MNRAS 194, 809

Larson, R. B., 1995, MNRAS 272, 213

Leisawitz, D., F. N. Bash, and P. Thaddeus, 1989, ApJS 70, 731

Léorat, J., T. Passot, and A. Pouquet, 1990, MNRAS 243, 293

Lesieur, M., 1997, *Turbulence in Fluids*, 3rd ed. (Kluwer, Dordrecht), p. 245

Lithwick, Y., and P. Goldreich, 2001, ApJ 562, 279

Mac Low, M.-M., 1999, ApJ 524, 169

Mac Low, M.-M., Klessen, R. S., 2003, Rev. Mod. Phys., in press (astro-ph/0301093)

Mac Low, M.-M., R. S. Klessen, A. Burkert, and M. D. Smith, 1998, Phys. Rev. Lett. 80, 2754

Mac Low, M.-M., C. F. McKee, R. I. Klein, J. M. Stone, and M. L. Norman, 1994, ApJ 433, 757

Matzner, C. D., 2002, ApJ 566, 302

Matzner, C. D., and C. F. McKee, 2000, ApJ 545, 364

McKee, C. F., 1999, in *NATO ASIC Proc. 540: The Origin of Stars and Planetary Systems*, edited by C. J. Lada and N. D. Kylafis (Kluwer Academic Publishers), p. 29

McKee, C. F., and J. P. Ostriker, 1977, ApJ 218, 148

McKee, C. F., and E. G. Zweibel, 1995, ApJ 440, 686

Mizuno, A., T. Onishi, Y. Yonekura, T. Nagahama, H. Ogawa, and Y. Fukui, 1995, ApJ 445, L161

Monaghan, J. J., 1992, ARAA 30, 543

Mouschovias, T. C., 1991a, in *The Physics of Star Formation and Early Stellar Evolution*, edited by C. J. Lada and N. D. Kylafis (Kluwer, Dordrecht), p. 61

Mouschovias, T. C., 1991b, in *The Physics of Star Formation and Early Stellar Evolution*, edited by C. J. Lada and N. D. Kylafis (Kluwer, Dordrecht), p. 449

Mouschovias, T. C., and L. Spitzer, Jr., 1976, ApJ 210, 326

Müller, W.-C., and D. Biskamp, 2000, Phys. Rev. Lett. 84, 475

Nakano, T., 1976, PASJ 28, 355

Nakano, T., 1998, ApJ 494, 587

Nakano, T., and T. Nakamura, 1978, PASJ 30, 681

Ng, C. S., and A. Bhattacharjee, 1996, ApJ 465, 845

Norman, C. A., and A. Ferrara, 1996, ApJ 467, 280

Norman, C. A., and J. Silk, 1980, ApJ 239, 968

Olmi, L., and L. Testi, 2002, A&A 392, 1053

Oort, J. H., & L. Spitzer, Jr., 1955, ApJ 121, 6

Ostriker, E. C., C. F. Gammie, and J. M. Stone, 1999, ApJ 513, 259

Ossenkopf V., and M.-M. Mac Low, 2002, A&A 390, 307

Padoan, P., 1995, MNRAS 277, 377

Padoan, P., and Å. Nordlund, 1999, ApJ 526, 279

Padoan, P., and Å. Nordlund, 2002, ApJ 576, 870

Palla, F., and S. W. Stahler, 2000, ApJ 540, 255

Passot, T., A. Pouquet, and P. R. Woodward, 1988, A&A 197, 392

Passot, T., E. Vázquez-Semadeni, and A. Pouquet, 1995, ApJ 455, 536

Porter, D. H., A. Pouquet, and P. R. Woodward, 1994, Phys. Fluids 6, 2133

Rodriguez-Gaspar, J. A., G. Tenorio-Tagle, and J. Franco, 1995, ApJ 451, 210

Salpeter, E. E., 1955, ApJ 121, 161

Scalo, J. M., E. Vázquez-Semadeni, D. Chappell, T. Passot, 1998, ApJ 504, 835

Schaye, J., 2002, ApJ, submitted (astro-ph/0205125)

She, Z., and E. Leveque, 1994, Phys. Rev. Lett. 72, 336

Shu, F. H., 1977, ApJ 214, 488

Shu, F. H., F. C. Adams, and S. Lizano, 1987, ARAA 25, 23

Sirianni, M., A. Nota, G. De Marchi, C. Leitherer, and M. Clampin, 2002, ApJ 579, 275

Smith, M. D., and M.-M. Mac Low, 1997, A&A 326, 801

Spaans, M., and J., Silk, 2000, ApJ 538, 115

Stone, J. M., and M. L. Norman, 1992a, ApJS 80, 753

Stone, J. M., and M. L. Norman, 1992b, ApJS 80, 791

Stone, J. M., E. C. Ostriker, and C. F. Gammie, 1998, ApJ 508, L99

Strom, K. M., S. E. Strom, and K. M. Merrill, 1993, ApJ 412, 233

Tafalla, M., D. Mardones, P. C. Myers, P. Caselli, R. Bachiller, and P. J. Benson, 1998, ApJ 504, 900

Vázquez-Semadeni, E., T. Passot, and A. Pouquet, 1995, ApJ 441, 702

Vázquez-Semadeni, E., T. Passot, and A. Pouquet, 1996, ApJ 473, 881

von Weizsäcker, C. F., 1943, Z. Astrophys. 22, 319

von Weizsäcker, C. F., 1951, ApJ 114, 165

Wada, K., and C. A. Norman, 1999, ApJ 516, L13

Wada, K., G. Meurer, and C. A. Norman, 2002, ApJ 577, 197

Walborn, N. R., R. H. Barbá, W. Brandner, M. Rubio, E. K. Grebel, and R. G. Probst, 1999, AJ 117, 225

Walder, R. and D. Folini, 2000, ApSS 274, 343

Whitworth, A. P., 1979, MNRAS 186, 59

Whitworth, A. P., A. S. Bhattal, N. Francis, and S. J. Watkins, 1996, MNRAS 283, 1061

Williams, J. P., L. Blitz, and C. F. McKee, 2000, in *Protostars and Planets IV*, edited by V. Mannings, A. P. Boss, and S. S. Russell (University of Arizona Press, Tucson), p. 97

Williams, J. P., P. C. Myers, D. J. Wilner, and J. di Francesco, 1999, ApJ 513, L61

Wolfire, M. G., D. Hollenbach, C. F. McKee, A. G. G. M. Tielens, and E. L. O. Bakes, 1995, ApJ 443, 152

Yorke, H. W., Tenorio-Tagle, G., Bodenheimer, P., and M. Różyczka, 1989, A&A 216, 207

Zinnecker, H., M. J. McCaughrean, and B. A. Wilking, 1993, in *Protostars and Planets III*, edited by E. H. Levy and J. F. Lunine (University of Arizona Press, Tucson), p. 429

Zweibel, E. G., and A. Brandenburg, 1997, ApJ 478, 563

Zweibel, E. G., and K. Josafatsson, 1983, ApJ 270, 511

Dynamics of Small Scale Motions in the Solar Photosphere

Arnold Hanslmeier

Institut für Geophysik, Astrophysik und Meteorologie
Universtäts-Platz 5, A-8010 Graz, Austria
arh@igam06ws.uni-graz.at

1 Introduction

The Sun is the star we live with at a distance of only 150×10^6 km and the only star where we can directly observe details on its surface. Sunspots have been observed since more than 2000 years because the largest are visible to the naked eye. Today the Sun is studied in all wavelengths, in short wavelengths from several space missions (e. g. SOHO, TRACE, YOHKOH, RHESSI), in the optical and near IR from ground based observatories and of course with ground based radiotelescopes. Thus a kind of tomography of the solar atmosphere can be made as it is shown in Table 1 illustrating that in order to observe higher layers of the solar atmosphere (chromosphere, corona), one has to go to UV, EUV, and X-rays.

Table 1: Examples of lines that are commonly used for solar observations

Line	Temperature in K	Remarks
visible	photosphere, $h = 0 \ldots 400$ km, $T = 6000$ K	from ground
Hα	lower chromosphere, $h = 2000$ km	from ground
He II	chromosphere, 60 000 K	in UV
Fe IX	1 million K, corona	extreme UV
Fe XII	1.5 million K, corona	extreme UV
soft X rays	> 2 million K, corona	

The great relevance of solar physics to space missions and space activities of all kind is the fact that our Sun is a variable star. In the visible light that is mainly emitted from the photosphere, this variation is very low (< 0.5 promille). However, the shorter the wavelength, the stronger the variation. The influences of solar activity to the Earth and near Earth space environment are called space weather and are described in the book of Hanslmeier (2002).

High energetic phenomena often are located in these higher layers but of course they must be related to the dynamics in the deeper layers. The photosphere of the Sun does not appear homogeneous but reveals the above mentioned sunspots and a cellular like pattern called granulation. The size of the granular elements is in the range of about 1000 km and thus about $1''$ as seen from the Earth.

Therefore, the study of small scale dynamics in the solar photosphere, granulation, fine structure of sunspots etc. requires spatially highly resolved images and spectrograms, ideally as time series that can only be obtained under excellent seeing conditions.

In the following review we will discuss first the standard solar model than some observational problems to get high resolution time series and then concentrate on the dynamics of granulation, sunspots and pores and some model simulations.

2 Standard Solar models

It is well known that the Solar interior can be divided into the following zones:

- solar core, where the nuclear reactions take place,
- radiation zone, where the energy is transported outward by radiation,
- tachocline, where shearing motions occur between the fluid motions of the upper lying convection zone and the stable radiative zone; this generates magnetic flux.
- convective zone, where because of the lower temperature atoms become only partially ionized which increases the opacity leading to convective motions.

In Table 2 we give the basic characteristics of these layers as well as of the photosphere.

Table 2: Basic characteristics of the main zones in the solar interior

Name	Extension in $R_\odot$	Temperature	density [g/cm^3]
Core	0 ... 0.25	$1.5 \times 10^7 \ldots 7 \times 10^6$	150 ... 20
radiative zone	0.25 ... 0.70	$7 \times 10^6 \ldots 2 \times 10^6$	20 ... 0.2
transition zone	thin		
convective zone	0.70 ... 1.0	$2 \times 10^6 \ldots 7 \times 10^3$	$0.2 \ldots 10^{-4} \rho_{\text{atm,SL}}^1$
photosphere	400 km	7000 ... 4500	$\sim 10^{-7}$

[1] density of Earth's atmosphere at sea level

By taking into account the results from helioseismology, standard solar models have been improved considerably in the recent years. From a great number of acoustic modes determined by several ground networks such as GONG (Global oscillation Network), IRIS (International research on the Interior of the Sun), BISON (Birmingham Solar Oscillation network), the Low-Degree (l) Oscillation Experiment

(LOWL) the base of the convection zone was determined as $0.713 \pm 0.0001 R_\odot$. Additionally, space helioseismic experiments aboard SOHO (Solar and Heliospheric Observatory), namely GOLF (Global Oscillations at Low Frequencies) and MDI (Michelson Doppler Imager) give new constraints on the transition between the radiative zone and the convection zone. A review and description of the various instruments was given e. g. by Fleck (2001).

The photospheric helium in mass fraction according to Brun et al. (1998, see this paper for further references) is 0.243 and they found for the sound-speed square difference between the Sun and the model a value of 1 %. Observing solar neutrinos provides a good means to probe the solar core where nuclear reactions take place. The standard model gives the following neutrino fluxes: 7.18 SNU for the chlorine experiment, 127.2 SNU for the gallium detector, and 4.82×10^6 cm^{-2}s^{-1} for the ^{8}B neutrino flux. Acoustic-mode predictions are also estimated in that paper. The initial solar helium abundance seems more and more constrained to values between 0.273 and 0.277 in mass fraction. Reduction of about 30 % in chlorine and water detectors, which is more than half the discrepancy with the experimental results can be noted. A direct test for the solar interior would be the observation of g-mode frequencies which may be accessible to SOHO.

A recent review about standard solar models was given by Christensen-Dalsgaard (1998). The computation of standard models depends on the basic equations of stellar structure. These are:

$$\frac{dp}{dr} = -\frac{Gm\rho}{r^2} \qquad \frac{dm}{dr} = 4\pi r^2 \rho \qquad (1)$$

$$\frac{dT}{dr} = -\frac{3\kappa L \rho}{16\pi a c r^2 T^3} \qquad \frac{dL}{dr} = 4\pi r^2 \rho \epsilon \qquad (2)$$

which are the hydrostatic equilibrium, equation for mass, luminosity and energy generation. r denotes the distance to the center, p, m, ρ pressure, mass and density inside, r, G the gravitational constant, L the luminosity at r, c the speed of light, κ the opacity, and ϵ the rate of energy generation per unit mass. In convective regions we must use equation

$$\frac{d\ln T}{d\ln p} \cong \left(\frac{\partial \ln T}{\partial \ln p}\right)_{\text{ad}} = \nabla_{\text{ad}} \qquad (3)$$

That means the temperature gradient is nearly adiabatic except near the very surface. In addition to these equations microphysics has to be defined by the equations of state, energy generation and opacity. In the convection zone the structure is independent on the opacity thus depending mainly on the equation of state and composition.

The composition is given by the abundance X (hydrogen), Y (helium) and Z (heavier elements) and $X + Y + Z = 1$. The composition changes because of the evolution of the Sun (nuclear reactions in the core) and because of diffusion. Heavier elements sink, lighter elements rise. The convective overshooting can be observed as granulation (see Fig. 1).

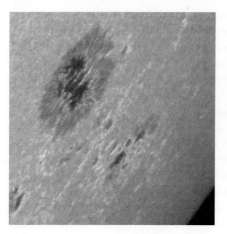 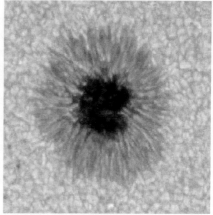

Figure 1: Solar granulation and sunspots. Near the disk center (right) the granulation is clearly visible, near the solar limb (left) it cannot be seen since because there we observe higher layers where these overshooting convective motions do not longer exist (center to limb variation). Courtesy: J. A. Bonet, M. Sobotka, M. Vazquez, and A. Hanslmeier, observed with the Swedish SVST at La Palma.

Convection in late type stars such as the Sun is important because
- it provides energy transport,
- influences stellar evolution,
- causes mixing of nuclear processed material,
- influences and is influenced by magnetic activity,
- contributes to chromospheric heating,
- contributes to pulsations, acoustic flux generation.

3 Theoretical constraints on convection

3.1 Introduction

The dynamics at small scales in the photosphere is dominated by overshooting convective motions. These motions can be modelled by HD-simulations. If ρ denotes the density, $\vec{v}$ the velocity, $\vec{g}$ the gravitational acceleration, p the pressure, e the internal energy, σ the viscous stress tensor, Q_{rad} the radiative heating/cooling rate and Q_{visc} the viscous dissipation, then the following set of equations describe these motions (Asplund et al., 2000)

$$\frac{\partial \ln \rho}{\partial t} = -\vec{v}.\nabla \ln \rho - \nabla.\vec{v} \tag{4}$$

$$\frac{\partial \vec{v}}{\partial t} = -\vec{v}.\nabla \vec{v} + \vec{g} - \frac{p}{\rho}\nabla \ln P + \frac{1}{\rho}\nabla.\vec{v} \tag{5}$$

$$\frac{\partial e}{\partial t} = -\vec{v}.\nabla e - \frac{p}{\rho}\nabla.\vec{v} + Q_{\text{rad}} + Q_{\text{visc}} \tag{6}$$

These convective motions leave signatures in spectral lines in the form of line shifts and asymmetries. A discussion about that can be found e. g. in Dravins and Nordlund (1990a, b). Today such 3D convection simulations are quite realistic and there is a good agreement with observed line shapes, shifts and asymmetries. For late type stars we observe a *C-shape* for the bisectors of unblended lines. The bisectors divide line profiles into two halves at equal intensity levels. The C-shape is the result of differences of area coverage and Doppler shifts between up- and downflows and differences in line strength and continuum level. Weak lines only show the upper part of the C-shape and have blue shifts (after subtraction of the effects of gravitational redshift). The cores of stronger lines are formed in or above the convective overshoot layers and therefore their blueshifts are reduced or even non existent.

3.2 Statistical measures of hydrodynamics

Since the dynamic equations for turbulent fluids generally show an unstable behavior, it is practical to use statistical measures of the velocity field rather than trying to make exact predictions.

Most theories for turbulence are based on Kolmogorov's definitions of "homogeneous" and "homogeneous isotropic" turbulence (Kolmogorov, 1941). These definitions are purely statistical, and basically impossible to test for an experiment, but Kolmogorov argues that the conditions typically are fulfilled in a small region far from the border of the system ("small" and "far" as compared to the length scale at which energy is pumped into the system). Assuming that a fluid system is homogeneous isotropic turbulent and that the average behavior only depends on the average energy dissipation in the system, it is possible to derive a simple relation between the average energy dissipation, ϵ, velocity differences between particles separated some distance, $\delta \vec{v}(R)$, and that distance, R.

Let us assume, for simplicity that the density is constant. The equation of continuity is reduced to

$$\nabla \cdot \vec{u} = -\frac{\partial \rho}{\partial t} \rightarrow \nabla \cdot \vec{u} = 0 \qquad (7)$$

The second viscosity that appears in the general form, is also removed and the system is described by a modified equation (5):

$$\frac{\partial \vec{u}}{\partial t} + (\vec{u} \cdot \nabla)\vec{u} = -\frac{1}{\rho}\nabla p + \nu \nabla^2 \vec{u} + \vec{F} \qquad (8)$$

where $\vec{u}$ denotes the velocity field.

Given a typical velocity and length of the system the dimensionless *Reynolds number* is:

$$N_{\text{Re}} = \frac{\text{velocity} \times \text{length}}{\nu} \qquad (9)$$

and we can make the distinction for
- low $N_{\text{Re}} \rightarrow$ laminar flow,
- high $N_{\text{Re}} \rightarrow$ chaotic flow.

Let us now define locally homogeneous and locally isotropic turbulence (see e. g. Kolmogorov, 1991). In short, the turbulence is called locally homogeneous if the probability distribution for the relative velocity of two particles in the fluid G depends only on the distance between the particles. Kolmogorov states that if you are looking at the system at length scales which are much greater than the length at which energy is pumped into the system and the length scale at which energy dissipates from the system (as heat due to viscosity) the only parameter to describe the probability distribution of the relative velocity field in a homogeneous isotropic turbulent fluid is the energy dissipation. Then there is a scaling relation:

$$< \delta v(R) > \sim (\epsilon R)^{1/3} \qquad (10)$$

According to Kolmogorov the following scenario holds:

1. Assume that energy is added to a system at largest scales L_0; this is called *outer scale*.

2. The energy cascades from larger to smaller scales, turbulent eddies break down into smaller and smaller structures.

3. The sizes where this takes place is called *inertial range*.

4. Finally, the eddy size becomes so small that it is subject to dissipation from viscosity, this is called the *inner scale l_0*.

A well known system that behaves like in the above scenario is the Earth's atmosphere where L_0 ranges from $10^{1 \ldots 2}$ m and l_0 is a few mm.

Let us examine this in more detail: $v \ldots$ velocity, $\epsilon \ldots$ energy dissipation rate per unit mass, $\nu \ldots$ viscosity, $l_0 \ldots$ inner scale, l the local scale. Then the energy/mass is given by:

$$v^2/2 \sim v^2 \qquad (11)$$

The energy dissipation rate per unit mass:

$$\epsilon \sim v^2/\tau = v^2/(l/v) = v^3/l \qquad (12)$$

Therefore,

$$v \sim (\epsilon l)^{1/3} \qquad (13)$$

and for the energy:

$$v^2 \sim \epsilon^{2/3} l^{2/3} \qquad (14)$$

For the relation between the inner and the outer scale one finds from the Navier Stokes equation:

$$l_0 \sim L_0 (N_{\text{Re}})^{-3/4} \qquad (15)$$

Let us now consider the Kolmogorov Power spectrum for the velocity fluctuations. If $k = 2\pi/l$, then

$$\Phi(k) \sim k^{-5/3} \tag{16}$$

and for three dimensions:

$$\Phi^{3D}(k) \sim k^{-11/3} \tag{17}$$

A more detailed discussion can be found e. g. in Tatarski (1961).

3.3 Routes to turbulence

For incompressible flows the Navier-Stokes equation is

$$\frac{\partial \vec{v}}{\partial t} + (\vec{v}\nabla\vec{v}) = \vec{F} - \frac{1}{\rho}\nabla p + \nu\nabla^2 \vec{v} \tag{18}$$

The Reynolds number N_{Re} defines the ratio between external buoyancy (u/l) and internal dissipation (ν/l^2). Flows with the same N_{Re} are similar.

3.3.1 Landau Scenario

This was investigated by Landau and Hopf (1944). They studied the stability of a stationary incompressible flow, given by $\vec{v}_0(\vec{x})$ and $p_0(\vec{x})$ dependent on N_{Re} under the influence of perturbations given by $\vec{v}_1(\vec{x},t)$ and $p_1(\vec{x},t)$. Thus:

$$(\vec{v}_0 \nabla)\vec{v}_0 = \vec{F} - \frac{1}{\rho}\nabla p_0 + \nu\nabla^2 \vec{v}_0 \qquad \nabla \vec{v}_0 = 0 \tag{19}$$

Now $\vec{v} = \vec{v}_0 + \vec{v}_1$ and $p = p_0 + p_1$ and under the assumption that the perturbations $\vec{v}_1, p_1$ are small:

$$\frac{\partial \vec{v}_1}{\partial t} + (\vec{v}_0 \nabla)\vec{v}_1 + (\vec{v}_1 \nabla)\vec{v}_0 = -\frac{1}{\rho}\nabla p_1 + \nu\nabla^2 \vec{v}_1 \qquad \nabla \vec{v}_1 = 0 \tag{20}$$

The result is that for higher $N_{\text{Re}} > N_{\text{Re}_1}$ the flow can be described as a superposition of a stationary and a periodic flow with a frequency $\omega_1(N_{\text{Re}})$, A is a finite amplitude:

$$\vec{v} = \vec{v}_0 + Ae^{i(\omega_1 t + \beta_1)}\vec{f}(\vec{x}) \tag{21}$$

For larger N_{Re} this superposition is no longer valid, the flow consists of a general periodic function with base frequency $\omega_1(N_{\text{Re}})$ and can be expanded into a Fourier-Series:

$$\vec{v}(\vec{x},t) = \sum_k A_k \vec{f}_k(\vec{x}) e^{ik(\omega_1 t + \beta_1)} \tag{22}$$

Thus we have the following scenario to produce turbulence:

- stationary → attractor
- periodic (after a so called Hopf bifurcation)
- 3-D Torus,
- ...

New incommensurable frequencies occur – the flow becomes rapidly very complex.

3.3.2 Ruelle Scenario

According to Landau and Hopf a complex phenomenon like turbulent flow requires a complex description with infinite number of degrees of freedom. Because Lorenz demonstrated that three ordinary linear differential equations can exhibit a very complex behavior, this model of Landau was seen critically. The motion on a strange attractor, which is typical for turbulent motions is characterized by a sensitive dependence on initial conditions, i. e. two arbitrary close trajectories will diverge exponentially (see next chapter). A quasiperiodic motion on an n-dimensional torus means that adjacent trajectories will stay in the neighborhood for all times. Quasiperiodic motions don't lead to a mixing of trajectories in phase space, but turbulent motion lead to such a mixing. This mixing is characterized by the fact that the averaged autocorrelation function approaches zero as $\tau \to \infty$:

$$\bar{a}(\tau) = \lim_{T \to \infty} \frac{1}{2T} \int_{-T}^{T} f(t+\tau)f(t)dt \to 0 \tag{23}$$

4 Chaotic dynamics

4.1 Introduction

The irregular and unpredictable time evolution of many nonlinear systems is called by the term *chaos*. Chaos occurs in mechanical oscillators (e. g. pendula or vibrating objects) in rotating or heated fluids and many other processes. The central characteristic of such a system is that it does not repeat its past behavior. This can also be seen by the fact that if we start a system twice, but from slightly different initial conditions then such error grows exponentially. For nonchaotic systems such an uncertainty leads to an error in prediction which grows linearly in time. Chaotic systems therefore are very sensitive to initial conditions. This was already recognized by Henri Poincaré in 1913: "it may happen that small differences in the initial conditions produce very great ones in the final phenomena. A small error in the former will produce an enormous error in the latter. Prediction becomes impossible, and we have a fortuitous phenomenon."

Because of this lack of predictability a chaotic system can resemble a stochastic system which is a system subject to random external forces. However the source of irregularity is quite different. For chaos, the irregularity is part of the intrinsic dynamics of the system not from unpredictable outside influences.

4.2 Tools of chaotic dynamics

4.2.1 Phase space

The *phase space* of a dynamical system is a mathematical space with orthogonal coordinate directions representing each of the variables needed to specify the instantaneous state of the system. Let us consider a particle moving in one dimension. Its state is characterized by its position $x(t)$ and velocity $v(t)$. Hence its phase space is a plane. A particle moving in three dimensions would have a six dimensional phase space with three position and three velocity directions. The orbit in phase space is called a *trajectory*. Two trajectories corresponding to similar energies will pass very close to each other, but the orbits will not cross each other. This noncrossing property derives from the fact that past and future states of a deterministic mechanical system are uniquely prescribed by a system state at a given time. A crossing of trajectories at time t would introduce ambiguity into past and future states.

The *preservation of areas* is another important property of the phase space of conservative systems i.e. systems with constant energy. Generally, dynamical systems can be classified into *conservative* or *dissipative* depending upon whether the phase space volumes stay constant or contract. Let us consider two examples. A linearized, undamped pendulum conserves energy and its trajectory preserves area, the trajectories of the linearized damped pendulum given by

$$\frac{d^2\Theta}{dt^2} + \frac{d\Theta}{dt} + \Theta = 0 \tag{24}$$

decay to a single point. Let us consider the *flux* out of a small region ΔS:

$$\vec{F}.\vec{n}\Delta S \tag{25}$$

It can be shown that it is sufficient to calculate $\nabla.\vec{F}$. If this is zero, the system is conservative, if the divergence of phase velocity is negative, the system is dissipative.

For illustration let us again consider the two examples:

- undamped pendulum:

$$\frac{d^2\Theta}{dt^2} + \Theta = 0 \tag{26}$$

The components of the phase velocity vector are then:

$$\frac{d\Theta}{dt} = \omega \qquad \frac{d\omega}{dt} = -\Theta \tag{27}$$

therefore $\vec{F} = (\omega, -\Theta)$ and $\nabla.\vec{F} = \partial\omega/\partial\Theta + \partial(-\Theta)/\partial\omega = 0$. The area is preserved.

- damped pendulum:

$$\frac{d^2\Theta}{dt^2} + \frac{d\Theta}{dt} + \Theta = 0 \tag{28}$$

and

$$\frac{d\Theta}{dt} = \omega \qquad \frac{d\omega}{dt} = -\omega - \Theta \tag{29}$$

therefore $\vec{F} = (\omega, -\omega - \Theta)$ and $\nabla.\vec{F} = \partial\omega/\partial\Theta + \partial(-\omega - \Theta)/\partial\omega = -1$. The phase area diminishes, the system is dissipative.

In many dynamical systems *attractors* occur. The stability of attractors can be tested by examining the dynamical behavior near critical points. A saddle point is an unstable point where the trajectories move away from it. Often the phase space can be separated into alternating regions. There are basins of attraction a *separatrix* divides one basin from another. Also limit cycles occur.

Poincaré introduced a method to simplify phase space diagrams of complicated systems. The construction is easy: the phase space diagram is viewed stroboscopically in such a way that the motion is observed periodically. For a driven pendulum the strobe period is the period of the forcing.

The time evolution of a dynamical system is represented by $f(t)$ and any function $f(t)$ may be represented as a superposition of periodic components. The determination of the relative strengths is called *spectral analysis*. Depending on the nature of $f(t)$ the spectrum may be given by:

- $f(t)$ periodic: the spectrum may be expressed as a linear combination of oscillations whose frequencies are integer multiples of a basic frequency. This leads to a Fourier series.

- $f(t)$ not periodic: the spectrum must be expressed in terms of oscillations with a continuum of frequencies. This is a *Fourier transform*.

The Fourier series representation of $f(t)$ may be written as:

$$f(t) = \sum_{n=-\infty}^{\infty} a_n e^{in\omega_0 t} \qquad a_n = \frac{\omega_0}{2\pi} \int_{-\pi/\omega_0}^{\pi/\omega_0} f(t) e^{-in\omega_0 t} dt \qquad (30)$$

A Fourier series follows from a Fourier transform by:

$$T \to \infty \qquad n\omega_0 \to \omega \qquad a_n \to a(\omega) d\omega$$

Here ω is a continuous variable. One gets:

$$f(t) = \int_{-\infty}^{\infty} a(\omega) d\omega e^{i\omega t} \qquad a(\omega) d\omega = \frac{d\omega}{2\pi} \int_{-\infty}^{\infty} f(t) e^{-i\omega t} dt \qquad (31)$$

$a(\omega)$ is complex. The real valued function

$$S(\omega) = |a(\omega)|^2 \qquad (32)$$

is called the *power spectrum*. How does noise appear in a power spectrum? For example Johnson noise, which results from thermal agitation in electric circuits is frequency independent (also called white noise). A $1/f$ noise is common in resistors and solid state devices.

4.2.2 Lyapunov Exponent

A. M. Lyapunov (1857–1918) introduced a measure of the sensitive dependence upon initial conditions that is characteristic of chaotic behavior. Consider for simplicity a one-dimensional map given by $x_{n+1} = f(x_n)$. The difference between two initially nearby states after the nth step can be written as:

$$f^n(x + \epsilon) - f^n(x) \sim \epsilon e^{n\lambda} \qquad (33)$$

or

$$\ln\left[\frac{f^n(x+\epsilon)-f^n(x)}{\epsilon}\right] \sim n\lambda \qquad (34)$$

For small ϵ, this expression becomes:

$$\lambda \sim \frac{1}{n}\ln\left[\frac{df^n}{dx}\right] \qquad (35)$$

Applying the chain rule for the derivative of the nth iterate and take the limit as n tends to infinity, we obtain

$$\lambda = \lim_{n\to\infty}\frac{1}{n}\sum_{i=0}^{n-1}\ln|f'(x_i)| \qquad (36)$$

The Lyapunov exponent gives the stretching rate per iteration, averaged over the trajectory.

For n-dimensional maps there are n Lyapunov exponents, since stretching can occur for each axis. An n-dimensional initial volume develops, on average, as

$$V = V_0 e^{(\lambda_1+\lambda_2+\cdots+\lambda_n)n} \qquad (37)$$

For a dissipative system, the sum of the exponents must be negative. If the system is chaotic then at least one of the exponents is positive. Lyapunov exponents are also defined for continuous time dynamical systems.

4.2.3 Entropy

Let us now apply the concept of entropy to dynamical systems. Consider a hypothetical statistical system for which the outcome of a certain measurement must be located on the unit interval. If the line is subdivided into N subintervals, we can associate a probability p_i with the ith subinterval containing a particular range of possible outcomes. The entropy of the system is then defined as:

$$S = -\sum_{i=1}^{N} p_i \ln p_i \qquad (38)$$

This may be interpreted as a measure of the amount of disorder in the system or as the information necessary to specify the state of the system. If the subintervals are equally probable so that $p_i = 1/N$ for all i, then the entropy reduces to $S = \ln N$.

4.3 Dimension

How can we define the dimension $d(A)$ of a given set A? One approach is the *capacity dimension* d_C. Let us consider a one-dimensional figure such as a straight line or curve of length L. Such a figure can be covered by $N(\epsilon)$ one-dimensional boxes of size ϵ on a side.

$$N(\epsilon) = L(1/\epsilon) \tag{39}$$

Similarly, a two-dimensional square of side L can be covered by $N(\epsilon) = L^2(1/\epsilon)^2$ boxes and for a three-dimensional cube the exponents would be 3. Thus, in general:

$$N(\epsilon) = L^d(1/\epsilon)^d \tag{40}$$

Taking the logarithms:

$$d = \frac{\log N(\epsilon)}{\log L + \log(1/\epsilon)} \tag{41}$$

In the limit of small ϵ the *capacity dimension* results as

$$d_{\mathrm{C}} = \lim_{\epsilon \to 0} \frac{\log N(\epsilon)}{\log(1/\epsilon)} \tag{42}$$

This is equivalent to regard d_C as the slope of the $\log N$ versus $\log(1/\epsilon)$ curve as $\epsilon \to 0$. The attractor for a dissipative system resides in an n-dimensional phase space but its dimension is less than n. Chaotic attractors generally have noninteger dimension and are called *strange attractors*.

For higher-dimensional systems another type of dimension is more efficient to compute, the *correlation dimension* d_G. Suppose that many points are scattered over a set. The typical number of neighbors of a given point will vary more rapidly with distance from that point if the set has high dimension than otherwise. Each point has a circle of radius R drawn around it. All points within all circles of size R contribute to $C(R)$. The correlation dimension is the slope of the graph $\log C(R)$ versus $\log R$ where $C(R)$ is given by:

$$C(R) = \lim_{N \to \infty} \left[\frac{1}{N^2} \sum_{i,j=1}^{N} H(R - |\vec{x}_i - x_j|) \right] \tag{43}$$

where $\vec{x}_i, \vec{x}_j$ are points on the attractor, $H(y)$ is the Heaviside function (1 if $y \geq 0$ and 0 if $y < 0$). N is the number of points randomly chosen from the entire data set. The Heaviside function counts the number of points within a radius R of the point denoted by $\vec{x}_i$ and $C(R)$ gives the average fraction of points within R.

Please take into account, that d_C and d_G are not equivalent. Small variations of density are ignored by the capacity dimension.

The *information dimension* is related to the entropy. It depends on the distribution of points on the attractor. Suppose the attractor is covered by a set of n boxes of size ϵ and let the probability that a point is in the ith box be p_i. Then the metric entropy or missing information is

$$I(\epsilon) = -\sum_{i=0}^{n} p_i \log p_i \tag{44}$$

and

$$d_I = \lim_{\epsilon \to 0} \left[-\frac{I(\epsilon)}{\log \epsilon} \right] \qquad (45)$$

The relation between these different types of dimension are:

$$d_C \leq d_I \leq d_G \qquad (46)$$

In recent studies the idea that complex sets may be described as *multifractals* emerged: sets consisting of many interwoven fractals, each with a different fractal dimension.

Kaplan and Yorke (1979) pointed out an interesting relationship between Lyapunov exponents and dimension. Let us assume that we have a stretching in phase space by a factor of $e^{\lambda_1 t}$, $\lambda_1 > 0$ and the other direction shrinks according to the factor $e^{\lambda_2 t}$, $\lambda_2 < 0$. Therefore, the area evolves as $A(t) = A_0 e^{(\lambda_1 + \lambda_2)t}$. By analogy with the capacity dimension, a Lyapunov dimension can be defined:

$$d_L = \lim_{\epsilon \to 0} \left[\frac{d(\log N(\epsilon))}{d(1/\epsilon)} \right] \qquad (47)$$

$N(\epsilon)$ is the number of squares with sides of length ϵ required to cover $A(t)$. $N(\epsilon)$ and ϵ depend on time as follows:

$$N(t) = \frac{A(t)}{\text{squarearea(t)}} = \frac{A_0 e^{(\lambda_1 + \lambda_2)t}}{A_0 e^{2\lambda_2 t}} = e^{(\lambda_1 - \lambda_2)t} \qquad (48)$$

and $\epsilon(t) = A_0^{1/2} e^{\lambda_2 t}$. This leads to the definition of d_L:

$$d_L = 1 - \frac{\lambda_1}{\lambda_2} \qquad (49)$$

which is known as Kaplan-Yorke relation. They also predicted that $d_L \leq d_G$ and that if the points on the fractal are approximately distributed uniformly then equality holds. Generalized to higher-dimensional spaces:

$$d_L = j + \frac{\lambda_1 + \lambda_2 + \lambda_3 + ... + \lambda_j}{|\lambda_{j+1}|} \qquad (50)$$

Here, the λ_i are ordered (λ_1 being the largest) and j is the index of the smallest non negative Lyapunov exponent.

4.3.1 Kolmogorov entropy, information change

As we have stressed, chaotic systems are very sensitive to initial conditions and initially adjacent dynamical states diverge. While an initial state may be known with a high degree of precision, the ability to predict later states diminishes because of trajectory divergence, information is lost, the entropy has increased.

For many systems the information function has a simple linear time dependence:

$$I(t) = I_0 + Kt \qquad (51)$$

The average rate, taken over long times is called Kolmogorov entropy:

$$K = \lim_{\epsilon \to 0} \lim_{T \to \infty} [I(\epsilon, T)/T] \qquad (52)$$

The Kolmogorov entropy for chaotic systems depends on the positive Lyapunov exponent and in systems with several positive exponents (j is the index of the smallest positive λ_i):

$$K \leq \sum_{i=1}^{j} \lambda_i \qquad (53)$$

Grassberger and Procaccia (1983) suggest equality. We can state that:
- positive but finite K implies chaotic behavior,
- $K = 0$ implies regular motion (periodic, quasiperiodic or stationary),
- $K > 0$ chaotic motion,
- $K = \infty$ random motion.

Another important feature of the Kolmogorov entropy is that it may be used to estimate the time for which future predictions of the state of a chaotic system are valid. After this prediction time, the system's uncertainty has diverged to the size of the phase space.

Suppose that the system is characterized by a positive Lyapunov exponent λ_+ and its initial state is defined to within a size ϵ. Then in a time T, the uncertainty in the coordinates will have expanded to the size L of the attractor:

$$L \sim \epsilon e^{\lambda_+ T} \qquad L \sim \epsilon e^{KT} \qquad (54)$$

and the *prediction time* becomes:

$$T \sim (1/\lambda_+) \log(L/\epsilon) \qquad T \sim (1/K) \log(L/\epsilon) \qquad (55)$$

The prediction time increases only logarithmically with the precision of the initial measurement. Chaotic states allow only short-term prediction.

4.3.2 Intermittency

Intermittency is defined by the occurrence of a signal that alternates randomly between long regular (laminar) phases (intermissions) and relatively short irregular bursts. Such a behavior occurs in many experiments.

4.4 Characterization of chaotic attractors

Possible criteria we have discussed for chaotic motion are:
- time dependence of the signal looks chaotic,
- power spectrum exhibits broadband noise,
- autocorrelation function decays rapidly,
- Poincaré map shows space-filling points,
- Non integer dimension.

4.5 Chaos in fluid dynamics

In a Rayleigh-Bénard convection experiment (see e.g. Schuster, 1969), a fluid is placed between two horizontally thermally conducting plates with the lower one warmer than the upper one. When the temperature difference ΔT exceeds a critical value δT_C, convection occurs (rolls resembling rotating parallel cylinders). Beyond a second threshold ΔT_{c_2} chaotic behavior occurs. The *Lorenz model* (1963) consists of three coupled ordinary differential equations and exhibited the first strange attractor to be studied numerically.

The Bénard experiment can be described by the following set of differential equations:

$$\dot{x} = -\sigma x + \sigma y \tag{56}$$
$$\dot{y} = rx - y - xz \tag{57}$$
$$\dot{z} = xy - bz \tag{58}$$

here σ and b are dimensionless constants which characterize the system and r is the control parameter which is proportional to δT. x is proportional to the circulatory fluid velocity, y characterizes the temperature difference between ascending and descending fluid elements and z is proportional to the deviations of the vertical temperature profile from its equilibrium value.

4.6 Applications

In this section we discuss several applications of the above mentioned concepts of non linear dynamics.

A short review about non linear dynamics in solar physics was given by Hanslmeier (1997).

Hanslmeier et al. (1994) made a first attempt to study the turbulent or non turbulent behavior of convective overshoot using Lyapunov exponents. The problem was however, that only 1-D data were available and no time series. They separated small scale from large scale motions using a filter to the data. At subgranular scales ($< 1''$) they found that the physics changes especially in the intergranulum (enhanced turbulence for temperature and velocity field).

Nesis et al. (2001) investigated the attractor in the three dimensional phase space spanned by intensity, Doppler velocity and turbulence (line broadening). They found that the attractor does not fill the entire phase space as one would have expected from high Reynolds numbers. The dimension of the attractor seems to be independent of the appearance of big granules and shear flow.

5 Observational constraints

5.1 General remarks

Let us consider an image formed in the x, y-plane by a telescope. Then because of the optics the observed intensity $I(x, y)$ has contributions from neighboring points.

Thus the intensity is given by the following convolution:

$$I(x,y) = \int\int_{-\infty}^{\infty} I_0(\xi,\eta) PSF(x,y,;\xi,\eta) d\xi d\eta \quad (59)$$

where PSF denotes the *point spread function*. Let $F_0(\vec{q})$ denote the Fourier transform of I_0 and $S(\vec{q})$ the Fourier transform of the PSF. $S(\vec{q})$ is called the transfer function and its modulus the *modulation transfer function* or MTF. In the Fourier representation the optical imperfections of the telescope and seeing can be written as a convolution:

$$MTF_{\text{total}} = MTF_{\text{telescope}} \bigotimes MTF_{\text{seeing}} \quad (60)$$

If the point spread function is rotationally symmetric in the x,y-plane, $PSF(x,y) = PSF(r)$ and $r = \sqrt{x^2+y^2}$, then

$$MTF(q) = 2\pi \int_0^{\infty} r PSF(r) J_0(2\pi q r) dr, \quad q = \sqrt{q_x^2 + q_y^2} \quad (61)$$

here J_0 is the Bessel function of order 0.

Let us consider a simple example: the so called *Airy image* of a point source formed by a diffraction-limited telescope with a circular unobstructed aperture (rotationally symmetric PSF):

$$PSF_D(r) = \frac{1}{\pi}[J_1(br)/r]^2 \quad (62)$$

J_1... Bessel function of order 1, $b = D\pi/\lambda f$; D... aperture, f... focal length. The first zero occurs at $r = r_1 = 3.832/b$, this is called the radius of the central Airy disc. The angle $\alpha_1 = r_1/f$ is considered as the resolution of a diffraction limited telescope:

$$\alpha_1 = 1.22\lambda/D \quad (63)$$

Generally one finds for the MTF:

$$MTF_D(q) = \frac{2}{\pi}\left[\arccos(q/q_m) - \frac{q}{q_m}\sqrt{1-(q/q_m)^2}\right] \quad (64)$$

Please note that spatial frequencies larger than $q_m = b/\pi$ are not transmitted by the telescope, $MTF_D = 0$ for $q \geq q_m$. The modulation transfer function is the factor by which a signal is reduced at each spatial frequency q. Thus e. g. an original intensity contrast of more than 10 % may be reduced to less than 1 % if a telescope just resolves that structure. This is extremely important for studying fine structures such as granulation which is of typical size of about $1''$, or even subgranular structures.

Averaged over periods of one second or longer, the seeing point spread function is rotationally symmetric and given by a Gaussian:

$$PSF_S(r) = \frac{1}{2\pi s_0^2} exp(-r^2/2s_0^2) \quad (65)$$

s_0 is a quantitative measure of the seeing. $s_0 = 1''$ is considered as good seeing. Since the Fourier transform of a Gaussian is another Gaussian, the MTF of seeing is given by:

$$MTF_s(q) = exp(-2\pi^2 s_0^2 q^2) \qquad (66)$$

One way to avoid seeing effects that are normally in the range of $1''$ is to choose a good observatory site. The evaluation of such sites can be done by measuring observable parameters.

For the evaluation of a good observatory site the Fried parameter is a good measure. The Fried parameter r_0 is defined as the aperture of an imaginary diffraction-limited telescope which has the same resolving power as a large optically perfect telescope used under the conditions of the actual seeing. Note that seeing is variable and therefore r_0 depends on the observatory site, time and varies with the wavelength:

$$r_0 \sim \lambda^{6/5} \qquad (67)$$

There exist several procedures how to estimate the seeing.

- In situ aircraft measurements with micro-thermal sensors.

- Radiosondes. These measurements give e. g. for the Canary Islands 20 cm $< r_0 <$ 30 cm.

- Spectral-ratio method (v. d. Lühe, 1984): If an object is structured at scales below the seeing limit, such structure will manifest in the Fourier transform.

- Seeing monitor (Brandt, 1970): attached to a telescope, intensity profiles across the solar limb are rapidly scanned and the gradient of the Gaussian profile $exp(-x^2/2\sigma^2)$ provides the blurring parameter.

Local turbulence can be minimized by a careful selection of the observatory site. A high building reduces ground-level turbulence. Local turbulence develops during the morning hours when the ground is heated by the Sun. E. g. at the Big Bear solar observatory (Ca, USA) or at the Huairou Solar Station (PR China) this effect is reduced because of the surrounding water (water has a high thermal inertia). Other possibilities to reduce ground-level turbulence are open structure of the telescope (e. g. the Dutch open telescope at La Palma) and a domeless design (70 cm VTT at Izana, Tenerife). Furthermore the telescope can be evacuated to minimize internal seeing due heating inside the telescope. A vacuum of a few mbar is sufficient. However one needs an entrance window and an exit window both have to withstand atmospheric pressure and their thickness must be in the order of about 1/10 of their diameter. Problems with such thick windows arise because of mechanical stresses, absorption and thus thermal effects. Instead of evacuation also tanks filled with He are used.

5.2 Techniques to improve the image quality

5.2.1 Adaptive optics

The principle of adaptive optics is very simple:

- measurements of wave-front perturbations,
- extraction of a correction signal,
- application of this signal to an active (tiltable or deformable) optical element,
- restoration of wavefront.

The telescope aperture is sampled by an array of lenslets, which in turn forms an array of images of the object. The subaperture size is e. g. 7 cm diffraction limits the rms contrast of the granulation images to 2.5–3 %, depending on the seeing conditions, compared to 6–8 % of the original value. Nevertheless, the image contrast is sufficient to compute cross-correlations between subaperture-images and a selected subaperture-image, which serves as reference. Cross-correlations are computed using:

$$corr(\Delta I) = \sum\sum I_M(\vec{x}) I_R(\vec{x} + \Delta_i) \tag{68}$$

where I_M denotes the subaperure image, I_R the reference image and Δ_i the pixel shift between subaperture image and reference image. Normally such a shift is in the range of ≤ 3 pixels.

The *Spot Tracker* was developed and tested at BBSO (Big Bear Solar Observatory) in 1982. The Spot Tracker used high-contrast solar features like small sunspots or pores to track solar images. Next generation of solar trackers was based on using low-contrast granulation patterns for tracking solar images. The main advantage of these trackers is that a granulation pattern is everywhere on the sun at least near the disk center and at any time (see e. g. Edwards et al. (1987) or von der Lühe et al. (1989), Rimmele et al. (1991), Ballesteros et al. (1996)).

5.2.2 Phase diversity

This technique is also very well suited to improve the image quality and reveal small scale structures. An object is measured in 2 different channels yielding the functions:

$$g_1(r) = f(r) * h_1(r) \qquad g_2(r) = f(r) * h_2(r) \tag{69}$$

Thus we have 2 measurements g_1, g_2 with three unknowns $f(r), h_1(r), h_2(r)$. The second measurement is made by a defocussing and therefore there exists a relation between $h_1(r)$ and $h_2(r)$, and written in the Fourier domain:

$$H_1(f) = |H_1(f)|e^{i\theta(f)} \qquad H_2(f) = |H_1(f)|e^{i\theta(f)+\alpha(f)} \tag{70}$$

And for the restoration:

$$F(f) = \frac{[G_1(f)H_1(f) + G_2(f)H_2(f)]}{[|G_1(f)|^2 + |G_2(f)|^2]} \tag{71}$$

A recent paper on phase diversity applied to photospheric fine structure was given by Tritschler and Schmidt (2002).

An implementation of a phase-diverse speckle imaging (PDS) technique for reducing the effects of aberrations (due to the earth's atmosphere and telescope optics) in solar images is described in the thesis of Löfdahl (1996). He demonstrated the application of this technique by a restoration and analysis of a $29'' \times 29''$ 70-minute time sequence of solar granulation and bright points.

5.3 How to observe fine structures?

In order to study the dynamics of solar photospheric fine structure 2-D observations are needed: from images we get information about the structures that are observed (granulation, intergranular lanes, pores ...), from spectrograms information about Doppler velocities, line broadening. The ideal case is simultaneous observation.

Birefringent filters were used by Lyot (1933) and Öhmnan (1938). The problems is the limited spectral resolution $R < 5 \times 10^4$ and the purity of available calcite elements. A tunable birefringent filter was used in the SOUP experiment (Title et al., 1986). At 5200 Å this has a passband of 0.05 Å. With the Multi-Channel Subtractive Double Pass spectrograph (MSDP) a 2-D field of view can be obtained simultaneously at a limited number of wavelengths (< 10). The spectral resolution is $R < 10^5$. A Fabry Perot Interferometer (FPI) has several advantages: it is capable of two-dimensional imaging, high spectral resolution ($R \sim 10^{5...6}$) which is comparable to those of high resolution grating spectrometers. For the Göttingen FPI a universal birefringent filter (UBF) which is tunable in the visible spectrum serves as an order sorter for the FPI. The instruments yield a passband of ≤ 30 m. An FPI consists of two partially transparent flat mirrors of high reflectivity $Re > 90\,\%$ which are parallel to each other and enclose a plane parallel 'plate' of air. Due to multiple-beam interference, a passband is produced with a very narrow FWHM which is given by the Airy formula for transmitted light. More details about the FPI can be found in e. g. Bendlin et al. (1992).

5.4 Techniques to obtain velocity vectors

In this paragraph we shortly describe methods that can be applied to obtain velocity maps from 2-D images of solar granulation. The cross correlation between two images separated by a sampling time delay τ is calculated. This time delay must be smaller than the lifetime of tracers in the image. Such a two-dimensional cross correlation is determined at each subregion (i. e. locally) by multiplying the intensity product with an apodizing window $W(x)$. The spatially localized cross correlation which is given by $C(\vec{\delta}, \vec{x})$ is a function of four dimensions:

- two-dimensional vector displacement $\vec{\delta}$ between the images.

- two-dimensional central location $\vec{x}$ of the window function.

Let $J_t(\vec{x})$ and $J_{t+\tau}(\vec{x})$ be two images, then

$$C(\vec{\delta}, \vec{x}) = \int J_t\left(\vec{\xi} - \frac{\vec{\delta}}{2}\right) J_{t+\tau}\left(\vec{\xi} + \frac{\vec{\delta}}{2}\right) W(\vec{x} - \vec{\xi}) d\vec{\xi} \qquad (72)$$

This integral is in principle over the full area of the images, in practice however it is limited to the size of $W(\vec{x})$. This window defines the spatial resolution of the vector displacement. It has the form of a Gaussian. The maximum of the cross correlation is the definition for the motion of the tracers because of their evolution. It must be stressed, that before forming the cross correlation, the images must be spatially filtered to remove large-scale components. This includes also e. g. intensity gradients which may bias the cross correlation. In order to reduce atmospheric seeing effects, the cross correlation can be averaged in time before locating its maximum. When the seeing is poor and the granulation contrast is low, the cross-correlation peak is broad and is a nearly constant function of displacement. For a more detailed description of this procedure see e. g. November and Simon (1988).

6 Solar granulation

The photosphere of the Sun is covered with a small cellular like pattern that is called granulation. In the bright granules hot matter rises whereas in the darker intergranular lanes cooler matter sinks down. The size of solar granulation is in the range of $1''$ and therefore spatially highly resolved observations require stable seeing conditions. In principle there are two observing techniques:

- taking images,

- doing spectroscopy.

A review about solar granulation was given by Muller (1999). Granulation can best be observed near solar disk center. Near the limb the phenomenon vanishes because of the decreased contrast since we observe at greater geometrical height near the limb and the convective overshooting of the granules vanishes there. Spectroscopic observations near the solar disk center yield e. g. vertical velocities and horizontal velocities as one approaches the limb. Measurements of displacements between successive images of time series yield horizontal velocities near disk center and vertical velocities near the limb. Doing a 2-D spectroscopy also 2-D velocity maps can be obtained near disk center. The dynamics with height may be studied by using spectral lines of different formation height or by observing in different wavelength regions.

The bisectors which divide line profiles into equal parts at the same intensity levels give information about the velocity gradients. A totally symmetric line profile gives a straight line as a bisector whereas due to convective motions the bisectors are not straight lines.

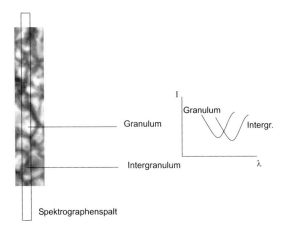

Figure 2: Observing solar granulation with a spectrograph. The entrance slit (spektrographenspalt) covers many granular/intergranular elements.

6.1 High resolution granulation observations

The Reynolds number in the lower solar photosphere is near to 10^{12}. Therefore, we have to expect that the granular flow field is highly turbulent.

High resolution granulation observations can be divided into white light images and spectroscopic observations. We give few examples of both kinds. Krieg et al. (2000) presented observations of granular velocities and their relation to the granular intensity pattern. They applied speckle methods to two-dimensional narrow band images in Na_2D from the Göttingen Fabry-Perot interferometer. The velocities refer to a geometric height of 50–200 km and velocity amplitudes of ± 2.2 km/s were found. High velocity regions are small-scale and the upflows coincide with bright borders of granules or with small-scale brightenings. They found also the –5/3 or –17/3 law of log (power) vs. log (wavenumber) which is expected for isotropic turbulence. Given isotropic turbulence, the different slopes stand for:

- –17/3 inertial convective range
- –5/3 inertial conductive range.

This was described e.g. in Espagnet et al. (1993). From considerations of fluid dynamics and from their numerical simulations Nordlund et al. (1997) stressed a possible laminar character of solar convection which is not compatible with isotropic turbulence.

All studies of solar granulation which is attributed to convection suffer from oscillations. To filter the 5 min oscillations it is necessary to go to the k-ω diagram – which can be done only in the case time series are available (see e.g. Espagnet et al., 1995). That means a filtering in time and space. If no time series are available only filtering in space is possible. Espagnet et al. (1996) have shown that the 5 min oscillations occur at wavelengths $> 4\ldots 8''$. Thus one can filter out these wavelengths.

Hirzberger et al. (1999) studied the evolution of solar granulation using an 80 min high resolution time series obtained at the Swedish La Palma telescope. To follow the evolution of individual granules, an automatic tracking algorithm was developed (more than 2600 individual granules were analyzed). The size of the individual images was about $30 \times 30''$ and the time interval between the individual images 18.9 s. Besides the usual reduction procedures a subsonic filter was applied. In each image granules are selected randomly and are tracked forward and backward in time. In order to obtain a stopping criterion, the correlation between boxes centered about the granules is calculated. Proper motions of the granules and residual shifts due to the distortion produce shifts smaller than 3 pixels in each direction (1 pixel $\sim 0.062''$. Therefore, the correlation was computed for ± 3 pixels. The process was stopped when the correlation was less than 0.2. When a granule splits the area changes dramatically and this serves as a further stopping condition for the tracking. They found a mean lifetime of about 6 min. The birth and death mechanism that were detected for granulation are:

- fragmentation – most frequent birth mechanism,

- merging – most frequent dying mechanism,

- spontaneous emergence from or dissolution into the background. Spontaneous emergence occurs very rarely, dissolution frequently.

The evolution of solar granulation deduced from 2-D simulations was investigated by Ploner et al. (1999). For death mechanism the authors consider the fragmentation and dissolution. Fragmentation is produced by buoyancy braking due to the stronger horizontal flow in larger granules. The expansion is due to pressure excess relative to neighboring granules (especially when these are dissolving). Dissolving granules are born small and shrink. Thus the fate of a granule depends on its properties of birth and its neighborhood. Kawaguchi (1980) detected that granules $> 2''$ predominantly fragment and granules $< 1''$ predominantly dissipate. Karpinsky and Pravdjuk (1998) detected different size ranges of granules with different fragmentation rates. Granules < 713 km rarely fragment, whereas granules > 1200 km show a large fragmentation rate.

Numerical granulation models (e. g. Steffen et al. 1989; Stein and Nordlund 1998; Gadun et al. 2000) show much less turbulence as expected or even almost laminar granular flow fields. Nordlund et al. (1997) interpreted this lack of turbulence as produced by the small pressure scale height in the solar photosphere and, thus, by the rapid expansion of upflowing matter when it enters into higher photospheric levels. Vice versa, increased turbulence should be visible in the intergranular lanes. This idea is supported by observations of e. g. Nesis et al. (1993, 1999) and by the numerical models of Rast (1995).

An ideal tool to study the fine structure of solar photospheric dynamics is to use a combination of two-dimensional high spatial resolution imaging and with spectroscopy. This was done e. g. by using a Fabry Perot Interferometer (FPI) at the VTT telescope at Tenerife.

In Hirzberger (2002) a 45 min time series of two-dimensional spectra was studied. The data were obtained with the Vacuum Tower Telescope at the Observatorio del Teide in Izaña, Tenerife. Scans over the non-magnetic ion Fe I 5576Å line of a quiet granular field at disk center were taken simultaneously with a time series of broad band images. From the spectra intensity and velocity maps have been calculated at different line-depths. It was found that intensities and velocities are well correlated at low photospheric levels. In the higher photosphere the intensity pattern dissolves but the velocity pattern shows almost no variation. Moreover, the intensity excess of small granules dissolves at lower heights than that of larger ones. The time evolution of the granular pattern shows a clear dependence of the lifetime of structures on the spatial wavenumber. The e-folding times of the temporal coherences decrease according to a power law with an exponent of $\beta = 3/2$ which is incompatible with the Kolmogorov energy spectrum of homogeneous and isotropic turbulence and might be taken as a hint against the overall turbulent character of granular motions.

In Fig. 3 an example of VTT observation with the Göttingen FPI is given. This image shows the edge of NOAA 9143, and the distance from disk center is $\cos \theta = 0.92$, the direction to the center about 75 degrees. The upper image is a speckle reconstruction of a broad bandwidth image (± 100 at 7090 Å). The speckle reconstruction was made of 150 individual images (exposure time 20 ms). The image in the middle gives Doppler velocities obtained from the bisectors at half value of the line intensity (Fe I 7090 Å). The scan through the line consists of 150 images which were obtained simultaneously with the broad band images. 15 positions (10 images/position) with a distance of 29.78 mÅ were measured. The FWHM of the FPI is 56 mÅ. The narrow band images were reconstructed following the procedures of Krieg et al. (1999). The spatial resolution in the broadband was $0.3''$, in the narrow band (Doppler) $0.4''$. Velocities directed towards the observer are blue. In the umbrae of pores the velocities are small, as well as in the light bridges and umbral dots – in the granules about ± 1.5 km/s. Above the pore in the middle a filamentary pattern of positive velocities can be seen which corresponds to a Evershed effect.

The lower image displays LCT velocities averaged from 50 broad band images at an interval of 47 s (i. e. 39 min of time series). The FWHM of the window function was $3.5''$, the maximum velocity was about 540 m/s. The left pore shows a slight inward motion which is well known for pores. The pore in the middle does not show a flow, at a distance from the pore a continuous flow away from the pore occurs. This is well known for spots but should be inverse for the pores (Roudier et al. 2002). The contours of the lower images are 75 % of the mean photospheric intensity obtained from an average over all 50 broad band images.

6.2 Granulation and magnetic fields

Magnetic flux tubes are a key to understand many solar phenomena. Flux tubes are believed to be associated with small network elements (about 0.25 arcsec) and the strength of the field is in the kG range (Stenflo, 1973). Convective motions sweep the diffuse magnetic field to the supergranule boundaries and flux tubes of moderate strength are formed. Parker (1978) suggested that the field is further compressed as a result of adiabatic cooling of the tube. Therefore, downdrafts associated with

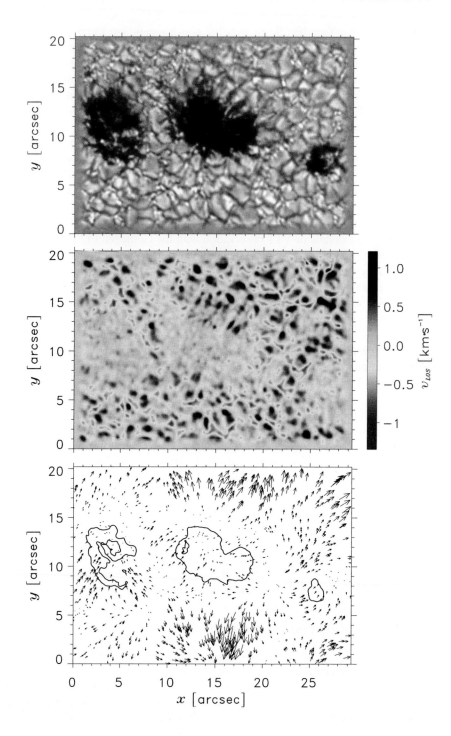

Figure 3: High resolution solar granulation observations. Courtesy: J. K. Hirzberger.

magnetic structures should appear. Dara et al. (1990) used a series of filtergrams in the magnetically sensitive Ca I line ($\lambda 6103$ obtained with the SPO Vacuum Tower Telescope and the universal filter and mapped the line of sight velocity and the longitudinal magnetic field. The effect of 5 min oscillation was eliminated and no association between magnetic field and line of sight velocity was found, only in a region with relatively low field strengths small downflows (300 m/s) were found.

Gurzinov et al. (1996) studied the generation of small scale dynamos. The generation of large-scale or mean magnetic fields is described by the kinematic dynamo theory of Steenbeck, Krause and Rädler. Small-scale magnetic fields can originate from the large-scale fields because of turbulent stretching and twisting plays an important role. In many cases the viscosity is much larger than the resistivity and the turbulent velocities cut off at a scale much larger than the scale at which resistivity destroys magnetic field. In high Reynolds number turbulence the SSF dynamo growth rate exceeds the growth rate of the mean field dynamo by a factor of $N_{Re}^{1/2}$.

A time series analysis of 2-D spectra of a sunspot at $\theta \sim 45°$ obtained with the FPI using the Fe I 7090.4 Å was made by Balthasar et al. (1996). They found that the gas plasma in the vicinity of the sunspot has a velocity of 500 m/s directed outwards, the Evershed flow shows a sharp decrease by about 1000 m/s at the outer boundary of the visible penumbra. The Evershed effect (Evershed, 1909) is a radial outward flow inside the penumbra reaching several km/s. The authors found Doppler shifts in the vicinity of the sunspot. These may be explained by granules moving away from the spot because the magnetic field widens with increasing height or by a continuation of the Evershed flow outside the sunspot.

6.3 Acoustic flux generation

What is the relation between the occurrence of the five minute oscillations and the granulation? This is still not solved and many authors tried to answer the question which is essential in order to understand the excitation mechanism and the internal properties. From theoretical studies Goldreich and Keeley (1977) or Goldreich et al. (1994) it is suggested that acoustic waves (that compromise the 5 minute oscillations) are stochastically generated by turbulent convection just beneath the photosphere. Brown (1991) predicted that their amplitude should increase with the amplitude of the turbulent motions. Therefore, the acoustic emission has to be expected from rather localized events. Goode et al. (1992) found these events in granules. Espagnet et al. (1996) found the most energetic oscillations in regions of expanding intergranular spaces with strong downflows. Rimmele et al. (1995) located the strongest acoustic flux above overshooting granules. Nesis et al. (1997) attributed enhanced full width at half maximum values to enhanced turbulence caused by enhanced acoustic flux. Khomenko et al. (2001) used a 30 min time series of CCD spectrograms of solar granulation with a spatial resolution of about $0\rlap{.}''5$ with the 70 cm VTT. The time resolution was 9.3 s. The Fe I line at $\lambda 5324$ was studied which has a line core formation height near the temperature minimum thus covering the whole range of the photosphere. The authors found amplitudes, phases and periods of the 5-min oscillations to be different above granules and intergranular lanes. Strong oscillations occurred well separated temporally and spatially. The oscillations above granules and intergranular lanes occur with different periods; the most energetic intensity oscillations occur above intergranular lanes; the most energetic velocity oscillations occur above granules and lanes with maximum contrast, i.e. above the regions with maximum convective velocities; velocity oscillations at the lower layers of the atmosphere lead oscillations at the upper layers in intergranular lanes. In granules the phase shift is nearly zero (Khomenko et al. 2001).

Rimmele et al. (1995) studied dark lanes in granulation and the excitation of oscillations using observations that were made at the Sacramento Peak Observatory. A 20 mÅ passband filter (universal birefringent filter and Fabry-Perot) was used to scan the profile of the magnetically insensitive Fe I $\lambda 5434$ line. The profile of that line was scanned at 14 wavelength positions. It took 32.5 s for each scan. In total 200 line scans were analyzed which correspond to more than 100 min. The images were stabilzed using a correlation tracker. From these data bisector line velocities were calculated at 10 different intensity levels in the line profile. Thus a 4-dimensional data set was obtained (x, y, z, t). The time sequence of velocity maps was processed by applying a Fourier-filter in the k-ω space. Thus contributions to the velocity signal that do not belong to the temporal-spatial frequency domain of the five minutes oscillations were eliminated. The authors used a bandpass for temporal frequencies, no spatial filter was applied. Using a Hilbert transform the data were filtered back to the time domain. Following Restaino et al. (1993) the mechanical flux can be derived from:

$$u \sim \frac{V^2}{\omega} \frac{\Delta \phi}{\Delta z} \tag{73}$$

here V is the velocity amplitude, $\Delta\phi/\Delta z$ is the observed phase gradient. The authors found that the formation or expansion of a dark lane precedes the appearance of an acoustic event in that lane. The most rapidly cooling regions generate the most acoustic noise. This is in accordance with models that acoustic noise is produced by rapid rapid cooling and collapse (Rast, 1993). The power needed for the generation of the acoustic modes is $\sim 10^{28}$ erg/s.

6.4 Fine structures inside a solar pore

A study of fine structures in solar pores was performed by Hirzberger et al. (2002). Using a 66 min time series of white light images ($\lambda = 5425 \pm 50$) of an active region obtained at the Swedish Vacuum Solar Telescope on la Palma a photometric analysis of the sub-structure of a granular light bridge in a large solar pore was performed. Due to the high quality of the data, the light bridge could be resolved into small grains embedded in a diffuse background with an intensity of about 85 % of the mean photospheric intensity. Proper motions up to 1.5 km/s were found and a lifetime distribution showing a maximum at 5 min and a second peak at 20 min was found. Processes typical for granulation such as fragmentation, merging and spontaneous origination and dissolution into the background were observed. Thus these motions can be interpreted as convective motions.

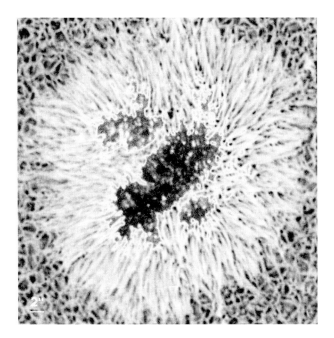

Figure 4: Umbral dots in sunspot NOAA 8580. The frame was taken on 13 June 1999 at the SVST, La Palma (observers P. N. Brandt, R. I. Bush, R. A. Shine, M. Sobotka). The penumbra and photosphere are shown in negative grey scale to increase the contrast of the umbra.

In this connection we can also discuss briefly the relation between magnetic fields and convection. Many authors have reported a reduced rms intensity fluctuation in abnormal granulation which is seen near strong fields (see e. g. Brandt and Solanki, 1990 and further references therein). The sizes of the convective cells are reduced and therefore the effect of lateral heating becomes stronger. Hirzberger et al. (2002) found that the rms fluctuations inside the light bridge are about 7.14 % and thus lower by more than 10 % than in quiet granulation. There seems to be some kind of transition between abnormal granulation, light bridges, penumbral grains and umbral dots. Rimmele (1997) found a strong correlation between the bright grains in the light bridges and upward velocities and in the darker lanes between the bright points downward velocities dominate. Sobotka et al. (1994) calculated 2-D power spectra and found a turbulent cascade in the intensities (which was found for normal granulation).

Dorotovic et al. (2002) made an analysis of an 11-hour series of high resolution white light observations of a large pore in an active region. The filamentary region attached to a pore was observed to change its structure back and forth between penumbra-like filaments and elongated granules. Moreover a clockwise rotation of this region around the center of the pore was detected during the whole observing period and this rotation had angular velocities decreasing with time from $7.6° \, h^{-1}$ to $2.7° \, h^{-1}$. Sobotka (1999) gave a review on the dynamics of motions inside sunspots and light bridges.

Acknowledgements

The author thanks the Austrian Fonds zur Förderung der Wissenschaftlichen Forschung for support. He also thanks J. A. Bonet, P. N. Brandt, A. Gadun, J. Hirzberger, A. Kucera, W. Mattig, R. Muller, A. Nesis, W. Poetzi, K. G. Puschmann, Th. Roudier, J. Rybak, M. Sobotka, M. Vazquez, and H. Wöhl for the many years of cooperation on solar granulation dynamics.

References

Asplund, M., Nordlund, Å., Trampedach, R., Allende Prieto, C., Stein, R.F. 2000, A&A 359, 729

Ballesteros, E., Collados, M., Bonet, J.A., Lorenzo, F., Viera, T., Reyes, M., Rodriguez Hidalgo, I. 1996, A&AS 115, 353

Balthasar, H., Schleicher, H., Bendlin, C., Volkmer, R. 1996, A&A 315, 603

Bendlin, C., Volkmer, R., Kneer, F. 1992, A&A 257, 817

Brandt, P.N. 1970, JOSO Ann. Rep. 1970, 50

Brandt, P.N., Solanki, S.K. 1990, A&A 231, 221

Brun, A.S., Turck-Chièze, S., Morel, P. 1998, ApJ 506, 913

Christensen-Dalsgaard, J. 1998, Space Sci. Rev. 85, 19

Dravins, D., Nordlund, Å 1990a, A&A 228, 184

Dravins, D., Nordlund, Å 1990b, A&A 228, 203

Dialetis, D., Macris, C., Prokakis, T., Sarris, E. 1986, A&A 168, 330

Edwards, C.G., Levay, M., Gilbreth, C.W., Tarbell, T.D., Title, A.M., Wolfson, C.J., Torgerson, D.D. 1987, BAAS 19, 929

Espagnet, O., Muller, R., Roudier, Th., Mein, N. 1993, A&A 271, 589

Espagnet, O., Muller, R., Roudier, Th., Mein, N., Mein, P. 1995, A&AS 109, 79

Espagnet, O., Muller, R., Roudier, Th., et al. 1996, A&A 313, 297

Evershed, J. 1909, MNRAS 69, 454

Fleck, B. 2001, in The Dynamic Sun, A. Hanslmeier, M. Messerotti and A. Veronig eds., Kluwer Dordrecht, 1

Gadun, A.S., Hanslmeier, A., Pikalov, K.N., et al. 2000, A&AS 146, 267

Goldreich, P., Keeley, D.A. 1977, ApJ 211, 934

Goldreich, P., Murray, N., Kumar, P. 1994, ApJ 424, 466

Goode, P.R., Gough, D., Kosovichev, A. 1992, ApJ 387, 707

Gruzinov, A., Cowly, S., Sudan, R. 1996, Phys. Rev. Lett. 77, 21, 4342

Hanslmeier, A. 1997, Hvar Obs. Bull. 21, 77

Hanslmeier, A., Nesis, A., Mattig, W. 1994, A&A 286, 263

Hanslmeier, A. 2002, The Sun and Space Weather, Kluwer, Dordrecht

Hirzberger, J., M., Bonet, J.A., Vazquez, M., Hanslmeier, A., Vazquez 1999, ApJ 515, 441

Hirzberger, J. Bonet, J.A., Sobotka, M., Vazquez, M., Hanslmeier, A. 2002, A&A 383, 275

Hirzberger, J. 2002, A&A 392, 1005

Kaplan, J.L., Yorke, J.A. 1979, Chaotic behavior of multidimensional difference equations, in Functional differential equations and the approximation of fixed points, eds H.O. Peitgen and H.O. Walther, Lect. Notes Math. 730, 204

Karpinsky, V.N., Pravdjuk, L.M. 1998, Kinematika i Fizika Nebesnukh Tel 14 (2), 119

Kawaguchi, I. 1980, Sol. Phys. 65, 207

Khomenko, E.V., Kostik, R.I., Shchukina, N.G. 2001, A&A 369, 660

Kolmogorov, A.N. 1941 Dokl. Akad. Nauk SSSR 30; 4:3201

Kolmogorov, A.N. (Translation by V. Levin) 1991, The local structure of turbulence in incompressible viscous fluid for very large reynolds numbers. Proc. R. Soc. Lond. A, 434:9-13

Krieg, J., Wunnenberg, M., Kneer, F., Koschinsky, M., Ritter, C. 1999, A&A 343, 983

Löfdahl, M.G. 1996, PHD Thesis

Lühe, O. von der, et al. 1989, A&A 224, 351

Lühe, O. von der 1984, J. Opt. Soc. America A 1, 510

Mehltretter, J.P. 1978, A&A 62, 311

Muller, R. 1999, in Motions in the Solar Atmosphere, A. Hanslmeier and M. Messerotti eds., Kluwer Dordrecht, 35

Nesis, A., Hanslmeier, A., Hammer, R., et al. 1993, A&A 279, 599

Nesis, A., Hammer, R., Hanslmeier, A., et al. 1997, A&A 326, 851

Nesis, A., Hammer, R., Kiefer, M., et al. 1999, A&A 345, 265

Nesis, A., Hammer, R., Roth, M., Schleicher, H. 2001, A&A 373, 307

Nordlund, Å, Spruit, H.C., Ludwig, H.-G., Trampedach, R. 1997, A&A 328, 229

November, L.J., Simon, G.W. 1988, ApJ 333, 427

Oda, N. 1984, Sol. Phys. 93, 243

Parker, E.N. 1978, ApJ 221, 328

Ploner, S.R.O., Solanki, S.K., Gadun, A.S. 1999, A&A 352, 679

Rast, M.P. 1993, ApJ 419, 240

Rast, M.P. 1995, ApJ 443, 863

Restaino, S.R., Stebbins, R.T., Goode, P.R. 1993, ApJ 408, L 57

Rimmele, T.R., Goode, P.R., Harold, E., Stebbins, R.T. 1995, ApJ 444, L119

Rimmele, T.R. 1997, ApJ 490, 458

Rimmele, Th., et al. 1991, ESO Tech. Rep

Rösch, J. 1962, in Trans. IAU 11B, 197

Roudier, Th., Bonet, J.A., Sobotka, M. 2002, A&A 395, 249

Schröter, E.H., Soltau, D., Wiehr, E. 1985, Vistas Astron. 28, 519

Schuster, H.G. 1969, Deterministic Chaos, VCH Weinheim

Sobotka, M., Bonet, J.A., Vazquez, M. 1994, ApJ 426, 404

Sobotka, M. 1999, in Motions in the Solar Atmosphere, A. Hanslmeier and M. Messerotti eds., Kluwer Dordrecht, 71

Steffen, M., Ludwig, H.-G., Krüß, A. 1989, A&A 213, 371

Stein, R.F., Nordlund, Å 1998, ApJ 499, 914

Stenflo, J.O. 1973, Solar Phys. 32, 41

Tatarski, V.I. 1961, Wave Propagation in a Turbulent Medium, Mc Graw-Hill New York

Title, A.M., Tarbell, T.D., Topka, K.P., Ferguson, S.H., Shine, R.A., & The SOUP Team. 1989, ApJ 336, 475

Tritschler, A., Schmidt, W. 2002, A&A 382, 1093

The Interstellar Medium and Star Formation: The Impact of Massive Stars

José Franco, Stan Kurtz, Guillermo García-Segura

Instituto de Astronomía, UNAM
Apdo. Postal 70-264, 04510 Mexico DF, Mexico
pepe@astroscu.unam.mx, s.kurtz@astrosmo.unam.mx, ggs@astrosen.unam.mx

Abstract

We discuss the interaction of massive stars with their parental molecular clouds. A summary of recent observational developments is presented, followed by a synopsis of the dynamical evolution of H II regions and wind-driven bubbles in high-pressure cloud cores. Both ultracompact H II regions and ultracompact wind-driven bubbles can reach pressure equilibrium with their surrounding medium. The structures stall their expansion and become static and, as long as the ionization sources and the ambient densities remain about constant, the resulting regions are stable and long lived. For cases with negative density gradients, and depending on the density distribution, some regions never reach the static equilibrium condition. For power-law density stratifications, $\rho \propto r^{-w}$, the properties of the evolution depend on a critical exponent, w_{crit}, above which the ionization front cannot be slowed down by recombinations or new ionizations, and the cloud becomes fully ionized. This critical exponent is $w_{crit} = 3/2$ during the expansion phase. We discuss three regimes, $w < 3/2$, $3/2 < w < 3$, and $w > 3$. We summarize numerical modeling that shows how the movement of the exciting star through a non-uniform medium can give rise to hyper and ultracompact H II regions, with the presence of extended emission.

1 Introduction

Young stars display vigorous activity and their energy output stirs and heats the gas in their vicinity. Low-mass stars have a small energy output and affect only small volumes, but their collective action can provide partial support against the collapse of their parental clouds and could regulate some aspects of cloud evolution (e. g., Norman & Silk 1980; Franco & Cox 1983; McKee 1989). In contrast, massive stars (> 8 M$_\odot$) inject large amounts of radiative and mechanical energy from their birth until their final explosion as supernovae. In the low-density interstellar medium of a gaseous galaxy, the combined effects of supernovae, stellar winds, and H II region expansion destroy star-forming clouds, produce the hottest gas phases, create large expanding bubbles, and are probably responsible for both stimulating and shutting off the star formation process at different scales (e. g., Cox & Smith 1974; Salpeter 1976; McKee & Ostriker 1977; Franco & Shore 1984; Cioffi & Shull 1991; Franco et al. 1994; Silich et al. 1996).

The strong UV radiation field from massive stars creates large H II regions. Young H II regions have large pressures and, when the pressure of the surrounding medium is low, they expand rapidly, driving a strong shock wave ahead of the ionization front. Expanding H II regions ionize and stir the parental cloud and, when the ionization front encounters a strong negative density gradient, they create "champagne" flows and can also generate cometary globules and elephant trunks (e. g., Yorke 1986; Franco et al. 1989,1990; Rodriguez-Gaspar et al. 1995; García-Segura & Franco 1996). These flows are generated by the pressure difference between the H II region and the ambient medium, and they are responsible for the disruption of the cloud environment (e. g., Whitworth 1979; Larson 1992; Franco et al. 1994).

Stellar winds and supernova explosions, on the other hand, generate overpressured regions which drive shock waves into the ambient medium (see review by Bisnovatyi-Kogan & Silich 1995). The resulting wind-driven bubbles and supernova remnants, either from a single progenitor or from an entire association, are believed to generate most of the structuring observed in a gaseous galactic disk (e. g., Heiles 1979; McCray & Snow 1979). Many observed structures in the Milky Way and in external galaxies have been ascribed to this stellar energy injection (e. g., Heiles 1979, 1984; Deul & den Hartog 1990; Palous et al. 1990, 1994). These bubbles may create fountains or winds at galactic scales (e. g., Shapiro & Field 1976; Chevalier & Oegerle 1979; Bregman 1980; Cox 1981; Heiles 1990; Houck & Bregman 1990), and their expanding shocks have also been suspected of inducing star formation (e. g., Herbst & Assousa 1977; Dopita et al. 1985). Thus, stellar activity creates a collection of cavities with different sizes, and can be viewed as an important element in defining the structure and activity of star-forming galaxies.

2 Observations

2.1 Classical H II Region Formation and Expansion

The classical picture of H II regions presents the idea of a newly-formed massive star that begins to produce large quantities of ultraviolet photons and hence to ionize the surrounding medium. If the star is still embedded within the molecular cloud from which it formed, then the molecular gas will dissociate and ionize, and an ionization front will form, expanding rapidly outward to form the initial Strömgren sphere. The ionization front expands very rapidly at first, but soon slows, and approaches the sound speed after several recombination times

The newly ionized gas will be greatly over-pressured with respect to the original molecular gas, and, in the classical picture, will expand toward pressure equilibrium with its environment. In pressure equilibrium, the final radius can be expressed in terms of the initial Strömgren radius, R_S, as

$$R_f = R_S \left(\frac{2T_e}{T_0}\right)^{2/3}. \qquad (1)$$

For typical values of T_e and T_0, the ionized and neutral temperatures, one expects the final radius to be nearly 100 times its initial value. The strong shock approximation for the expansion gives the Strömgren sphere radius as a function of time (eqn. 13). Adopting $c_i = 10$ km s^{-1} as the sound speed, and assuming a factor of 100 expansion, implies that the H II region will reach pressure equilibrium in about 3×10^6 years. At this time, it will have a radius of about one parsec. The basic physical processes relevant for the detailed structure and expansion of H II regions are discussed extensively in Osterbrock (1989). Other important considerations driving the formation and expansion phases are discussed in Franco et al. (1990).

The salient point here is that the general scheme outlined above for the expansion of recently formed H II regions is problematic, in that it predicts that the region will rapidly expand out of its initial, small, dense (ultracompact) state. In particular, adopting 0.1 pc as the nominal size for an ultracompact (UC) H II region, one expects the lifetime of this phase to be of order R/c_i or about 10^4 yr for the parameters used here. Many more UC H II regions are seen than this naive view predicts, hence the conclusion that their lifetimes must be longer than anticipated by the simple theory.

During the past decade seven distinct attempts have been made to address this so-called "lifetime problem" (see Garay & Lizano 1999, Kurtz et al. 2000, and Churchwell 2002 for reviews). The various attempts fall into two categories: those which prolong the life by providing an additional confining force, and those which extend the lifetime by replenishing the ionized gas. A key result of the former attempts is the realization that the "compactness" (i. e., the small diameter and high density) of the H II region may not be a good indicator of its youthfulness.

As indicated in Table 1, the physical properties of H II regions span orders of magnitude. The classes most closely linked to star formation are the smallest, densest, and possibly youngest stages: the compact, ultracompact, and hypercompact H II regions. Compact H II regions were first identified as such by Mezger et al. (1967), who characterized them with sizes from 0.06 to 0.4 pc and electron densities $\sim 10^4$ cm^{-3}. The smallest and densest of these came to be known as ultracompact H II regions. By the mid-1980s around a dozen of this regions were known. The situation changed dramatically with the survey of Wood & Churchwell (1989) which identified over 70 UC H II regions. Further surveys by Becker et al. (1994), Kurtz et al. (1994), and Walsh et al. (1998) identified many more UC H II regions, and the total number of probable ultracompacts currently stands at more than 500. As indicated in Table 1, these regions are (observationally) defined as having sizes $\lesssim 0.1$ pc and densities $\gtrsim 10^4$ cm^{-3}. Although not generally mentioned, it is implicitly assumed that the ionization is maintained by one or more early-type stars, and not by some other mechanism (e. g., accretion shocks).

Table 1: Physical Parameters of H II Regions

Class of Region	Size (pc)	Density (cm^{-3})	EM (pc cm^{-6})	Ionized Mass (M$_\odot$)
Hypercompact	~ 0.003	$\gtrsim 10^6$	$\gtrsim 10^9$	$\gtrsim 10^{-2}$
Ultracompact	$\lesssim 0.1$	$\gtrsim 10^4$	$\gtrsim 10^7$	$\sim 10^{-2}$
Compact	$\lesssim 0.5$	$\gtrsim 5 \times 10^3$	$\gtrsim 10^7$	~ 1
Classical	~ 10	~ 100	$\sim 10^2$	$\sim 10^5$
Giant	~ 100	~ 30	$\sim 5 \times 10^5$	$10^3 - 10^6$
Supergiant	>100	~ 10	$\sim 10^5$	$10^6 - 10^8$

2.2 Beyond Ultracompact H II Regions

Most of the radio surveys mentioned above suffered from two observational biases: they were insensitive to larger regions ($\gtrsim 30''$) and to higher density regions (emission measures $\gtrsim 10^8$ pc cm^{-6}). The former bias arises from the nature of interferometric observations, which are sensitive to only a certain range of angular sizes, determined by the range of baseline lengths. The latter bias results from optical depth effects, resulting in substantially lower flux densities for high density regions at the wavelengths used in the surveys. We discuss each of these biases in turn, and summarize more recent data which are less sensitive to these observational restrictions.

2.2.1 Ultracompact H II regions with extended emission

The original UC H II region surveys were typically made with arc second or better angular resolution. For the radio interferometers used in the surveys (the VLA and the ATCA) this implies observations that are insensitive to structures larger than roughly $30''$.

To date, two efforts to search for larger-scale structures associated with UC H II regions have been made: Kurtz et al. (1999) and Kim & Koo (2001). Kim & Koo selected 16 sources that had large single-dish to interferometer flux density ratios. Their low-resolution VLA observations found extended continuum emission in all cases. Kurtz et al. observed 15 sources selected randomly from published UC H II region survey data, and found extended emission in 12. In eight of these, they suggest that a direct physical relationship may exist between the ultracompact and extended components. It must be noted that at present the physical relationship between the two components is unconfirmed. In general, the evidence for a relationship is strictly morphological (see Figure 1). Kim & Koo report radio recombination line observations which support the hypothesis of a relation based on close agreement of the line velocities between the ultracompact and extended components. Although highly suggestive, this is not a foolproof test, as physically distinct H II regions at nearly the same distance could present very similar LSR velocities.

Further observational work is needed in this area, to confirm the relationship between the compact, high density component and the extended, low density compo-

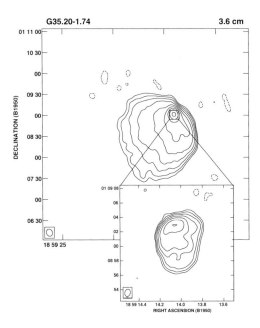

Figure 1: Two views of the H II region G35.20−1.74. The original view (inset), made with sub-arcsecond resolution, was sensitive only to structures smaller than about 20″ (Wood & Churchwell 1989). Subsequent lower resolution observations, sensitive to structures up to 3′ in size, show the full extent of the ionized gas in the region. The region has a cometary morphology at both arcsecond and arcminute scales.

nent. If confirmed, there will be a number of ramifications for our understanding of UC H II regions. First, we may have substantially underestimated the stellar content of the star forming regions by missing a large fraction of the free-free continuum emission, and hence the UV photons required to maintain the ionization. The extended emission may also require further theoretical developments. For example, the presence of ultracompact gas within a more diffuse envelope cannot be easily explained by the majority of the currently popular models. Although this appears to be the case for Orion (which has ionized components ranging from n_e of about 300 cm^{-3} to around 10^5 cm^{-3}), none of the currently proposed models can easily explain the co-existence of these two components of distinctly different densities. A schematic outline has been suggested by Franco et al. (2000a) and by Kim & Koo (2001), and will be discussed below (see Figure 4).

2.2.2 Hypercompact H II regions

The second observational bias of the existing UC H II region surveys was their use of centimeter waves ($\lambda = 2$–6 cm or $\nu = 5$–15 GHz). The problem is clear from the expression for the optical depth of thermal free-free emission:

$$\tau_\nu \approx 0.082\, T^{-1.35} \nu^{-2.1} \int n_e\, ds. \qquad (2)$$

As seen at once from this expression, the transition from optically thick to optically thin emission (at $\tau \approx 1$) rises to higher frequencies with increasing emission measure. Indeed, for emission measures greater than 4×10^9 pc cm^{-6}, the turnover frequency will be > 30 GHz ($\lambda < 1$ cm): very high emission measure H II regions are optically thick at centimeter wavelengths. The UC H II region surveys were typically performed at mid-centimeter wavelengths, and hence would be in the optically thick portion of the spectrum (where flux density nominally falls as ν^2) for very high emission measure (i.e., very dense) H II regions. A population of very small, very dense H II regions, optically thick in the centimeter regime and with peak millimeter flux densities of order 100 mJy, would have been largely missed by the existing surveys.

In retrospect, the first hint that this population might exist was reported by Gaume et al. (1995) in their 1.3 cm VLA observations of NGC7538-IRS1. They reported a "hypercompact core" of size 1100 AU and note that the optical depth must be of order one at their observing frequency, although a few individual clumps "may have an optical depth somewhat greater". They found an unusually broad (160 km s^{-1} FWZI) radio recombination line toward the source, and reported electron densities in the range of 10^4–10^5 cm^{-3}.

More recently, the 7 mm receivers at the VLA have offered an even better opportunity to probe this potential population of very small, very dense H II regions. Carral et al. (1997) found 7 mm continuum emission coincident with a water maser clump where no 6 cm emission had been found by Wood & Churchwell (1989; see Figure 2). De Pree, Goss & Gaume (1998) report 7 mm observations of the Sgr B2 Main complex, and find 22 candidate hypercompact H II regions, with average emission measures exceeding 10^9 pc cm^{-6}. A handful of other candidate hypercompact regions has been reported in other star forming regions (see Kurtz 2002 and references therein).

These regions are typically an order of magnitude smaller and two orders of magnitude denser than traditional UC H II regions. As such, they are even more deserving of a new designation than the UC H II regions, which were roughly classified as being five times smaller and five times denser than compact H II regions (Mezger et al. 1967, Wood & Churchwell 1989). We adopt the designation hypercompact or HC H II regions, defined by sizes of tens of milliparsecs and densities greater than 10^6 cm^{-3}, or emission measures in excess of 10^9 pc cm^{-6} (see Table 1).

A precise definition for "hypercompact" H II regions is not entirely clear. When the phrase has been used in the literature it has generally been taken to mean "very small". As indicated by the parameters listed in Table 1, we adopt a definition based on both size ($\lesssim 30$ mpc) and density ($\gtrsim 10^6$ cm^{-3}). Roughly equivalent is to require an emission measure $\gtrsim 10^9$ pc cm^{-6}. Ultimately, it may be desirable to classify hypercompact H II regions on the basis of the nature or evolutionary state of the object. The presence of broad radio recombination lines and the presence of hot molecular cores are possible criteria for such a classification. But at present, our understanding of hypercompact regions is not sufficiently advanced to allow a phenomenological definition.

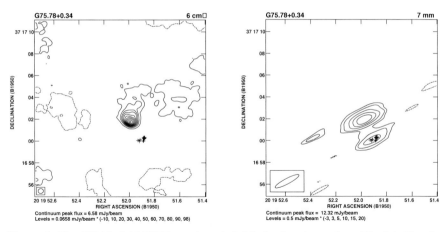

Figure 2: The G75.78+0.34 UC H II region. At left is the 6 cm image from Wood & Churchwell (1989). The crosses indicate water masers, from Hofner & Churchwell (1996). At right is the same sky area, imaged at 7 mm. Continuum emission is detected at the water maser position; from Carral et al. (1997). The spectrum of the millimeter-wave source has been extended to 3 and 1 mm points with OVRO, and is consistent with free-free rather than dust emission.

2.3 Density Gradients in UC and HC H II Regions

Clearly the simple Strömgren sphere model of H II regions is a very great idealization. One of the assumptions that is probably not well-met is of a homogeneous medium surrounding the newly-formed star. Studies too numerous to mention have shown that molecular clouds have a great deal of structure. Hot molecular cores (currently a favored spot for massive star formation) are expected to show distinct density gradients (see Kurtz et al. 2000 and references therein). As such, UC or HC H II region are likely to form in the presence of density gradients, and the observed properties of the H II regions may provide valuable information about cloud density structure.

The radio-continuum spectral index of a non-uniform H II region depends on the exponent of the density gradient – provided that the central density is high enough to remain optically thick. The spectral index behavior for a number of different density structures was worked out a quarter of a century ago by Olnon (1975) and Panagia & Felli (1975). The main result of these studies is that the radio spectral index, α, for unresolved sources with power-law forms $n_e \propto r^{-\omega}$, is in the range $-0.1 \leq \alpha \leq +2$. For a pure power-law, which has a singularity at the center, the plasma in the central regions is optically thick at all frequencies and the spectral index is given by the simple formula $\alpha = (2\omega - 3.1)/(\omega - 0.5)$; see Olnon (1975). For a flattened power-law, which has a finite, constant-density core (or any other density distribution that is bounded at $r = 0$), the optical depth toward the center depends on frequency, and the spectral index has a more complicated functional form. At frequencies below a critical value where the central core becomes optically thick, the spectral index for both the flattened and pure power-laws tends to the same value and the expression above applies.

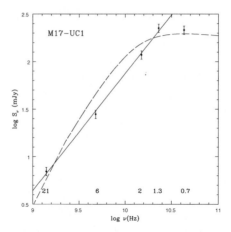

Figure 3: The flux density distribution for M17-UC1. The straight line is a least squares fit to the data from 21 to 1.3 cm, which yields a spectral index of +1.2. The dashed line is the best fit assuming a uniform, spherical model.

Despite the utility of these results, they have not been applied to the study of H II region densities until rather recently (Franco et al. 2000b). Figure 3 shows the application of the technique to the candidate HC H II region M17-UC1. Note that the continuum emission remains optically thick into the millimeter range, as expected for hypercompact H II regions. Absent, however, is the nominal +2 spectral index of a homogeneous, spherical H II region. Rather, M17-UC1 has a spectral index of +1.2. Assuming a power-law density distribution, the expression above yields a density gradient exponent of 3. This is a rather steep gradient; significantly higher than values found from molecular line studies of star-forming clouds which more typically show density gradient exponents of about two. Franco et al. (2000b) find power-law exponents of -2, -2.5, and -4 for the three regions they study, and discuss several alternative explanations for why the free-free continuum spectral index might over-estimate the density gradient.

High ambient pressures (De Pree, Rodríguez & Goss 1995; Xie et al. 1996) may be able to confine the ultracompact component. If these pressures occur within a hot molecular core, it could give rise to a hypercompact H II region. Density structure within the cloud (but outside the high density core) could provide the ultracompact component, with a power-law density gradient, and also extended emission. A conceptual picture of this scheme is presented in Fig. 4, and we discuss numerical simulations of such a model in §4.

3 Evolution: Analytical Results

3.1 The Expansion Phase

The evolution can be derived by assuming the existence of a thin shell with mass M, containing all the swept-up ambient gas. This approximation can be applied

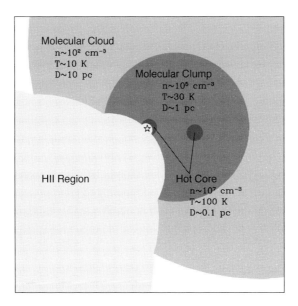

Figure 4: Schematic representation of the model proposed by Kim & Koo to explain the extended emission around UC H II regions. The figure is not to scale; taken from Kim & Koo (2000), their Figure 8. Franco et al. (2000a) also suggest density structure within the parent molecular cloud as a possible cause of the extended emission.

to H II regions when the fraction of mass eroded by the ionization front from the shell is small. This is true for constant, increasing, or mildly decreasing density stratifications, but is not applicable for strongly decreasing gradients because the shell is easily eroded by photoionization (Franco et al. 1990). Here we consider the constant density case and the thin shell approximation can be used without restrictions. Neglecting magnetic fields and self-gravity, the equation of motion of the shell, located at a distant R from the central star, is

$$4\pi R^2 (P_i - P_0) = \frac{d}{dt}(Mv), \qquad (3)$$

where P_i and P_0 are the internal and external pressures, and v is the shell velocity. Assuming that the shell radius can be written as a power-law in time, $R = R_0 \zeta^\beta$ (where R_0 is the initial radius of the region, $\zeta = (t_0 + t)/t_0$, and t_0 is a reference initial time), with constant R_0 or β (i.e., neglecting the existence of terms with $\dot{R}_0$ and $\dot{\beta}$), the right hand side of the equation becomes

$$\frac{d}{dt}(Mv) = \left(\frac{4\beta - 1}{3\beta}\right) 4\pi \rho_0 R^2 v^2. \qquad (4)$$

The assumption of a constant β, as we will see below, does not hold for cases with $P_0 > 0$, but the present approximations can still be applied in a piece-wise fashion (i.e., one can use them in segments, changing the values for R_0 and t_0).

The equation of motion can then be written as

$$P_i - P_0 = \rho_0 v^2 \frac{(4\beta - 1)}{3\beta},\tag{5}$$

and the shell evolution is given by

$$\frac{dR}{dt} = \left[\frac{3\beta}{(4\beta - 1)}\frac{(P_i - P_0)}{\rho_0}\right]^{1/2}.\tag{6}$$

Thus, one can write the formal solution to the equation of motion simply as

$$R - R_0 = \int_1^\zeta \left[\frac{3\beta}{(4\beta - 1)\rho_0}(P_i - P_0)\right]^{1/2} t_0 \, d\zeta.\tag{7}$$

This formal solution can be applied to H II regions, wind-driven bubbles, and SN remnants (see below). Obviously, the integration is not straightforward unless $P_0 = 0$, or the pressure difference, $P_i - P_0$, can be written as an explicit function of either R or t. In general this is not possible, but one can *always* check the behavior at early and late times, when the external pressure can be neglected and when the internal and external pressures become approximately equal. Here we illustrate the behavior in both cases.

3.2 Pressure Equilibrium

For the case of expanding H II regions, using $R_0 = R_S$ as the initial Strömgren radius, we can set the solution for the expansion simply as

$$R_{HII}(t) = R_S \zeta^\beta.\tag{8}$$

The average ion density inside the H II region at any time is given by

$$<n_i> = \left[\frac{3 F_\star}{4\pi \alpha_B}\right]^{1/2} R_{HII}^{-3/2},\tag{9}$$

and the corresponding internal pressure is

$$P_i = \left(\frac{3 k^2 T_i^2 F_\star}{\pi \alpha_B}\right)^{1/2} R_{HII}^{-3/2}.\tag{10}$$

The region, then, evolves as

$$R_{HII} = R_S + \int_1^\zeta \left(\frac{3\beta}{(4\beta - 1)\rho_0}\left[\left(\frac{3 k^2 T_i^2 F_\star}{\pi \alpha_B R_S^3 \zeta^{3\beta}}\right)^{1/2} - P_0\right]\right)^{1/2} t_0 \, d\zeta.\tag{11}$$

At early times, when $P_0/P_i \ll 1$, the integration with constant β gives $R_{HII} \propto \zeta^{1-3\beta/4}$. Thus, our initial assumption for the power-law implies that $\beta = 1 - 3\beta/4$, and one gets $\beta = 4/7$. The solution, then, is written as

$$R_{HII} = R_S \left[1 + \frac{7}{4}\left(\frac{8 k T_i}{3\mu_H}\right)^{1/2} \frac{t_0}{R_S}(\zeta^{4/7} - 1)\right].\tag{12}$$

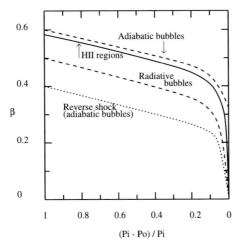

Figure 5: The evolution of the exponent β as a function of the normalized pressure difference $(P_i - P_0)/P_i$.

Defining $c_i \simeq (8\,k\,T_i/3\mu_H)^{1/2}$, and recalling the initial definition of the power-law $\zeta^{4/7} = R_{HII}/R_S$, the reference time is given by $t_0 = (4/7)(R_S/c_i)$. As expected, one recovers the well-known law for H II region expansion in a pressureless medium with a constant density

$$R_{HII} \simeq R_S \left(1 + \frac{7}{4}\frac{c_i t}{R_S}\right)^{4/7}. \qquad (13)$$

Using this explicit time dependence, the internal pressure decreases as

$$P_i = P_{i,0} \left(1 + \frac{7}{4}\frac{c_i t}{R_S}\right)^{-6/7}, \qquad (14)$$

where $P_{i,0}$ is the pressure at $t = 0$.

The expansion continues until $P_i \to P_0$, and the internal pressure tends to a constant value. In this limit, the time dependence in the formal solution vanishes, giving $3\beta/2 \to 0$. These limits show that, for a constant density medium, β evolves as a function of the pressure difference from 4/7 to zero. Figure 5 shows the evolution of β, as a function of the normalized pressure difference $(P_i - P_0)/P_i$, for H II regions and wind-driven bubbles (García-Segura & Franco 1996). The exponents were derived from high-resolution numerical simulations performed in one dimension for the evolution inside high density cores.

When pressure equilibrium is reached, the ion density is simply given by

$$n_{i,\text{eq}} = \left(\frac{P_0}{2kT_i}\right) \simeq 3.6 \times 10^4 P_7 T_{\text{HII},4}^{-1} \text{ cm}^{-3}, \qquad (15)$$

where $P_7 = P_0/10^{-7}$ dyn cm^{-2}, and $T_{\text{HII},4} = T_i/10^4$ K. The equilibrium radius of the H II region, then, corresponds to a Strömgren radius at this equilibrium density

$$R_{S,\mathrm{eq}} \approx 2.9 \times 10^{-2} \, F_{48}^{1/3} \, T_{\mathrm{HII},4}^{2/3} \, P_7^{-2/3} \quad \mathrm{pc}, \tag{16}$$

where $F_{48} = F_\star/10^{48} \, \mathrm{s}^{-1}$.

For high-pressure cores with $r_c \sim 0.1$ pc, the photoionized regions can reach pressure equilibrium without breaking out of the core. The resulting sizes are similar to those of the ultracompact class and indicate that UC H IIs can be explained by simple pressure equilibrium. This is in agreement with Xie et al. (1996), who show that smaller UC H IIs are embedded in higher pressure cores. Also note that the equilibrium values with $P_7 = 1$, $T_{\mathrm{HII},4} \sim 1$, and $F_{48} \sim 1$, are *very similar* to the average sizes and electron densities in UC H IIs (see Figure 151 of Kurtz et al. 1994). The apparent longevity problem of UC H IIs is rooted in the notion that young H II regions should grow fast and reach the expanded state on a relatively short time-scale. This statement is false, however, if the external pressure is large and halts the expansion at a small radius: *in pressure equilibrium, UC H IIs are stable and long lived.*

3.3 H II Evolution in Decreasing Density Gradients

If a star is born at a distance smaller than $R_{S,\mathrm{eq}}$ from the core boundary, the speeds of the ionization and shock fronts are modified by the negative density gradient. The gas of the H II regions is accelerated in supersonic flows and no static solution in pressure equilibrium exists (Tenorio-Tagle 1982; Franco et al. 1989, 1990). Under these conditions the ultracompact stage is indeed a transient phase. Here we assume isothermal clouds with $\rho \propto r^{-2}$, but one can easily find the solutions for the general power-law case, $\rho \propto r^{-w}$ (Franco et al. 1990). For $R_s \geq r_c$, the initial ionization front reaches the core radius with a speed

$$U_c \simeq 90 \, \alpha_0 \, n_3 \, r_{17} \left[\left(\frac{R_s}{r_c} \right)^3 - 1 \right] \, \mathrm{km \, s^{-1}}, \tag{17}$$

and in a time scale

$$t_c \simeq 130 \, \alpha_0^{-1} n_3^{-1} \ln \left[\frac{1}{1 - (r_c/R_s)^3} \right] \, \mathrm{yr}, \tag{18}$$

where $r_{17} = r_c/10^{17}$ cm. Afterwards, the ionization front enters the density gradient and its speed becomes

$$U_{if} = \frac{U_c}{(R_s/r_c)^3 - 1} u(w), \tag{19}$$

with

$$u(w) = \begin{cases} (r_c/r_i)^{2-w} \left[(R_s/r_c)^3 + 2w\beta - 3\beta (r_i/r_c)^{1/\beta} \right] & \text{for } w \neq 3/2, \\ (r_c/r_i)^{1/2} \left[(R_s/r_c)^3 - 1 - 3\ln(r_i/r_c) \right] & \text{for } w = 3/2, \end{cases} \tag{20}$$

where r_i is the location of the front, and $\beta = (3 - 2w)^{-1}$. This defines a critical exponent, corresponding to the maximum density gradient that is able to stop the ionization front,

$$w_f = \frac{3}{2}\left[1 - \left(\frac{r_c}{R_s}\right)^3\right]^{-1}, \qquad (21)$$

and above which no initial H II radius exists. Note that for $R_s/r_c > 2$ the critical value becomes $w_f \simeq 3/2$. Obviously, the concept of a critical exponent is not restricted to power-law stratifications and can also be applied to other types of density distributions, for instance, exponential, gaussian, and sech^2 profiles (see Franco et al. 1989).

For $w \neq 3/2$, the initial H II region radius can be written as

$$R_w = g(w)R_s, \qquad (22)$$

with

$$g(w) = \left[\frac{3 - 2w}{3} + \frac{2w}{3}\left(\frac{r_c}{R_s}\right)^3\right]^{\beta}\left(\frac{R_s}{r_c}\right)^{2w\beta}, \qquad (23)$$

where R_s is the Strömgren radius for the density n_c. The solution for $w = 3/2$ is

$$R_{3/2} = r_c \exp\left\{\frac{1}{3}\left[\left(\frac{R_s}{r_c}\right)^3 - 1\right]\right\}. \qquad (24)$$

After the formation phase has been completed in clouds with $w \leq w_f$, the H II region begins its expansion phase. For simplicity, we assume that the shock evolution starts at $t = 0$ when R_w is achieved. Given that the expansion is subsonic with respect to the ionized gas, the density structure inside the H II region can be regarded as uniform and its average ion density at time t is

$$\rho_i(t) \simeq \mu_i \frac{(9 - 6w)^{1/2}}{3 - w}(2n_c)R_s^{3/2}R^{-3/2}(t), \qquad (25)$$

where μ_i is the mass per ion, and $R(t)$ is the radius of the H II region at the time t. For $w \leq 3/2$, the radius can be approximated by

$$R(t) \simeq R_w\left[1 + \frac{7 - 2w}{4}\left(\frac{12}{9 - 4w}\right)^{1/2}\frac{c_i t}{R_w}\right]^{4/(7-2w)}, \qquad (26)$$

where c_i is the sound speed in the ionized gas. The ratio of total mass (neutral plus ionized), $M_s(t)$, to ionized mass, $M_i(t)$, contained within the expanded radius evolves as

$$\frac{M_s(t)}{M_i(t)} \simeq \left[\frac{R(t)}{R_w}\right]^{(3-2w)/2}. \qquad (27)$$

This equation indicates: i) for $w < 3/2$, the interface between the ionization front and the leading shock accumulates neutral gas and its mass grows with time to exceed even the mass of ionized gas, and ii) for $w = 3/2 = w_{crit}$, the two fronts move together without allowing the formation and growth of a neutral interface. Note that the decreasing ratio predicted by equation (27) for $w > 3/2$ is physically meaningless and it only indicates that the ionization front overtakes the shock front (and proceeds to ionize the whole cloud). Thus, regardless of the value of the critical exponent for the formation phase, w_f, the expansion phase is characterized by a critical exponent with a well defined value, $w_{crit} = 3/2$, which is independent of the initial conditions.

For $3/2 < w < w_f$, the ionization front overtakes the shock and the whole cloud becomes ionized. In this case, the pressure gradient simply follows the density gradient. The ionized cloud is set into motion, but the expanded core (now with a radius identical to the position of the overtaken shock) is the densest region and feels the strongest outwards acceleration. Then, superimposed on the general gas expansion, there is a wave driven by the fast-growing core (the wave location defines the size of the expanded core), and the cloud experiences the so-called "champagne" phase. This core expansion tends to accelerate with time and two different regimes, separated by $w = 3$, are apparent: a *slow* regime with almost constant expansion velocities, and a *fast* regime with strongly accelerating shocks. The slow regime corresponds to $3/2 < w < 3$ and the core grows approximately as

$$r(t) \simeq r_c + \left[1 + \left(\frac{3}{3-w}\right)^{1/2}\right] c_i t, \tag{28}$$

where for simplicity the initial radius of the denser part of the cloud has been set equal to r_c, the initial size of the core. For $w = 3$ the isothermal growth is approximated by

$$r(t) \simeq 3.2 r_c \left[\frac{c_i t}{r_c}\right]^{1.1}. \tag{29}$$

For $w > 3$, the fast regime, the shock acceleration increases with increasing values of the exponent and the core expansion is approximated by

$$r(t) \simeq r_c \left[1 + \left(\frac{4}{w-3}\right)^{1/2} \left(\frac{\delta + 2 - w}{2}\right) \frac{c_i t}{r_c}\right]^{2/(\delta+2-w)}, \tag{30}$$

where $\delta \simeq 0.55(w-3) + 2.8$.

3.4 Wind-driven Bubbles

The evolution of the cavity created by a stellar wind, i.e., a wind-driven bubble, can also be derived with the thin shell approximation described above. The thermalization of the wind creates a hot shocked region enclosed by two shocks: a reverse shock that stops the supersonic wind, and an outer shock that penetrates the ambient gas.

The gas processed by each shock is separated by a contact surface: the contact discontinuity. The kinetic energy of the wind is transformed into thermal energy at the reverse shock producing a hot gas (e. g., Weaver et al. 1977). The cooling properties are poorly known for a strong wind evolving in a high-density, dusty molecular core, but the shocked ambient gas cools very quickly and a thin external shell is formed on time-scales of a few years. Thus, the ambient gas is collected in a thin shell by the outer shock during most of the evolution. The case of the shocked stellar wind is less clear because the cooling time can be substantially longer than in the shocked ambient gas and it is difficult to define when the thin shell is formed behind the reverse shock. Thus, one can derive the limits for the evolution of the reverse shock in both the adiabatic and radiative modes.

The density in a steady wind, with a constant mass loss, decreases as

$$\rho_w = \frac{\dot{M}}{4\pi r^2 v_\infty}, \tag{31}$$

where $\dot{M}$ is the stellar mass-loss rate, and v_∞ is the wind speed. The pressure in the shocked wind region is defined by the wind ram pressure, $\rho_w v_\infty^2$, at the location of the reverse shock and is given by

$$P_i = \frac{\dot{M} v_\infty}{4\pi R_{rs}^2}, \tag{32}$$

where R_{rs} is the radius of the reverse shock.

3.4.1 Adiabatic Case

For an adiabatic bubble evolving in a constant density medium and powered by a constant mechanical luminosity, L_w, the thermal energy of the shocked wind region grows linearly with time, $E_{th} = 5 L_w t/11$ (Weaver et al. 1977). The shocked ambient medium is concentrated in the external thin shell and the the bubble radius, R_b is the radius of the contact discontinuity. The thermal pressure of the bubble interior changes as $P_i = (5 L_w t)/(22\pi R_b^3)$. This pressure is equal to that given by equation (32), and the locations of the reverse and outer shocks are related by mass conservation

$$R_{rs} = R_b^{3/2} \left(\frac{11 \dot{M} v_\infty}{10 L_w\, t} \right)^{1/2}. \tag{33}$$

The solution for R_b, then, also provides the evolution of the reverse shock. The initial radius in this case is very small, and we simply set $R_b = R_0(t/t_0)^\beta$, where R_0 now represents the bubble radius at some reference time t_0. Again, as in the H II region case, the formal solution is

$$R_b = \int_0^t \left[\frac{3\beta}{(4\beta - 1)\rho_0} \left(\frac{5 L_w t^{1-3\beta} t_0^{3\beta}}{22\pi R_0^3} - P_0 \right) \right]^{1/2} dt. \tag{34}$$

At early times, $P_i \gg P_0$ and P_0 can be neglected in the above equation. The exponent is then defined by $\beta = 1 + (1 - 3\beta)/2$, giving the well known adiabatic expansion in a medium with constant density $R_b \propto L_w^{1/5} \rho_0^{-1/5} t^{3/5}$. The position of the reverse shock (eqn. 33), then, evolves as $R_{rs} \propto t^{2/5}$. With this time dependence, the internal pressure drops as $P_i \propto t^{-4/5}$. At later times, when $P_i \to P_0$, the bubble reaches quasi-equilibrium with the ambient gas and the growth is $R_{b,eq} \propto (L_w/P_0)^{1/3} t^{1/3}$. In pressure equilibrium, the radius of an *adiabatic* bubble grows at a slow rate, but *no steady state* solution exists during this stage. The final radius of the reverse shock is simply given by the balance between the external and the wind ram pressures. The wind density at equilibrium is $\rho_{w,eq} = P_0/v_\infty^2$, and the location of the reverse shock is

$$R_{rs,eq} = \left[\frac{\dot{M} v_\infty}{4\pi P_0}\right]^{1/2} \simeq 2.3 \times 10^{-2} \left[\frac{\dot{M}_6 \, v_{\infty,8}}{P_7}\right]^{1/2} \text{ pc,} \qquad (35)$$

where $\dot{M}_6 = \dot{M}/10^{-6} \, M_\odot \, \text{yr}^{-1}$, and $v_{\infty,8} = v_\infty/10^8 \, \text{cm s}^{-1}$. Using mass conservation in the shocked wind region, $M_{sw} = \dot{M} t$, the quasi-equilibrium radius of an *ultracompact* wind-driven bubble is given by

$$R_{b,eq} = R_{rs,eq} \left(1 + \frac{3 L_w t}{8\pi P_0 R_{rs,eq}^3}\right)^{1/3}. \qquad (36)$$

Clearly, for adiabatic bubbles, β goes from $3/5$ at early times to $1/3$ at late times.

3.4.2 Radiative Case

Once radiative losses become important, the hot gas loses pressure and the bubble collapses into a simple structure: the free-expanding wind collides with the cold shell and the gas is thermalized and cools down to low temperatures in a cooling length. At this moment, the shell becomes static at the radius $R_{rs,eq}$.

If the bubble becomes radiative before the final radius, $R_{rs,eq}$, is reached, the shell is pushed directly by the wind pressure. For this radiative bubble case, the formal solution is

$$R_b = \int_0^t \left[\frac{3\beta}{(4\beta - 1)\rho_0} \left(\frac{\dot{M} v_\infty t_0^{2\beta}}{4\pi R_0^2 t^{2\beta}} - P_0\right)\right]^{1/2} dt. \qquad (37)$$

The early times solution, with $P_i \gg P_0$, gives the relation $\beta = 1 - \beta$, and one recovers the well-known solution for radiative bubbles $R_b \propto t^{1/2}$. The ram pressure now evolves as $P_i \propto t^{-1}$. At late times, one gets $\beta \to 0$ and the shell reaches the final equilibrium radius (eqn. 35). Thus, the exponent in this case evolves from $1/2$ to 0. Note that if the bubble starts on an adiabatic track and becomes radiative before pressure equilibrium is achieved, the exponent varies from $3/5$ to $1/2$, and then to zero.

4 Evolution: Numerical Models

4.1 Initial considerations

For a spherically symmetric, isothermal, and self-gravitating cloud, the density structure in equilibrium is proportional to r^{-2}. For our 1-Dimensional study, we set $r = 0$ at the cloud center, and assume a central core with constant mass density ρ_c and radius r_c, giving a density structure for $r \geq r_c$ of $\rho = \rho_c(r/r_c)^{-2}$.

The gravitational acceleration at locations inside the constant density core, $r \leq r_c$, is given by

$$g_r = -\frac{4\pi G}{r^2}\rho_c \int_0^r r^2 dr = -\frac{4\pi G}{3}\rho_c r, \tag{38}$$

where the minus sign indicates that the force is directed toward the origin. For positions outside the core, $r > r_c$, the acceleration grows as

$$g_r = -\frac{4\pi G}{r^2}\rho_c \left(\int_0^{r_c} r^2 dr + r_c^2 \int_{r_c}^r dr \right) = -4\pi G \rho \left(r - \frac{2}{3}r_c \right). \tag{39}$$

In hydrostatic equilibrium, the pressure difference between two positions located at radii r_1 and r_2 is given by $\Delta P = -\int_{r_1}^{r_2} \rho g_r dr$. Assuming a vanishing pressure at $r \to \infty$, the pressure at the core boundary, r_c, is

$$P(r_c) = \frac{10\pi G}{9}\rho_c^2 r_c^2, \tag{40}$$

and the total pressure at the center is

$$P(0) = P_0 = \frac{2\pi G}{3}\rho_c^2 r_c^2 + P(r_c) = \frac{8}{5}P(r_c) \simeq 2 \times 10^{-7}\, n_6^2 r_{0.1}^2 \quad \text{dyn cm}^{-2}, \tag{41}$$

where $n_6 = n_c/10^6$ cm^{-3}, and $r_{0.1} = r_c/0.1$ pc. Hence, the pressure inside the core varies less than a factor of two between $r = 0$ and $r = r_c$. The central pressure is about two-thirds the average pressure in a logatropic (i.e., $\rho \propto r^{-1}$) cloud with a similar mass column density (see McKee 1989).

Using $r_{0.1} = 1$ and the maximum core density value derived from several cloud complexes, $n_c \sim 5 \times 10^6$ cm^{-3} (e.g., Bergin et al. 1996), one gets a rough upper bound for the expected core pressures

$$P_0 \simeq 5 \times 10^{-6} \quad \text{dyn cm}^{-2}. \tag{42}$$

This value for the central pressure is larger than the *thermal* values inferred from observations, but is very similar to the expected *maximum* total value (magnetic + turbulent), 5×10^{-6} dyn cm^{-2}. The agreement is fortuitous, and merely shows that self-gravity is capable of producing such high core pressures.

Given the wide range in observed cloud properties, it is not possible (and is probably meaningless) to define an "average" pressure value. Here we are interested in the effects appearing at high core pressures, and simply use $P_0 = 10^{-7}$ dyn cm^{-2} as a reference value for a massive star forming core. Note that the corresponding core mass is

$$M_c \simeq \left(\frac{\pi P_0}{G}\right)^{1/2} r_c^2 \sim 10^2 \, P_7^{1/2} r_{0.1}^2 \, M_\odot, \qquad (43)$$

which, for $P_7 \sim 1$ and $r_{0.1} \sim 1$, is well within the range of observationally derived core masses (i.e., from 10 to about 300 $M_\odot$).

For our 2-Dimensional study we adopt an improved scheme: an isothermal cloud divided into an internal core of radius r_c and an external zone, for which $P = \rho \, c_s^2$ (c_s = constant) in hydrostatic equilibrium,

$$\vec{\nabla} P = -\rho \, \vec{g} \Rightarrow \vec{\nabla} \rho = -\frac{\rho}{c_s^2} \, \vec{g}. \qquad (44)$$

Assuming that the cloud is spherically symmetric, $\vec{\nabla} \longrightarrow \frac{d}{dr}$; $\vec{g} \longrightarrow g_r$

$$\frac{d\rho}{dr} = -\frac{\rho}{c_s^2} \, g_r. \qquad (45)$$

We first assume that the density in the external zone falls off as a power law r^{-2}, which is the case of a self-gravitating isothermal cloud:

$$\rho(r) = \rho_c \, (r/r_c)^{-2} \quad \text{for } r \geq r_c. \qquad (46)$$

Solving for g_r using (39) and (40) we find

$$g_r = \frac{2 \, c_s^2}{r} = \frac{2 \, c_s^2}{r_c} \, (r/r_c)^{-1} \quad \text{for } r \geq r_c. \qquad (47)$$

Inside the core, g_r must grow from zero up to $2 \, c_s^2/r_c$. Then, a simple manner is assuming that g_r grows linearly in this zone, such as $g_r = A \, (r/r_c)$. Thus, in order to join the external solution (41), we require that $A = 2 \, c_s^2/r_c$.

This implies that the density distribution inside the core is found by integrating (39) with the new g_r, giving

$$\rho(r) = \rho_0 \exp\left[-(r/r_c)^2\right] \quad \text{for } r \leq r_c, \qquad (48)$$

where ρ_0 is the central density at $r = 0$. Equating (40) and (42) at $r = r_c \Rightarrow \rho_c = \rho_0/e$.

In summary, the final solution for the density distribution of the cloud is given by

$$\rho(r) = \begin{cases} \rho_0 \exp\left[-(r/r_c)^2\right] & \text{for } r \leq r_c \\ \rho_0/e \, (r/r_c)^{-2} & \text{for } r \geq r_c \end{cases} \qquad (49)$$

and the gravitational acceleration

$$g_r = \begin{cases} 2 \, c_s^2/r_c \, (r/r_c) & \text{for } r \leq r_c \\ 2 \, c_s^2/r_c \, (r/r_c)^{-1} & \text{for } r \geq r_c \end{cases} \qquad (50)$$

Our second assumption, for the next set of simulations, is that the density structure in the external zone falls off as a power law r^{-3}, and using the above approach, we find

$$\rho(r) = \begin{cases} \rho_0 \exp\left[-3/2 \ (r/r_c)^2\right] & \text{for } r \leq r_c \\ \exp(-3/2) \ \rho_0 \ (r/r_c)^{-3} & \text{for } r \geq r_c \end{cases} \quad (51)$$

and the gravitational acceleration

$$g_r = \begin{cases} 3 \ c_s^2/r_c \ (r/r_c) & \text{for } r \leq r_c \\ 3 \ c_s^2/r_c \ (r/r_c)^{-1} & \text{for } r \geq r_c \end{cases} \quad (52)$$

We also assume that the modeled stars were born in situ in their parental cloud, i.e., we are not modeling 'run-away' stars. The largest expected stellar velocity that such a cloud can produce is given by the dispersion velocity. Noting that $P = 16/9 \ \pi \ G \ \rho_c^2 \ r_c^2 = \rho_c \ c^2$, and solving for c, we find $c = 4.07 \ r_{0.1} \ n_6^{1/2}$ km s^{-1} where $r_{0.1}$ is the core radius in units of 0.1 pc, and n_6 the density in units of 10^6 cm^{-3}. Thus, stellar velocities up to ~ 13 km s^{-1} can be considered in this study, for core densities of 10^7 cm^{-3} and for $r_{0.1} = 1$.

4.2 Hydrodynamical Experiments

The 1-D and 2-D hydrodynamical computations were done with the magneto-hydrodynamic fluid solver ZEUS-3D. This code is the updated, 3-D version (3.4) of the two-dimensional code ZEUS-2D (Stone & Norman 1992), an Eulerian explicit code which integrates the equations of hydrodynamics for a magnetized ideal gas. ZEUS-3D *does not* include radiation transfer, but we have implemented a simple approximation to derive the location of the ionization front for arbitrary density distributions (see Bodenheimer et al. 1979 and Franco et al. 1989, 1990). This is done by assuming that ionization equilibrium holds at all times, and that the gas is fully ionized inside the H II region. We perform the 1-D simulations in spherical polar coordinates, and the position of the ionization front in any given direction (θ, ϕ) from the photoionizing source is given by $\int n^2(r, \theta, \phi) r^2 dr \approx F_\star/4\pi\alpha_B$. The stellar wind is set as usual: several computational zones at the origin are set with the wind conditions; $\rho = \dot{M}/4\pi r^2 v_\infty$, $T = $ constant, and $v_\infty = $ constant. The models include the Raymond & Smith (1977) cooling curve above 10^4 K. For temperatures below 10^4 K, the unperturbed gas is treated adiabatically but the shocked gas region is allowed to cool with the radiative cooling curves given by Dalgarno & McCray (1972) and MacDonald & Bailey (1981). Finally, the photoionized gas is always kept at 10^4 K, so no cooling curve is applied to the H II regions (unless, of course, there is a shock inside the photoionized region).

4.2.1 Ultracompact Regions: 1-D Gas Dynamical Simulations in the Core

The calculations including the effects of a large external pressure were performed with three different 1-D models called "**R**", "**W**", and "**RW**" (for "radiation" only,

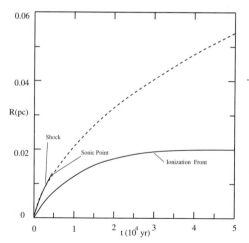

Figure 6: The time evolution of an ultracompact H II region in 1-D (Model R). The solid lines show the locations of the ionization and shock fronts. For reference, the analytical expansion law for H II regions evolving in a constant density medium is also shown (dashed line). The medium has $n_0 = 10^7$ cm^{-3}, $T_0 = 100$ K, and the stellar ionizing flux is $F_\star = 10^{48}$ s^{-1}.

"wind" only, and a combination of "radiation plus wind", respectively). Model R, shown in Figure 6, includes only the effects of a steady photoionizing radiation field with $F_\star = 10^{48}$ s^{-1}. As stated before, the photoionized gas is kept at 10^4 K. Model W is displayed in Figure 7, and has only an input stellar wind with $\dot{M} = 2.48 \times 10^{-8}$ M$_\odot$ yr^{-1} and $v_8 = 2$. Finally, model RW (Fig. 8) includes both energy inputs, using the previous values. This choice of parameters is arbitrary, but they are within the range of values expected for massive stars embedded in dense cores. Also, given that the evolutionary features have a moderate dependence on most parameters, the results are not particularly parameter-dependent. The simulations do not include magnetic fields or turbulent motions. To partially compensate for the magnetic and turbulent pressure, we use large temperatures and densities: the ambient ISM in all models is set to $n_0 = 10^7$ cm^{-3} and $T_0 = 100$ K.

Figure 6 summarizes the expansion of the UC H II defined as Model R. The ionization front (lower solid curve) decelerates with time until, at pressure equilibrium, it becomes a steady front at the final radius (~ 0.02 pc in this case). Its speed is always subsonic with respect to the ionized medium, but it becomes supersonic, and later turns subsonic, with respect to the ambient medium. The advance of the ionization front maintains a shock during a time $t \leq 5 \times 10^3$ yr, and then the expansion becomes subsonic (upper solid line). At this stage, the compressed shocked gas (which is accumulated in the swept-up neutral shell) tends to re-expand to its original density and drives a compression wave into the unperturbed medium. This wave continues to expand even after the ionization front has reached its final equilibrium value. For reference, the early times analytical solution (i.e., $P_0 \ll P_i$) for a constant density medium is displayed as the dashed line. This solution is meaningless for $P_0 \sim P_i$, and departures from the numerical results in the presence of the external thermal pressure are obvious.

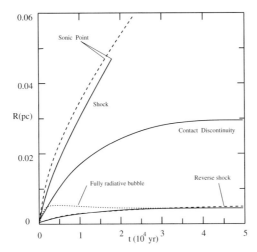

Figure 7: The time evolution of an ultracompact stellar wind bubble in 1-D (Model W). Radiative cooling (above 10^4 K) in the shocked gas is computed with the Raymond & Smith cooling curve. The solid lines show the locations of the (reverse and leading) shocks and contact discontinuity. For reference, the analytical solutions (reverse and leading shocks) for adiabatic wind-driven bubbles evolving in a constant density medium are also shown (dashed lines). The evolution of an isothermal bubble model (i.e., fully radiative) is also displayed: the positions of the contact discontinuity and the reverse shock are given by the dotted line. The ambient medium has $n_0 = 10^7$ cm^{-3}, and $T_0 = 100$ K. The wind has $\dot{M} = 2.48 \times 10^{-8}$ M$_\odot$ yr^{-1} and $v_8 = 2$.

Figure 7 displays the expansion of the ultracompact wind-driven bubble defined as Model W. The reverse shock (lower solid curve) moves outwards until it becomes a steady shock at ~ 0.004 pc. The evolution of the location of the contact discontinuity (middle solid curve) has an early $t^{3/5}$ behavior, up to about 10^4 yr. Then the expansion decelerates to a $t^{1/3}$ mode, and finally stalls due to radiative cooling. The structure at pressure equilibrium is just a hot cavity surrounded by an unperturbed ISM. The final collapse of the hot region occurs after a cooling time for the shocked wind gas, and the final radius is given by the stagnation radius of the reverse shock. This occurs at $\sim 10^6$ yr, much later than the final time-scale shown in Figure 7. The evolution of the swept-up shell (upper solid line) has the same behavior as that described in Figure 6: the compressed gas moves supersonically at early times, but, re-expands and drives an outward compression wave when the shell tends to the subsonic regime. As noted before, the wave continues its expansion after the reverse shock has reached its final equilibrium value. The early times (i.e., $P_0 \ll P_i$) solution for an adiabatic shocked wind (Weaver et al. 1977) is displayed by the dashed line. Note that the external shock and the contact discontinuity are assumed to be located at the same place in the analytical solution. This occurs in the numerical simulations only at the very early stages, but the shell becomes thicker with time, and the departure of the contact discontinuity evolution from the analytical solution is extreme.

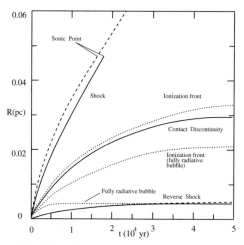

Figure 8: Same as Fig. 7, but for an ultracompact composite bubble, Model RW. The stellar flux is $F_\star = 10^{48}$ s^{-1}. Two runs are shown: one model is done with the Raymond & Smith cooling curve (above 10^4 K), the other one is isothermal (fully radiative).

Figure 8 shows the evolution for the combined radiative and mechanical energy input of Model RW. The bulk properties of the outer and reverse shocks are similar to those already described for the pure wind case of Figure 7. The only appreciable difference is in the behavior of the contact discontinuity at early times, which now is the inner boundary of the H II region. In this case, the ionized part of the swept-up shell (dotted line) has a temperature of $\sim 10^4$ K, and the expansion of the shocked wind region is affected by the enhanced thermal pressure of the photoionized shell. This causes a slower expansion velocity of the hot bubble at the beginning. Note that the position of the contact discontinuity at early times remains below the corresponding curve in Figure 7. The pressure in the photoionized region, however, decreases with time until pressure equilibrium within the whole structure is achieved, and the later evolution for the contact discontinuity is similar to that of Figure 6. In summary, the main difference between models RW and W is that the expansion in Model W occurs at constant external pressure while Model RW expands in a time-variable external thermal pressure.

4.3 2-D gas dynamical simulations: Blister Formation

We have chosen Cartesian coordinates with the Y-axis being the symmetry axis (slab geometry) in this study. Thus, the star can be placed at any location along the X-Z plane. This particular choice is very safe, since it does not include any axis artifact in the computations (as would occur in cylindrical coordinates).

The setup in all models is similar and is summarized in Table 2. The star is fixed on the computational mesh and it remains static during the computation. The stellar motion (for models A2 and B2) is then translated into a stream of gas in the whole computational mesh. With time, the outer boundary of the Z axis is updated with the oncoming gas at the stellar velocity. This update depends on the nature of the

Table 2: Model attributes

Model	Stellar Velocity km s^{-1}	Density Law ($r \geq r_c$)	Linear Scale pc
A0	0	r^{-2}	0.4
B0	0	r^{-3}	0.4
A2	2	r^{-2}	0.5
B2	2	r^{-3}	0.5

problem, i.e., it takes into account the slope of the ISM density distribution. All models have the same numerical resolution of 250 zones along the X and Z axes.

We use the same modeling approach as above, i.e., we solve a radial integral to find the position of the Strömgren radius (this time in Cartesian coordinates). Then, the minimum temperature inside the H II region is set to the photoionization equilibrium temperature ($\sim 10^4$ K).

The multidimensional modeling of a realistic UC H II includes the hypersonic stellar wind (i.e., an ultracompact bubble) and is computationally expensive because the wind velocities from massive main sequence stars are of order 10^3 km s^{-1}, when their bubble sizes are of the order of 10^{-2} pc. This forces the Courant condition to calculate very small time-steps during the simulations. Thus, 2-D stellar wind simulations of order 10^3 yr are feasible, but simulations of order 10^5 yr are not. For this reason, we do not include effects of the stellar wind in the models. However, we showed above that the size of the computed UC H II is mainly determined by the thermal pressure of the ionized gas at these high densities. Thus, models A and B are qualitatively correct.

We began our 2-D study by computing several cases (not shown here) where the star is located at the center of the core. For a core radius of 0.1 pc, a central density of 10^7 cm^{-3} and $F_\star = 10^{48}$ s^{-1}, the models behaved very much like the 1-D solutions. This is understandable, because the gaussian core of (48) produces a plateau at the center, which resembles the uniform medium of the 1-D case. Thus, pressure equilibrium was achieved for all models at timescales which matched the one calculated above. The novel aspect is that the final density structure is affected by gravity. This important feature will be discussed below. The first set of models (A0 and B0) are calculated with the stars fixed at the core edge (Figs. 9, 10). The purpose of these models is to study the expansion and dynamics of the H II regions in the density ramps given by (49) and (51) for $r \geq r_c$. The ionized gas acquires a flat density distribution during the first forty thousand years in both cases (Fig. 9), however, as time passes, the ionized gas notices the cloud gravity and makes the necessary re-adjustments to come into hydrostatic equilibrium. This feature is clearly observable in model A0, where pressure equilibrium is achieved and the final density distribution follows the initial density ramp. Although model A0 reaches pressure equilibrium and the H II region does not expand after 10^5 yr, the gas is not stationary. Rather, gas is continuously injected into the ISM by photoevaporation. This is visible in the lower left panel of Figure 10, where a wave pattern appears to the left of the H II region, at the neutral ramp. Model B0, on the other hand, does not

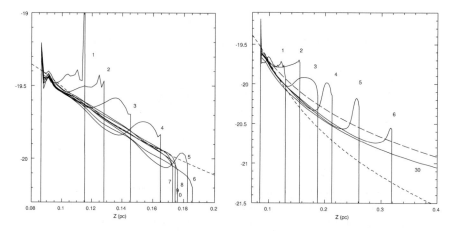

Figure 9: (left) Evolution of the photoionized gas density of model A0. Labels are in units of 2×10^4 yr. The star is located at $z = 0.1$. The short dashed line represents the initial atmosphere divided by 200, which is the expected value at equilibrium (power law r^{-2} for $r \geq r_c$). (right) Same as on the left, but for the B0 model. The short dashed line represents the initial atmosphere divided by 200, which is the expected value at equilibrium (power law r^{-3} $r \geq r_c$). The long dashed line is a power law r^{-2} with the same density value at the stellar location.

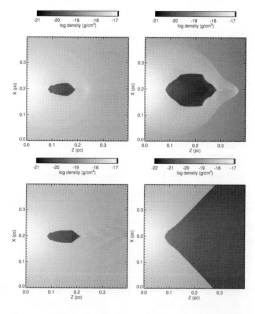

Figure 10: Gas density, snapshots of the two stationary models A0 (left) and B0 (right). The star is located 0.1 pc from the cloud center, just at the edge of the core $[(x,z)_\star = (0.2, 0.1)]$. Times in years are $t = 1 \times 10^5$ (top), $t = 6 \times 10^5$ (bottom). Model A0 reaches pressure equilibrium, while model B0 produces a blister.

The Interstellar Medium and Star Formation: The Impact of Massive Stars

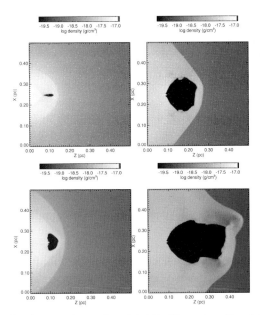

Figure 11: Gas density snapshots of model A2. Times in units of 10^5 years are (top to bottom) 0.8, 1.6 (left), 2.4, 3.2 (right). The star departs from the cloud center with $v_\star = 2$ km s^{-1}. The last stellar position is 0.65 pc from the center; X and Z coordinates are fixed to the star $[(x,z)_\star = (0.25, 0.1)]$.

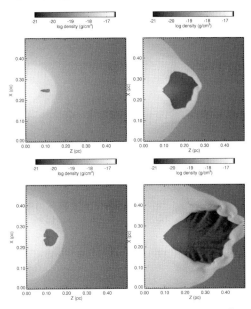

Figure 12: Gas density snapshots of model B2. Times in units of 10^5 years are (top to bottom) 0.8, 1.2 (left), 1.66, 2 (right). The star departs from the cloud center with $v_\star = 2$ km s^{-1}. The last stellar position is 0.4 pc from the center; X and Z coordinates are $[(x,z)_\star = (0.25, 0.1)]$.

reach pressure equilibrium in the computational domain, and a blister-like region is produced. Gas is permanently photoevaporated from the cloud, producing a steady state mass loss rate from the cloud. Due to photoevaporation, the ionized gas does not follow the imposed ramp ($\rho \propto r^{-3}$); instead, the final solution is closer to the r^{-2} ramp.

Note that the density ramps in these cases are off-center from the stellar locations, i.e., the origin of the power laws do not coincide with the stellar coordinates, so the solutions given by Franco et al. (1990), and references therein, are not applicable.

The second set of models (A2 and B2) include stellar motion. The stars depart from the cloud center with $v_\star = 2$ km s^{-1}, and after crossing the gaussian cores at $\sim 49,000$ yr, they encounter the density ramp described above. During the early stages inside the core, these models have a cometary shape due to a leading bow-shock, with sizes of the order of hypercompact H II regions (see Figs. 11 and 12). Once they enter into the density ramp, their sizes grow with time and the initially cometary shapes turn into blister type regions. This is because the expansion velocity of the H II region is larger than the stellar velocity.

To summarize, these models tell us that hypercompact and ultracompact H II regions are long-lived astronomical objects while they are inside their parental cloud cores. When stars escape from these cores (models A2 and B2), transitions to larger structures such as classical H II regions are expected. The stars can also fall back into the core, forming smaller structures again. Thus, the lifetime of an ultracompact or a hypercompact H II region is just defined by the crossing time of the cloud core, given by $\tau_{UCHII} = r_c/v_\star$. In cases where the stellar orbits are always smaller than the core radius, the lifetime will be the stellar lifetime.

5 Summary

The history of the pressure in a star forming region follows a simple scheme. The initial properties and pressure of the gas are defined by self-gravity. Once young stars appear, their energy input modifies the structure and evolution of the cloud. This is particularly true for massive stars: their radiative and mechanical energy inputs are even able to disrupt their parental clouds. In the case of dense cloud cores, the sizes of either H II regions or wind-driven bubbles are severely reduced by the large ambient pressure (García-Segura & Franco 1996). In fact, the pressure equilibrium radii of ultracompact H II regions are actually indistinguishable from those of ultracompact wind-driven bubbles, and they could be stable and long-lived.

The situation is completely different when the stars are located near the edge of a cloud core. The resulting H II regions (and also wind-driven bubbles) generate supersonic outflows. Cases with $w > w_{crit}$ lead to the champagne phase: once the cloud is fully ionized, the expansion becomes supersonic. For spherical clouds with a small, constant-density core and a power-law density distribution, r^{-w}, outside the core, there is a critical exponent ($w_{crit} = 3/2$) above which the cloud becomes completely ionized. This represents an efficient mechanism for cloud destruction and, once the parental molecular cloud is completely ionized, can limit the number of massive stars and the star formation rate (Franco et al. 1994). Photoionization from OB stars can destroy the parental cloud on relatively short time scales, and defines the limiting number of newly formed stars. The fastest and most effective destruction mechanism is due to peripheral, blister, H II regions, and they can limit the star forming efficiency at galactic scales.

For a cloud of mass M_{GMC}, with only 10% of this mass concentrated in star-forming dense cores, the number of newly-formed OB stars required for complete cloud destruction is

$$N_{OB} \sim 30 \frac{M_{GMC,5} n_3^{1/5}}{F_{48}^{3/5} (c_{i,15} t_{MS,7})^{6/5}}. \tag{53}$$

where $M_{GMC,5} = M_c/10^5 \, M_\odot$, $n_3 = n_0/10^3 \, \text{cm}^{-3}$, $c_{i,15} = c_i/15 \, \text{km s}^{-1}$, and $t_{MS,7}$ is the main sequence lifetime in units of 10^7 yr. This corresponds to a total star forming efficiency of about 5%. Larger average densities and cloud masses can result in higher star formation efficiencies. Internal H II regions at high cloud pressures can also result in large star forming efficiencies and they may be the main limiting mechanism in star forming bursts and at early galactic evolutionary stages.

The hydrodynamical modeling shows that when stellar motion is considered, stars can move from the central core to the edge of the cloud. There are transitions from hypercompact to ultracompact to extended H II regions as the star moves into lower density regions. However, the opposite behavior is expected when stars fall into the cloud core. Thus, UC H II regions are pressure-confined entities as long as they remain embedded within dense cores. The pressure confinement comes from the ram and thermal pressures. Their survival only depends on the stellar lifetime and on the crossing time of the cloud core. This mechanism, of stellar motion through density structures within the cloud, may contribute to the range of sizes and densities observed for hypercompact, ultracompact and compact H II regions with extended emission.

References

Becker, R.H., White, R.L., Helfand, D.J., & Zoonematkermani, S. 1994, ApJS 91, 347

Bergin, E., Snell, R., & Goldsmith, P. 1996, ApJ 460, 343

Bisnovatyi-Kogan, G.S. & Silich, S.A. 1995, RvMP 67, 661

Bodenheimer, P., Tenorio-Tagle, G., & Yorke, H.W. 1979, ApJ 233, 85

Bregman, J.N. 1980, ApJ 236, 577

Carral, P., Kurtz, S.E., Rodríguez, L.F., De Pree, C.G., & Hofner, P. 1997, ApJ 486, L103

Chevalier, R.A. & Oegerle, W.R. 1979, ApJ 227, 398

Churchwell, E. 2002, ARAA 40, 27

Cioffi, D.F. & Shull, J.M. 1991, ApJ 367, 96

Cox, D.P. 1981, ApJ 245, 534

Cox, D.P. & Smith, B.W. 1974, ApJ 189, L105

Dalgarno, A. & McCray, R.A. 1972, ARAA 10, 375

De Pree, C.G., Goss, W.M., & Gaume, R.A. 1998, ApJ 500, 847

De Pree, C.G., Rodríguez, L.F., & Goss, W.M. 1995, RevMexAA 31, 39

Deul, E.R. & den Hartog, R.H. 1990, A&A 229, 362

Dopita, M.A., Mathewson, D.S., & Ford, V.L. 1985, ApJ 297, 599

Franco, J. & Cox, D.P. 1983, ApJ 273, 243

Franco, J. & Shore, S. 1984, ApJ 285, 813

Franco, J., Tenorio-Tagle, G., & Bodenheimer, P. 1989, RMxAA 18, 65

Franco, J., Tenorio-Tagle, G., & Bodenheimer, P. 1990, ApJ 349, 126

Franco, J., Kurtz, S., García-Segura, G., & Hofner, P. 2000a, ApSS 272, 169

Franco, J., Kurtz, S., Hofner, P., Testi, L., García-Segura, G., & Martos, M. 2000b, ApJ 542, L143

Franco, J., Shore, S.N., & Tenorio-Tagle, G. 1994, ApJ 436, 795

Franco, J., Tenorio-Tagle, G., & Bodenheimer, P. 1990, ApJ 349, 126

Garay, G. & Lizano, S. 1999, PASP 111, 1049

García-Segura, G. & Franco, J. 1996, ApJ 469, 171

Gaume, R.A., Goss, W.M., Dickel, H.R., Wilson, T.L., & Johnston, K.J. 1995, ApJ 438, 776

Heiles, C. 1979, ApJ 229, 533

Heiles, C. 1984, ApJS 55, 585

Heiles, C. 1990, ApJ 354, 483

Herbst, W. & Assousa, G.E. 1977, ApJ 217, 473

Hofner, P. & Churchwell, E. 1996, AAS 120, 283

Houck, J.C. & Bregman, J.N. 1990, ApJ 352, 506

Kim, K.-T. & Koo, B.-C. 2001, ApJ 549, 979

Kurtz, S., Cesaroni, R., Churchwell, E., Hofner, P., & Walmsley, C.M. 2000, in "Protstars & Planets IV", ed. V. Mannings, A. Boss, & S. Russell, (Tucson: Univ. of Arizona Press), 299

Kurtz, S., Churchwell, E., & Wood, D.O.S. 1994, ApJS 91, 659

Kurtz, S.E., Watson, A.M., Hofner, P., & Otte, B. 1999, ApJ 514, 232

Kurtz, S. 2002, ASP Conf. Proc. 267, ed. P. Crowther, (San Francisco: ASP), 81

Larson, R.B. 1992, MNRAS 256, 641

MacDonald, J. & Bailey, M.E. 1981, MNRAS 197, 995

McCray, R. & Snow, T.P. 1979, ARAA 17, 213

McKee, C.F. 1989, ApJ 345, 782

McKee, C.F. & Ostriker, J.P. 1977, ApJ 218, 148

Mezger, P.G., Altenhoff, W., Schraml, J., Burke, B.F., Reifenstein, E.C. III, & Wilson, T.L. 1967, ApJ 150, L157

Norman, C. & Silk, J. 1980, ApJ 238, 158

Olnon, F.M. 1975, A&A 39, 217

Palous, J., Franco, J., & Tenorio-Tagle, G. 1990, A&A 227, 175

Palous, J, Tenorio-Tagle, G., & Franco, J. 1994, MNRAS 270, 75

Panagia, N. & Felli, M. 1975, A&A 39, 1

Raymond, J.C. & Smith, B.W. 1977, ApJS 35, 419

Rodriguez-Gaspar, J.A., Tenorio-Tagle, G., & Franco, J. 1995, ApJ 451, 210

Salpeter, E.E. 1974, RvMP 46, 433

Shapiro, P.R. & Field, G.B. 1976, ApJ 205, 762

Silich, S.A., Franco, J., Palous, J., & Tenorio-Tagle, G. 1996, ApJ 468, 722

Stone, J.M. & Norman, M.L. 1992, ApJS 80, 753

Tenorio-Tagle, G. 1982, in "Regions of Recent Star Formation", ed. R.S. Roger & P.E. Dewdney (Dordrecht: Reidel), 1

Walsh, A.J., Burton, M.G., Hyland, A.R., & Robinson, G. 1998, MNRAS 301, 640

Weaver, R., McCray, R., Castor, J., Shapiro, P., & Moore, R. 1977, ApJ 218, 377

Whitworth, A. 1979, MNRAS 186, 59

Wood, D.O.S. & Churchwell, F. 1989, ApJS 69, 831

Xie, T., Mundy, L.G., Vogel, S.N., & Hofner, P. 1996, ApJ 473, L131

Yorke, H.W. 1986, ARAA 24, 49

Circuit of Dust in Substellar Objects

Christiane Helling

Zentrum für Astronomie und Astrophysik, TU Berlin
Hardenbergstraße 36, 10623 Berlin, Germany

Konrad-Zuse-Zentrum für Informationstechnik Berlin
Takustraße 7, 14195 Berlin, Germany

chris@astro.physik.tu-berlin.de
http://astro.physik.tu-berlin.de/~chris/

Abstract

Substellar atmospheres are cool and dense enough that dust forms very efficiently. As soon as these particles are formed, they sizedependently precipitate due to the large gravity of the objects. Arriving in hot atmospheric layers, the dust evaporates and enriches the gas by those elements from which it has formed. The upper atmospheric layers are depleted by the same elements. Non-continuous and spatially inhomogeneous convective element replenishment, generating a turbulent fluid field, completes the circuit of dust.

The formation of dust in substellar atmosphere is described by extending the classical theory of Gail and Sedlmayr for the case of different gas and dust velocities. Turbulence is modeled in different scale regimes which reveals turbulence as trigger for dust formation in hot environments. Both mechanisms cause the dust to be present in else wise dust-hostile region: precipitation transports the dust into hot regions, and turbulence allows the formation of dust in there.

1 Introduction

The classical problem of stellar atmospheres had been considered as almost solved: observed spectra are reproduced by solving hydrostatic equilibrium equation coupled with a frequency-dependent radiative transfer involving an extensive body of opacity data and a gas phase equilibrium chemistry. Typical examples are A-type stars (e. g. Allard et al. 1998, Castelli and Kurucz 2001), F and G type stars (e. g. Fuhrmann 1998) or the carbon-rich AGB stars (e. g. Loidl et al. 2001). Due to the appropriately high precision of such reproductions, synthetic spectra fits can be fully automated as it has been demonstrated in (Bailer-Jones 2000).

This situation changed when the first brown dwarf Gliese 226B was discovered in 1994 and the first spectra became available. At that time, none of the classical model atmosphere simulations were able to reproduce satisfactorily such a substellar atmosphere spectrum. Tsuji et al. (1996) were the first who communicated that important physical effects, very likely correlated with the presence of dust, had not been taken into account in the modeling of substellar objects.

Substellar objects, i.e., brown dwarfs and planets, are high gravity objects with considerable compact atmospheres. Their atmospheres are cool enough that complex molecules are present and phase transitions (*dust formation*) can take place. The carbon is no longer locked into the CO molecule because of the high density of the atmosphere, therefore, molecules like CH_4, CO_2, H_2O, ... appear simultaneously thereby challenging opacity calculations. As soon as solid (or fluid) particles (*dust*) have formed, they will sizedependently precipitate due to the large gravity of the objects. The dust grains will then arrive in an atmospheric layer which is hot enough to evaporate them thermally. The lower atmospheric regions will be enriched by those elements from which the dust grains have formed and the upper atmospheric layers will be depleted by the same. The *circuit of dust in a substellar atmosphere* (Fig. 1) is completed by convective element replenishment. Brown dwarfs are nearly fully convective (see e.g. Chabrier and Baraffe 1997 but Tsuji 2002) which creates a turbulent environment in their atmospheres. Therefore, a non-continuous and spatially inhomogeneous replenishment of the upper atmosphere from deeper layers with those elements previously preticipated can be expected. The dust formation process in hot layers is nevertheless triggered by mixing events due to convective motion. Since the vertical extension of a brown dwarf atmosphere is quite small – typical density scale heights are as small as $H_\rho \approx 10^6$ cm – all these processes occur in a thin layer, similar to the "weather" in the Earth's atmosphere.

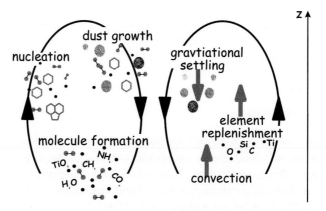

Figure 1: The circuit of dust in substellar atmospheres: dust formation (nucleation, growth) → gravitational settling (drift) → element replenishment by convection → dust formation.

As a consequence, the classical theory of stellar atmospheres misses two major points: i) dust formation with its strong feedbacks on the atmosphere, and ii) dust formation in turbulent gases. Both will be discussed in the following.

2 Key processes of dust formation

According to the classical theory of Gail and Sedlmayr (1984, 1986, 1999), dust formation is to be considered as a two step process: 1) small seed particles (*clusters*) form out of the gas phase (*nucleation*) and 2) subsequently grow to macroscopic sizes (*growth*). A third process influences the dust complex in the gravitationally dominated atmosphere of a substellar object: gravitational settling (*drift*)[1].

Nucleation: The formation of *seed particles* can in the most simple case be described as a linear reaction chain where the same monomer (e. g. TiO_2, Jeong 2000) is added during each reaction step:

$$A_1 \rightleftharpoons A_2 \rightleftharpoons \cdots \rightleftharpoons A_{N-1} \underset{\tau_{ev}^N}{\overset{\tau_{gr}^{N-1}}{\rightleftharpoons}} A_N \underset{\tau_{ev}^{N+1}}{\overset{\tau_{gr}^N}{\rightleftharpoons}} A_{N+1} \rightleftharpoons \cdots \rightleftharpoons A_{N_*} \rightleftharpoons \cdots.$$

The first stable cluster (N_*, *critical cluster*) forms if the growth rate τ_{gr} exceeds the evaporation rate τ_{ev}. Now, the growth of subsequently larger clusters can be considered as a stationary flux through the cluster space from which a *stationary nucleation rate* J_* results.

Growth: Once seed particles are formed, they subsequently grow via chemical surface reactions but can only sustain in the gas phase if they are thermally stable. Thermal stability can be visualized by the stability sequence[2] in Fig. 2 where those $(T, n_{<H>})$ are plotted for which the supersaturation ratio $S = 1$. The compounds are stable below $S = 1$ and evaporate at temperatures/densities above such a curve.

Seed particles, which provide the necessary surface for the formation of macroscopic dust grains, are located in the stable growth regime (see Fig. 2) of any stability sequence because a considerable supersaturation is necessary that non-planar particles form from the gas phase. Therefore, many compounds are already thermally stable and rather *heterogeneous core-mantle grains* will form in the atmosphere of a brown dwarf and maybe also in planetary atmospheres.

Gravitational settling: The dust formation process is influenced by drift, i. e., dust and gas move with the different velocities ($\boldsymbol{v}_{\rm dust} = \boldsymbol{v}_{\rm dr} + \boldsymbol{v}_{\rm gas}$) because of the high gravity of substellar objects. It was shown in Woitke and Helling (2002) that the time needed by a particle to reach its final velocity, $\tau_{\rm acc}$, is much smaller than any other time scales involved, i.e. $\min\{\tau_{\rm hyd}, \tau_{\rm gr}, \tau_{\rm sink}\} \gg \tau_{\rm acc}$. Therefore, the *equilibrium drift* concept can be applied where the equilibrium drift velocity, $\boldsymbol{v}_{\rm dr} = \overset{\circ}{\boldsymbol{v}}_{\rm dr}$, is implicitly given by the force equilibrium,

$$m_{\rm d}\, \boldsymbol{g}(\boldsymbol{x}) + \boldsymbol{F}_{\rm fric}(\boldsymbol{x}, a, \overset{\circ}{\boldsymbol{v}}_{\rm dr}) = 0, \qquad (1)$$

---

[1] Drift occurs if some force on a particle counter balances friction and causes the grains to move with a different velocity than the gas. Examples are the radiation force in red giants or the gravity in substellar objects.

[2] Stability sequences are not to be confused with condensation sequences.

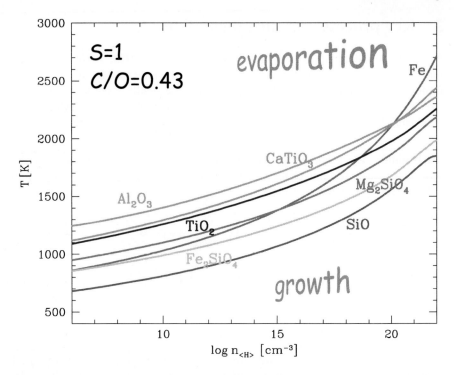

Figure 2: Stability sequence for an oxygen-rich gas (C/O = 0.43), typical for a brown dwarf atmosphere.

where the force of gravity is balanced by the force of friction $\boldsymbol{F}_{\rm fric}(\boldsymbol{x}, a, \mathring{\boldsymbol{v}}_{\rm dr})$. Depending on the hydrodynamic situation, characterized by the Knudsen number $Kn = \bar{l}/(2a)$ ($\bar{l}$ – mean free path, a – particle size), different cases need to be considered for the force of friction and also for the dust volume accretion rate dV/dt (compare Eqs. 4, 6):

free molecular flow $Kn \gg 1$		viscous case $Kn \ll 1$
subsonic		laminar
	force of friction	
supersonic		turbulent
	volume accretion rate	
freely impinging		diffusive

The volume accretion rate, which defines the particle's growth time scale $\tau_{\rm gr} = \frac{4\pi a^3/3}{dV/dt}$, is either determined by molecules freely impinging on the grains surface or by diffusion towards the surface. Aiming at a theoretical description of the dust formation, a typical time which a particle of size a needs to cross a density scale hight H_ρ with a velocity $\mathring{\boldsymbol{v}}_{\rm dr}$, $\tau_{\rm sink} = \frac{H_\rho}{\mathring{v}_{\rm dr}}$, is compared with $\tau_{\rm gr}$ in Fig. 3. Herein, the (a, ρ) plane in Fig. 3 (ρ – gas density) is subdivided by a dashed line where $\tau_{\rm sink} = \tau_{\rm gr}$: Above this line, i.e., where $a > a_{\tau_{\rm sink}=\tau_{\rm gr}}$, the grains would be removed

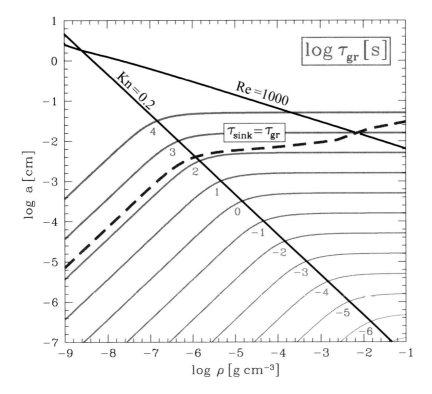

Figure 3: Contour plot of the growth time scale $\log \tau_{gr}$ [s] as function of the grain size a and the gas density ρ at constant temperature $T = 1500$ K for quartz (SiO$_2$, $\rho_d = 2.65$ g cm^{-3}). We assume growth by accretion of the key species SiO with maximum particle density $n_{SiO} = 10^{(7.55-12)} n_{\langle H \rangle}$ and extreme supersaturation ($S \to \infty$; for more details see Woitke and Helling 2002).

by gravitational settling before they can form, hence, such particle sizes can not exist. Applying a gas density range typically for a brown dwarf atmosphere, the maximum grain size $a_{\max}$ to be expected depends on the gas density and is 1 μm in the cold and thin outer layers and 100 μm in the hot and dense layers inside the atmosphere. Furthermore, $\tau_{\text{sink}} = \tau_{\text{gr}}$ limits the number of hydrodynamical regimes to be expected in a brown dwarf atmosphere: *the subsonic free molecular flow and the laminar viscous flow need to be considered.*

2.1 Theory of dust formation for substellar atmospheres

A theoretical description of the formation of dust in compact, substellar atmosphere, i.e., in gravitationally dominated gases, can be derived by utilizing the key idea of Gail and Sedlmayr (1988) of defining dust moments, $\rho L_j(\boldsymbol{x}, t)$ [cm^j/cm^3],

$$\rho L_j(\boldsymbol{x}, t) = \int_\infty^{V_l} V^{j/3} f(V, \boldsymbol{x}, t) dV, \qquad (2)$$

of the grain size distribution function $f(V, \boldsymbol{x}, t)$ and the grain volume V. Allowing – in contrast to Gail and Sedlmayr – for $v_\text{dust} \neq v_\text{gas}$, systems of dust moment equations result for the two hydrodynamic cases relevant for a compact atmosphere (Woitke and Helling 2002):
- **subsonic free molecular flow** ($\text{Kn} \gg 1 \wedge \overset{\circ}{\boldsymbol{v}}_\text{dr} \ll c_T$):

$$\frac{\partial}{\partial t}(\rho L_j) + \nabla(\boldsymbol{v}_\text{gas}\, \rho L_j) = [\text{Da}_\text{d}^\text{nuc} \cdot \text{Se}_j]\, J(V_l) + \left[\text{Da}_\text{d,lKn}^\text{gr}\right] \frac{j}{3} \chi_\text{lKn}^\text{net}\, \rho L_{j-1}$$
$$+ \left[\left(\frac{\pi \gamma}{32}\right)^{1/2} \frac{\text{M} \cdot \text{Dr}}{\text{Kn}^\text{HD} \cdot \text{Fr}}\right] \xi_\text{lKn} \nabla\left(\frac{L_{j+1}}{c_T} \vec{e}_r\right) \quad (3)$$

$$\chi_\text{sKn}^\text{net} = \sqrt[3]{48\pi^2} \sum_r \Delta V\, n_r\, D_r \left(1 - \frac{1}{S_r}\right) \qquad \xi_\text{sKn} = \frac{2}{9}\left(\frac{3}{4\pi}\right)^{2/3} g\, \rho_d \quad (4)$$

- **laminar viscous flow** ($\text{Kn} \ll 1 \wedge \text{Re}_\text{d} < 1000$):

$$\frac{\partial}{\partial t}(\rho L_j) + \nabla(\boldsymbol{v}_\text{gas}\, \rho L_j) = [\text{Da}_\text{d}^\text{nuc} \cdot \text{Se}_j]\, J(V_l) + \left[\text{Da}_\text{d,sKn}^\text{gr}\right] \frac{j}{3} \chi_\text{sKn}^\text{net}\, \rho L_{j-2}$$
$$+ \left[\left(\frac{\pi \gamma}{288}\right)^{1/2} \frac{\text{M} \cdot \text{Dr}}{\text{Kn} \cdot \text{Kn}^\text{HD} \cdot \text{Fr}}\right] \xi_\text{sKn} \nabla\left(\frac{\rho L_{j+2}}{\mu_\text{kin}} \vec{e}_r\right) \quad (5)$$

$$\chi_\text{lKn}^\text{net} = \sqrt[3]{36\pi} \sum_r \Delta V\, n_r\, v_r^\text{rel}\, \alpha_r \left(1 - \frac{1}{S_r}\right) \qquad \xi_\text{lKn} = \frac{\sqrt{\pi}}{2}\left(\frac{3}{4\pi}\right)^{1/3} g\, \rho_d, \quad (6)$$

with $j = 0, 1, 2, \ldots$. The equations reveal the same conservative form with source terms as those in the classical Gail and Sedlmayr case: the first source term on the r.h.s. of Eqs. (3) and (5) represents the nucleation since it contains the nucleation rate J_*. The second source term describes the heterogeneous dust growth with χ^net being the growth speed which depends on the hydrodynamical regime considered (l.h.s. of Eqs. (4) and (6)). By allowing for $v_\text{dust} \neq v_\text{gas}$, a third – advective – source term appears which transports already existing particles in regions with possibly still supersaturated gas where they can continue to grow. ξ is the gravitational force density for each of the regimes (r.h.s. of Eqs. (4) and (6)).

The equations (3) and (5) are written *dimensionless*, therefore, characteristic numbers appear in front of the source terms being put in brackets $[\cdot]$ for lucidity[3]. Evaluating the characteristic numbers for a typical brown dwarf model atmosphere, a *hierarchical appearance of the source terms* become evident. If the thermodynamic conditions are appropriate for nucleation, the nucleation source term will determine the solution of the equations. The growth source term becomes most dominant if nucleation becomes inefficient. The advective drift source term determines the solution of the equations only if dust nucleation and growth are negligibly small.

---
[3] A detailed description of these characteristic numbers and all other quantities is given in Helling et al. (2002) and Woitke and Helling (2002).

One may, however, observe that the systems of equations (3) and (5) are not closed. Here, further work is necessary but, nevertheless, the closure is possible in the case of a static, stationary subsonic free molecular flow to be demonstrated in the next section. For the general case, one might take notice of the work of Deuflhard and Wulkow (1989) and Wulkow (1992) who adopted orthogonal polynomials of a discrete variable to construct closure terms.

2.2 Formation and structure of a quasi-static cloud layer

The formation and the structure of a quasi-static cloud layer can be investigated by solving the extended dust moment equations (Eqs. 3–5). As example, the static ($v_{gas} = 0$), stationary ($\partial/\partial t = 0$) case of Eqs. (3, 4) for a subsonic ($\mathring{v}_{dr} \ll c_T$) free molecular flow ($Kn \gg 1$) is presented. The equations have been supplemented with element consumption and a homogeneous gas mixing, and are solved on top of a prescribed brown dwarf atmosphere structure kindly provided by Tsuji (2002). For simplicity, the formation of pure TiO_2-grains is considered where first $(TiO_2)_N$ seed particles form and second, continue to grow a TiO_2 mantle. The homogeneous mixing occurs with a time scale of $\tau_{mix} = 10^4$s in the example presented in Fig. 4.

Five major zones can be distinguished. In *Zone I*, nucleation takes place efficiently due to a very large supersaturation of the gas. Most of the Ti is now locked in the dust which immediately precipitates into deeper layers. Therefore, a depth-dependent depletion of those molecules consumed by dust formation results. In our case, a large depletion of TiO and TiO_2 occurs in the upper zones. In *Zone II*, nucleation is considerably less efficient and this region is determined by the growth of already existing particles. The mean particles size $\langle a \rangle$ increases steeply, and therewith also the mean drift velocity $\langle \mathring{v}_{dr} \rangle^4$. In the Zones III – V, no dust formation is possible since $J_* = 0$ because of the temperature being too high. Therefore, the dust contained in these regions must have formed in the upper layers (Zone I and II). Therefore, *Zone III* is dominated by the drift process which transports dust in regions which are still supersaturated and, therefore, the dust can continue to grow. However, $\langle a \rangle$ increases only slightly and $\langle \mathring{v}_{dr} \rangle$ reaches a terminal value due to the competition between the increasing grain size and the increasing gas density which hinders the dust to move faster inwards. In *Zone IV*, no effective growth takes place since $S \approx 1$. The rain is saturated in these layers. In *Zone V*, temperatures are high enough to evaporate the grains. Consequently, $\langle a \rangle$ and $\langle \mathring{v}_{dr} \rangle$ decrease. Note, that in Zone V the element abundance in the gas phase, ε_{gas}, exceeds the number of elements locked in the dust grains ε_{dust}, and even the initial solar abundance of the gas phase. This shows that precipitating dust grains have elementally enriched the deepest atmospheric layers causing the molecular abundance to be larger than without precipitating dust.

The consistent description of dust formation, drift and element depletion has provided inside into the formation and the structure of a quasi-static cloud layer which exhibits

---

[4]The mean drift velocity is calculated according to Eq. (66) in Woitke and Helling (2002) adopting the mean particles size $\langle a \rangle$, $\langle \mathring{v}_{dr} \rangle = \frac{\sqrt{\pi} g \rho_d}{2\rho c} \langle a \rangle$ (ρ_d – grain material density, c – speed of sound).

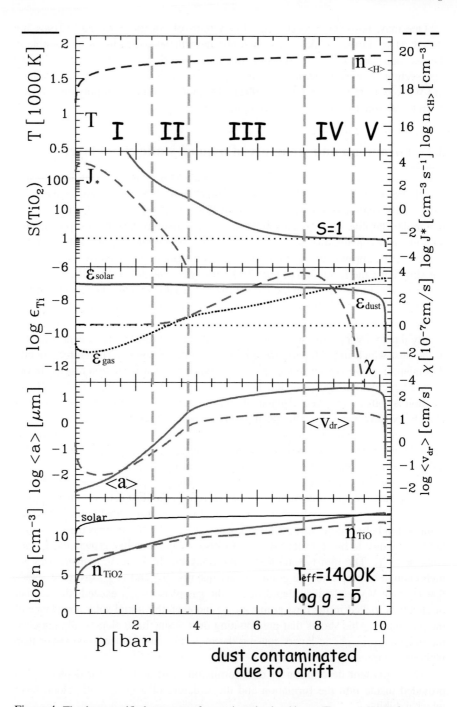

Figure 4: The dust-stratified structure of a quasi-static cloud layer ($T_{eff} = 1400$ K, $\log g = 5$, $\tau_{mix} = 10^4$ s): *Zone I* – nucleation dominated, *Zone II* – growth dominated, *Zone III* – drift dominated, *Zone IV* – saturated rain, *Zone V* – evaporation dominated.

- a dust stratified large scale structure where only in the outer regions dust formation is possible. The inner regions are dust contaminated due to drift.

- a depth dependent gas depletion in the upper atmosphere and an enrichment of the deeper atmosphere by heavy elements.

3   Dust formation in turbulent gases

The large scale structure of a dust forming substellar, brown dwarf atmosphere has been investigated by supplying a consistent theoretical description of dust formation in gravitationally dominated media. On such large scales, the atmosphere of a brown dwarf is largely convective. Convection provides a disturbance which is large enough to cause the atmospheric fluid to be highly turbulent because the inertia of the fluid is large compared to its friction ($\Rightarrow$ large Reynolds number).

Figure 5: Leonardo da Vinci (c. 1513, *Old Man with Water Studies*).

Turbulence is a multi-scale phenomenon as it was already observed and painted by Leonardo da Vinci (Fig. 5) 500 years ago and as for example observations of Jupiter's surface reveal it today. However, only the largest scales of a turbulent fluid

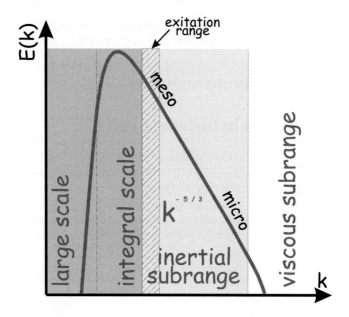

Figure 6: Scale regimes involved in a turbulent fluid field. Driven turbulent is exited in a small wavenumber interval where $k_{\min} = 2\pi/(3h)$ and $k_{\max} = 2\pi/l_{\text{ref}}$ (h – spatial grid resolution).

are accessible by any kind of observation (compare Fig. 6). Therefore, observations of brown dwarf variability, e. g., by Bailer-Jones and Mundt (2001), Eislöffel and Scholz (2001), and those of planetary atmospheres reveal merely the large scale structures of the fluid.

Numerical experiments of moderately large Reynolds number flows have shown that the simulation of the largest scales requires only 2 % of the whole computing time. The remaining 98 % are necessary to resolve the small scale structures down to the viscous subrange where the energy is dissipated by frictional forces (Dubois, Janberteau and Temam 1999). Therefore, for hydrodynamic simulations one naturally desires to model these scales. In addition, chemical and structure formation processes, like e. g. combustion or dust formation, are seeded in the small scale regime. While the smallest scales may be not considerably influenced by the large scale flow structure (though they are energized by them), the structure formation on the small scales may seed large scale structure formation. Only the latter will be observable but the previous are, nevertheless, needed to reach a detailed understanding of the largest scales and thereby of any observation.

Therefore, the different scale regimes have to be studied in order to reveal the basic physical mechanisms involved in the dust formation in turbulent media.

3.1 Dust formation in the microscopic scale regime

The study of the microscopic regime ($l_{\text{ref}} \ll H_\rho$) provides an unique possibility to gain insight in governing processes involved in structure formation and in particular

Circuit of Dust in Substellar Objects 125

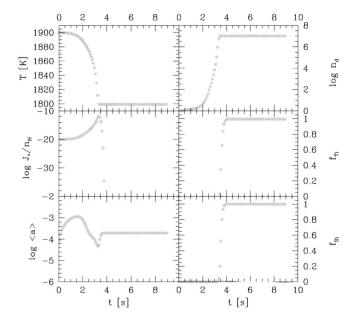

Figure 7: Time evolution of a gas box after an expansion wave superposition staring from the following initial values: $T(x,0) = 1900\,\text{K}$, $\rho(x,0) = 10^{-4}\,\text{g cm}^{-3}$, $\rho(x,0)L_0(x,0) = 1\,\text{cm}^{-3}$, $v_{\text{ref,micro}} = 10^5\,\text{cm s}^{-1}$, $l_{\text{ref,micro}} = 10^4\,\text{cm}$; see also Helling et al. (2001) Fig. 3.
1st row: T [K] – gas temperature (l.h.s.), n_d [cm^{-3}] – dust particle density (r.h.s);
2nd row: $J_*/n_{<H>}$ [s^{-1}] – nucleation rate (l.h.s.), f_{Ti} – degree of condensation of Ti (r.h.s.);
3rd row: $\langle a \rangle$ [10^{-4} cm] – mean grain radius (l.h.s.), f_{Si} – degree of condensation of Si (l.h.s.).

in the formation of dust in a turbulent fluid since direct numerical simulations of hydrodynamic processes can be carried out. The turbulent fluid field is considered to consist of interacting acoustic waves carrying temperature fluctuations as characteristic of the large wavenumber end of the inertial subrange (compare Fig. 6)[5]. Thereby, the superposition of expansion waves causes the local temperature to drop below the nucleation threshold for a short time interval. During this time, seed particles form, persist in the gas phase and continue to grow to macroscopic sizes (Helling et al. 2001). The principal mechanisms can nicely be studied in a 0D model of a gas box after such a superposition since a long term study of the dust complex is possible.

A fully time-dependent solution of the hydrodynamic equations, the energy equation with radiative cooling, the classical Gail and Sedlmayr dust moment equations consistently coupled to the element conservation equation has been obtained by applying a multi-dimensional Euler-solver for compressible fluids (Smiljanovski et al. 1997) and the operator splitting method for the source terms. The coupled complex of classical dust formation, the element consumption and the radiative cooling is solved by the LIMEX DAE solver (Deuflhard and Nowak 1987) for each hydrody-

---

[5]Note that the characteristic velocity of a regime is inverse proportional to its characteristic length. Therefore, $v_{\text{ref,micro}} = c_s$ in the microscopic regime does not contradict the findings of the mixing length theory since it concerns the macroscopic scale regime.

namic time step because of the strong coupling between these equations. LIMEX solves reliably the steady state of the dust complex after the transition from a dynamic (Fig. 7: $t < 3.2$ s) to an equilibrium situation (Fig. 7: $t > 3.2$ s). The simulation now concerns the formation of core-mantel grains made of a TiO_2-seed on which SiO and TiO_2 grow a heterogeneous mantle.

micro-Results: The simulation starts from a temperature much too high for nucleation. But, the small amount of grains formed during a wave superposition event will grow ($\Rightarrow$ increase of $\langle a \rangle$; Fig. 7: 3rd row, l.h.s.). Soon, the particles are large enough that radiative cooling sets in, therefore, the temperature starts to decrease. The larger the initial particles grow, the faster the temperature drops. Finally, the temperature is low enough that nucleation is re-initiated ($\Rightarrow$ increase of J_*; 2nd row, l.h.s.) and the number of dust particles n_d increases (1st row, r.h.s.). Now, a *feedback loop* establishes (compare Fig. 10) since more dust intensifies the radiative cooling which causes the temperature to increase even further. Conditions for most efficient nucleation are met, therefore, even more particles form. The process stops if all material has been consumed, hence, the degree of condensation of the elements involved reaches one ($f_{Ti} = 1$, $f_{Si} = 1$; 2nd and 3rd row, r.h.s.).

3.2 Dust formation in the mesoscopic scale regime

Based on the knowledge gained from the microscale investigations, the scale regime can be extended to larger scales by involving the small wavenumber end of the Kolmogoroff inertial subrange (compare Fig. 6). In this *mesoscopic* regime ($l_{\text{ref}} < H_\rho$) *driven turbulence* is supposed to model a constantly occurring energy input from some convectivly active zone outside a test volume.

In order to fulfill the conservation equations inside the test volume, the stochastic, dust-free velocity, pressure and entropy fields are prescribed on ghost cells located outside the test volume. The stochastic disturbance of the velocity field is, thereby, calculated according to the Kolmogoroff spectrum in the wavenumber space. By solving the hydro-/thermodynamic equations (see Sect. 3.1), the stochastically created waves continuously enter the model volume and, thereby, a turbulent fluid field is generated.

Stochastic boundary conditions driving turbulence: Turbulence is modeled by boundary conditions for our test volume: A disturbance $\delta\alpha(\boldsymbol{x}, t)$ is added to a homogeneous background field $\alpha_0(\boldsymbol{x}, t)$,

$$\alpha(\boldsymbol{x}, t) = \alpha_0(\boldsymbol{x}, t) + \delta\alpha(\boldsymbol{x}, t). \tag{7}$$

The present pseudo-spectral model for driven turbulence[6] comprises:

---

[6]For examinations concerning the forcing in direct numerical simulations of turbulence see Eswaran and Pope (1988).

- **a stochastic distribution of velocity amplitudes $\delta v(x,t)$:**

$$e(k) = C_K \varepsilon^{2/3} k^{-5/3} \tag{8}$$

$$E_{\text{turb}}^i = \int_{k_{i+1}}^{k_i} e(k)\, dk = \frac{3}{2} C_K \varepsilon^{2/3} [k_i^{-2/3} - k_{i+1}^{-2/3}] \tag{9}$$

$$A_v(k_i) = \sqrt{2 z_3 E_{\text{turb}}^i} \tag{10}$$

$$\delta v(x) = \Sigma_i A_v(k_i) \cos(k_i \hat{k}_i x + \omega_i t + \varphi_i) \hat{k}_i \tag{11}$$

The turbulent energy $E_{\text{turb}}^i = A_v(k_i)^2/2$ according to the Kolmogoroff spectrum is assumed to be the most likely value around which a stochastic fluctuation is generated by a Gaussian distributed number z_3 according to the Box-Müller formula $z_3 = \sqrt{-2 \log z_1} \sin(\pi z_2)$. z_1 and z_2 are equally distributed random numbers. The random numbers are chosen once at the beginning of the simulation. ε is the energy dissipation rate, $C_K = 1.5$ the Kolmogoroff constant (Dubois, Janberteau and Temam 1999), $k_i = |k_i \hat{k}_i|$ are N equidistantly distributed wavenumbers in Fourier space ($i = 1, \ldots N$ with N the number of modes), and $\varphi_i = 2\pi z_4$ are the equally distributed random phase variations. From dimensional arguments the *dispersion relation* $\omega_i = (2\pi k_i^2 \varepsilon)^{1/3}$ was derived.

- **a stochastic distribution of pressure amplitudes $\delta P(x,t)$:**

The pressure amplitude is determined depending on the wavenumber of the velocity amplitude $A_v(k_i)$ such that the compressible and the incompressible limits are matched for the smallest and the largest wavenumber:

$$A_P(k_i) = \frac{[k_{\max} - k_i]\, \rho A_v(k_i)^2 + [k_i - k_{\min}]\, \rho c_s A_v(k_i)}{[k_{\max} - k_{\min}]}. \tag{12}$$

The Fourier cosine transform (as Eq. 9) provides $\delta P(x,t)$ in ordinary space. $k_{\max} = 2\pi/(3h)$ is the maximum wavenumber with h the spatial grid resolution, $k_{\min} = 2\pi/l_{\text{ref}}$ is the minimum wavenumber with l_{ref} the size of test volume.

- **a stochastic distribution of the entropy $S(x,t)$:**

The entropy $S(x,t)$ is a purely thermodynamic quantity and a distribution can in principle be chosen independently on the distribution of the hydrodynamic quantities. For a given entropy $S(x,t)$ and given $P(x,t)$ the gas temperature $T(x,t)$ is given by:

$$\log T(x,t) = \frac{S(x,t) + R \log P(x,t) - R \log R}{c_V + R}. \tag{13}$$

So far, $S(x,t)$ has been kept constant. R is the ideal gas constant, $c_V = 3\,k/(2\mu)$ the specific heat capacity for an ideal gas, k the Boltzmann constant, and $\mu = 2.3\,m_H$ the mean molecular weight.

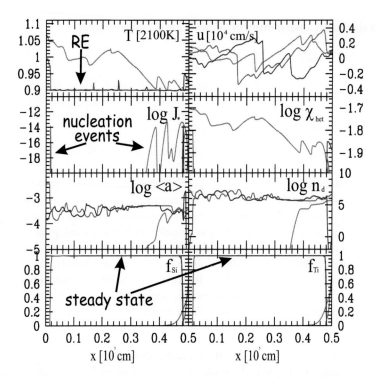

Figure 8: The evolution of the dust complex in space under the conditions of driven turbulence for three time steps. ($T(\boldsymbol{x},0) = 2100$ K, $\rho(\boldsymbol{x},0) = 10^{-4}$ g cm^{-3}, $v(\boldsymbol{x},0)_{\rm ref,meso} = 10^3$ cm s^{-1}, $l_{\rm ref,meso} = 10^5$ cm, 50-k mode driven).
1st row: T [K] – gas temperature (l.h.s.), u [10^4 cm s^{-1}] – hydrodynamic velocity (r.h.s);
2nd row: $J_*/n_{<H>}$ [s^{-1}] – nucleation rate (l.h.s.), $\chi_{\rm het}$ [cm s^{-1}] heterogeneous growth velocity (r.h.s);
3rd row: $\langle a \rangle$ [10^{-4} cm] – mean grain radius (l.h.s.), $n_{\rm d}$ [cm^{-3}] – dust particle density (r.h.s);
4th row: $f_{\rm Si}$ – degree of condensation of Si (l.h.s.), $f_{\rm Ti}$ – degree of condensation of Ti (r.h.s.)).

meso-Results: Figure 8 exhibits the principal behavior of the dust complex under the conditions of driven turbulence as function of the horizontal position for three different time steps. The turbulent fluid field creates a strongly fluctuating velocity field and singular temperature decreases which are low enough to cause singular nucleation events. The feedback loop described in Sect. 3.1 causes the temperature to reach its radiative equilibrium quite soon and only small spikes indicate a short deviation due to interacting compression waves. The stochastic appearance of nucleation events results in an rather *inhomogeneous number and mean particle size as function of horizontal position and time* (3rd row). If all condensible material has been consumed, the dust complex has reached a steady state. Further incoming waves will only transport but not modify the dust grains.

In the long term run, the dust complex is characterized by small fluctuation in the mean particles size $\langle a \rangle$ ($\log \langle a \rangle \approx -3.5 \pm 0.25$) and number density $n_{\rm d}$ ($\log n_{\rm d} \approx 6 \pm 1$).

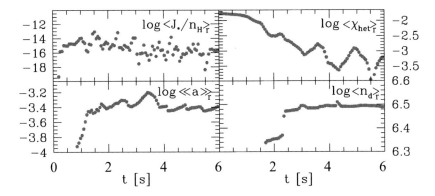

Figure 9: Space mean values of the dust properties as function of time from the simulations depicted in Fig. 8.
1st row: $J_*/n_{<H>}$ [s^{-1}] – nucleation rate (l.h.s.),
χ_{het} [cm s^{-1}] heterogeneous growth velocity (r.h.s);
2nd row: $\langle a \rangle$ [10^{-4} cm] – mean grain radius (l.h.s.), n_{d} [cm^{-3}] – dust particle density (r.h.s)).

It is, however, possible to gain some inside by calculating mean quantities as for instance in Fig. 9 the space means of all the time steps available for the simulation in Fig. 8. Figure 9 depicts the resulting quantities for the dust complex. One observes large variations in the means of the mean particle radius $\langle\langle a \rangle\rangle$ and the grain number density $\langle n_{\mathrm{d}} \rangle$ in the beginning of the calculation but some kind of saturation with only slight variations in the long term run. Only the nucleation rate $\langle J_* \rangle$ and the heterogeneous growth velocity $\langle \chi_{\mathrm{het}} \rangle$ fluctuate largely in time which is correlated to the inflow of dust free material from the boundaries into the test volume.

4 Summary

The dust formation in substellar objects is presently approximated quite roughly in the context of the classical stellar atmosphere and synthetic spectrum calculations. Often, a fixed particle size or the ISM grain size distribution are assumed. More elaborate treatments apply time scale arguments based on the work of Rossow (1978). Authors assume the presence of homogeneous dust particles made of, e. g., pure iron, and dust condensation is considered independently from gravitational settling.

We have argued, that the **dust in brown dwarf atmospheres**

- consists of **heterogeneous core-mantel grains** because of the necessity of the creation of a first seed particle out of the gas phase. Many compounds are then already stable and grow therefore efficiently on such first surfaces.

- has a **maximum possible mean grain size** between 1 μm outside and 100 μm inside the atmosphere caused by the equilibrium between gravity and friction.

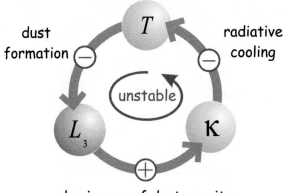

Figure 10: Unstable feedback loop.

- **precipitates into dust-hostile region**. Therefore, the site of dust formation does not coincide with the site of maximum dust content in the atmosphere.

- has to be treated consistently with element consumption and precipitation (drift). A **depth-dependent depletion** of molecules results in the upper atmospheric regions while an enrichment results for the innermost atmosphere.

In extension to the classical theory of dust formation developed by Gail and Sedlmayr, the **theory of dust formation in gravitationally dominated gases** allows for different velocities of dust and gas, and thereby

- can rely on the equilibrium drift concept, since the acceleration time scale $\tau_{\rm acc}$ of the dust particles is very small.

- provides a simultaneous description of nucleation, growth, drift, and evaporation by means of systems of moment equations.

Only the largest structures of a turbulent fluid field are observationally accessible but structure formation processes are seeded in the regime of the small, microscopic scales (but $l_{\rm ref} > \eta$, η dissipation length). Dust formation in a turbulent compact atmosphere

- proceeds via a **run away process** which is governed by an unstable **feedback loop** (Fig. 10): The more dust is present, the more efficient the radiative cooling decreases the temperature. More dust forms which again intensifies the radiative cooling (Fig. 10).

- occurs via **stochastic nucleation events** due to stochastically superimposed waves.

- results in an **inhomogeneous distribution** of dust in spaces and time, size and number.

5 Conclusions

The circuit of dust in substellar atmospheres is influenced on the macroscopic scales by convective upwelling which provides an efficient mechanism to elementally enrich the upper atmospheric regions. These regions have been depleted by precipitating dust which simultaneously continues to grow on its way into the deeper atmosphere. On the small scales, a turbulent fluid field influences the dust formation process creating random dust formation events. Both mechanisms cause the dust to be present in elsewise dust-hostile regions: precipitation transports the dust into hot regions, and turbulence allows the formation of dust in there.

Acknowledgement:

The work on dust formation is carried out in collaboration with P. Woitke, the work on turbulence models mainly with R. Klein. I thank both for their supporting interest and the many enlightning and guiding discussions. I thank E. Sedlmayr for his inspiring questions and criticism. B. Patzer is thanked for careful reading the manuscript. This work has been supported by the *DFG* (grands SE 420/19-1, 19-2; Kl 611/7-1, 9-1). Most of the literature search has been performed with the ADS system.

References

Allard, N.F., Drira, I., Gerbaldi, M., Kielkopf, J., Spielfiedel, A. 1998, A&A 335, 1124
Bailer-Jones, C.A.L., Mundt, R. 2001, A&A 367, 218
Bailer-Jones, C.A.L., 2000, A&A 357, 197
Castelli, F., Kurucz, R. 2001, A&A 372, 260
Chabrier, G., Baraffe, I. 1997, A&A 327, 1039
Deuflhard, P., Wulkow, M. 1989, IMPACT Comp. Sci. Eng. 1, 269
Deuflhard, P., Nowak, U. 1987, in: Deuflhard, P., Engquist, B. (eds.), Large Scale Scientific Computing. Progress in Scientific Computing 7, 37. Birkhäuser
Dubois, T., Janberteau, F., Temam, R. 1999, Dynamic Multilevel Methods and the Numerical Simulaion of Turbulence, Cambridge Univ. Press
Eislöffel, J., Scholz, A. 2001, in: Alves, J. (ed.), The Origins of Stars and Planets: The VLT View, in press
Fuhrmann, K. 1998, A&A 338, 161
Gail, H.-P., Sedlmayr, E. 1986, A&A 166, 225
Gail, H.-P., Sedlmayr, E. 1988, A&A 206, 153
Gail, H.-P., Sedlmayr, E. 1999, A&A 347, 594
Gail, H.-P., Keller, R., Sedlmayr, E. 1984, A&A 133, 320
Helling, Ch., Oevermann, M., Lüttke, M., Klein, R., Sedlmayr, E. 2001, A&A 376, 194
Jeong, K.S. 2000, Dust shells around oxygen-rich Miras and long-period variables, PhD thesis, Technische Universität, Berlin, Germany
Loidl, R., Lançon, A., Jørgensen, U. 2001, A&A 371, 1065
Rossow, W.V. 1978, Icarus 36, 1
Smiljanovski, V., Moser, V., Klein, R. 1997, Combustion Theory & Modelling 1, 183
Tsuji, T. 2002, ApJ 575, 264
Tsuji, T., Ohnaka, K., Aoki, W. 1996, A&A 305, L1
Woitke, P., Helling Ch. 2002, A&A, in press
Wulkow, M. 1992, IMPACT Comp. Sci. Eng. 4, 153

Hot Stars:
Old-Fashioned or Trendy?

A. W. A. Pauldrach

Institut für Astronomie und Astrophysik der Universität München
Scheinerstraße 1, 81679 München, Germany
UH10107@usm.uni-muenchen.de,
http://www.usm.uni-muenchen.de/people/adi/adi.html

Abstract

Spectroscopic analyses with the intention of the interpretation of the UV-spectra of the brightest stars as individuals – supernovae – or as components of star-forming regions – massive O stars – provide a powerful tool with great astrophysical potential for the determination of extragalactic distances and of the chemical composition of star-forming galaxies even at high redshifts.

The perspectives of already initiated work with the new generation of tools for quantitative UV-spectroscopy of Hot Stars that have been developed during the last two decades are presented and the status of the continuing effort to construct corresponding models for Hot Star atmospheres is reviewed.

Since the physics of the atmospheres of Hot Stars are strongly affected by velocity expansion dominating the spectra at all wavelength ranges, hydrodynamic model atmospheres for O-type stars and explosion models for Supernovae of Type Ia are necessary as basis for the synthesis and analysis of the spectra. It is shown that stellar parameters, abundances, and stellar wind properties can be determined by the methods of spectral diagnostics already developed. Additionally, it will be demonstrated that models and synthetic spectra of Type Ia Supernovae of required quality are already available. These will make it possible to tackle the question of whether Supernovae Ia are standard candles in a cosmological sense, confirming or disproving that the current SN-luminosity distances indicate accelerated expansion of the universe.

In detail we discuss applications of the diagnostic techniques by example of two of the most luminous O supergiants in the Galaxy and a standard Supernova of Type Ia. Furthermore, it is demonstrated that the spectral energy distributions provided by state-of-the-art models of massive O stars lead to considerably better agreement with observations if used for the analysis of H II regions. Thus, an excellent way of determining extragalactic abundances and population histories is offered.

Moreover, the importance of Hot Stars in a broad astrophysical context will be discussed. As they dominate the physical conditions of their local environments and the life cycle of gas and dust of their host galaxies, special emphasis will be given to the corresponding diagnostic perspectives. Beyond that, the relevance of Hot Stars to cosmological issues will be considered.

1 Introduction

It is well known that Hot Stars are not a single group of objects but comprise sub-groups of objects in different parts of the HR diagram and at different evolutionary stages. The most important sub-groups are massive O/B stars, Central Stars of Planetary Nebulae, and Supernovae of Type Ia and II. All these sub-groups have in common that they are characterized by high radiation energy densities and expanding atmospheres, and due to this the state of the outermost parts of these objects is characterized by non-equilibrium thermodynamics. In order to cover the best-known fundamental stages of the evolution of Hot Stars in sufficient depth this review will be restricted to the discussion of O stars and Supernovae of Type Ia; we will not discuss objects like Wolf-Rayet stars, Luminous Blue Variables, Be-stars, Supernovae of Type II, and others. Furthermore, Hot Stars play an important role in a broad astrophysical context. This implies that a complete review covering all aspects in theory and observation is not only impossible, but also beyond the scope of this review. Thus, this review will focus on a special part of the overall topic with the intent to concentrate on just one subject; the subject chosen is UV spectral diagnostics. It will be shown, however, that this subject has important implications for astrophysical topics which are presently regarded as being "trendy".

But first of all we have to clarify what UV spectral diagnostics means. This is best illustrated by the really old-fashioned (1977, Morton and Underhill) UV spectrum of one of the brightest massive O stars, the O4 I(f) supergiant ζ Puppis. As can be seen in Figure 1, expanding atmospheres have a pronounced effect on the emergent spectra of hot stars – especially in the UV-part. The signatures of outflow are clearly recognized by the blue-shifted absorption and red-shifted emission in the form of the well-known P Cygni profiles. It is quite obvious that these kind of spectra contain information not only about stellar and wind parameters, but also about abundances. Thus, in principle, all fundamental parameters of a hot star can be deduced from a comparison of observed and synthetic spectra.

Although spectra of the quality shown in Figure 1 have been available for more than 25 years, most of the work done to date has concentrated on qualitative results and arguments. In view of the effort put into the development of modern – state-of-the-art – instruments, it is certainly not sufficient to restrict the analyses to simple line identifications and qualitative estimates of the physical properties. The primary objective must be to extract the complete physical stellar information from these spectra. Such diagnostic issues principally have been made possible by the superb quality and spectral resolution of the spectra available.

For this objective, the key is to produce realistic synthetic UV spectra for Hot Stars. Such powerful tools, however, are still in development and not yet widely used. They offer the opportunity to determine the stellar parameters, the abundances of the light elements – He, C, N, O, Si – and of the heavy elements like Fe and Ni, quantitatively, as is indicated not only by IUE, but also more recent HST, ORFEUS, and FUSE observations of hot stars in the Galaxy and Local Group galaxies. All these observations show that the spectra in the UV spectral range are dominated by a dense forest of slightly wind-affected pseudo-photospheric metal absorption lines

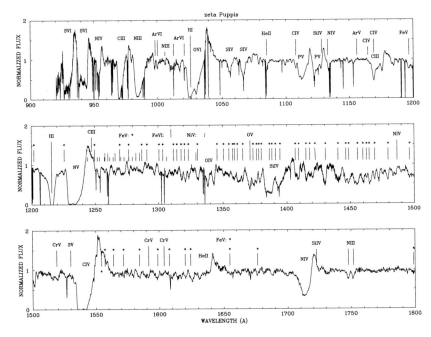

Figure 1: Merged spectrum of Copernicus and IUE UV high-resolution observations of the O4 I(f) supergiant ζ Puppis (900–1500 Å: Morton and Underhill 1977; 1500–1800 Å: Walborn et al. 1985). The most important wind lines of the light elements are identified and marked. Also marked are the large number of wind-contaminated lines of the iron group elements (e.g., Fe V) which are especially present between 1250 and 1500 Å. (Figure from Pauldrach et al. 1994b).

overlaid by broad P Cygni line profiles of strong, mostly resonance lines, formed in different parts of the expanding atmosphere. Thus, the obvious objective is to investigate the importance of these lines with respect to the structure of the expanding atmospheres that are characterized by the strength and the velocity of the outflow, and through which the shape of the spectral lines is mainly determined. Over the last 30 years it turned out that the achievement of this objective, which will also be the primary aim of this paper to review, remains a difficult task.

Before we discuss in detail the status quo of the diagnostic tool required (Section 6), we will first examine whether such a tool has been made obsolete by the general development in astrophysics or whether it is still relevant to current astronomical research.

For this purpose we first discuss the diagnostic perspectives of galaxies with pronounced current star formation. Due to the impact of massive stars on their environment the physics underlying the spectral appearance of starburst galaxies are rooted in the atmospheric expansion of massive O stars which dominate the UV wavelength range in star-forming galaxies. Therefore, the UV-spectral features of massive O stars can be used as tracers of age and chemical composition of starburst galaxies even at high redshift (Section 2).

With respect to the present cosmological question of the reionization of the universe – which appeared to have happened at a redshift of about $z \sim 6$ – the ionization efficiency of a top-heavy Initial Mass Function for the first generations of stars is discussed (Section 3).

Starting from the impact of massive stars on their environment it is demonstrated that the spectral energy distributions provided by state-of-the-art models of massive O stars lead to considerable improvements if used for the analysis of H II regions. Thus, the corresponding methods for determining extragalactic abundances and population histories are promising (Section 4).

Regarding diagnostic issues, the role of Supernovae of Type Ia as distance indicators is discussed. The context of this discussion concerns the current and rather surprising surprising result that distant SNe Ia appear fainter than standard candles in an empty Friedmann model of the universe (Section 5).

Finally, we discuss applications of the diagnostic techniques by example of two of the most luminous O supergiants in the Galaxy; additionally, basic steps towards realistic synthetic spectra for Supernovae of Type Ia are presented (Section 7).

2 The impact of massive stars on their Environment – UV Spectral Analysis of Starburst Galaxies

The impact of massive stars on their environment in the present phase of the universe is of major importance for the evolution of most galaxies. Although rare by number, massive stars dominate the life cycle of gas and dust in star forming regions and are responsible for the chemical enrichment of the ISM, which in turn has a significant impact on the chemical evolution of the host galaxy. This is mainly due to the short lifetimes of massive stars, which favours the recycling of heavy elements in an extremely efficient way. Furthermore, the large amount of momentum and energy input of these objects into the ISM controls the dynamical evolution of the ISM. This takes place in an extreme way, because massive stars mostly group in young clusters, producing void regions around themselves and wind- and supernova-blown superbubbles around the clusters. These superbubbles are ideal places for further star formation, as numerous Hubble Space Telescope images show. Investigation of these superbubbles will finally yield the required information to understand the various processes leading to continuous star formation regions (cf. Oey and Massey 1995). The creation of superbubbles is also responsible for the phenomenon of galactic energetic outflows observed in starbursts (Kunth et al. 1998) and starburst galaxies even at high redshift (Pettini et al. 1998). It is thus not surprising that spectroscopic studies of galaxies with pronounced current star formation reveal the specific spectral signatures of massive stars, demonstrating in this way that the underlying physics for the spectral appearance of starburst galaxies is not only rooted in the atmospheric expansion of massive O stars, but also dominated by these objects (cf. Figure 2 from Steidel et al. 1996, for star-forming galaxies at high redshift see also Pettini et al. 2000, and for UV line spectra of local star-forming galaxies see Conti et al. 1996; note that the similarity of the spectra at none/low and high redshifts suggests a similar stellar content).

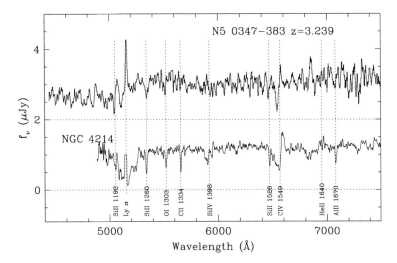

Figure 2: UV spectrum of a $z > 3$ galaxy (upper part). For comparison, a recent HST spectrum of the central starburst region in the Wolf-Rayet galaxy NGC 4214 is also shown (lower part). Note the characteristic P-Cygni lines, especially of C IV and Si IV, pointing to the dominating influence of massive O stars. Figure from Steidel et al. 1996.

The characteristic P-Cygni lines observed as broad stellar wind lines, especially those of the resonance lines of C IV and Si IV, integrated over the stellar populations in the spectra of starbursting galaxies, allow quantitative spectroscopic studies of the most luminous stellar objects in distant galaxies even at high redshift. Thus, in principle, we are able to obtain important quantitative information about the host galaxies of these objects, but diagnostic issues of these spectra require among other things *synthetic UV spectra of O-type stars* as input for the population synthesis calculations needed for a comparison with the observed integrated spectra.

The potential of these spectra for astrophysical diagnostics can nevertheless be investigated in a first step by using *observed* UV spectra of nearby O-type stars as input for the corresponding population synthesis calculations instead. In the frame of this method stars are simulated to form according to a specified star-formation history and initial mass function and then follow predefined tracks in the HR-diagram. The integrated spectra are then built up from a library of observed UV spectra of hot stars in the Galaxy and the Magellanic Clouds. The output of this procedure are semi-empirical UV spectra between 1000 and 1800 Å at 0.1 to 0.7 Å resolution for populations of arbitrary age, star-formation histories, and initial mass function. The computational technique of this method is described comprehensively in the literature and we refer the reader to one of the latest papers of a series (Leitherer et al. 2001).

As an example of the analyses performed in this way a comparison of the average spectrum of 8 clusters in NGC 5253 to synthetic models at solar (top) and 1/4 solar (bottom) metallicity is shown in Figure 3 (Leitherer et al. 2001). The Figure shows clearly that a representative value of the overall metallicity of this starburst galaxy can be determined, since the model spectrum in the lower part fits the observation

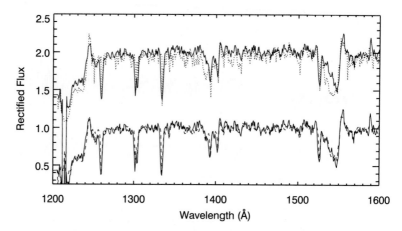

Figure 3: Average spectrum of 8 clusters in NGC 5253 compared to that of a population model at solar (top) and 1/4 solar metallicity (bottom). It is clearly shown that a representative value of the overall metallicity of this starburst galaxy can be determined. Figure from Leitherer et al., 2001.

almost perfectly. This result is especially convincing as both models (dashed lines) are based on the same parameters, except for the metallicity. A standard Salpeter IMF between 1 and 100 $M_\odot$ was used and the starburst has been assumed to last 6 Myr, with stars forming continuously during this time. The worse fit to the observations produced by the solar metallicity model spectrum – particularly discrepant are the blue absorption wings in Si IV and C IV which are too strong in the models – could be improved by reducing the number of the most massive stars with a steeper IMF, but at the cost of the fit quality in the emission components. As an important result the ratio of the absorption to the emission strengths is therefore a sensitive indicator for the metallicity.

Moreover, the strong sensitivity of the emission parts of the P-Cygni lines of N V, Si IV, and C IV on the evolutionary stage of the O stars makes these spectra quite suitable as age tracers, which is shown in Figure 4, where the time evolution of the integrated spectrum following an instantaneous starburst is presented. The line profiles gradually strengthen from a main-sequence-dominated population at 0–1 Myr to a population with luminous O supergiants at 3–5 Myr; and the lines weaken again at later age due to termination of the supergiant phase – this behavior is especially visible in the shape of the C IV $\lambda 1550$ resonance line.

The conclusion from these examples is that this kind of analysis is very promising, but relies on observed UV spectra, which are just available for a small number of metallicity values, namely those of the Galaxy and the Magellanic Clouds. In order to make progress in the direction outlined before, *realistic synthetic UV Spectra of O-type stars* are needed. This is in particular the case for high-redshift galaxies (which are observable spectroscopically when the flux is amplified by gravitational lensing through foreground galaxy clusters) since in these cases the expected metallicities of starbursting galaxies in the early universe (cf. Pettini et al. 2000) are most probably different from local ones.

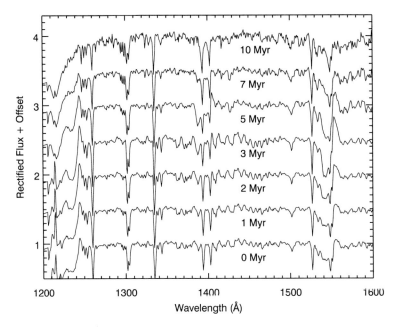

Figure 4: Time evolution of the integrated spectrum for the 10 Myr following an instantaneous starburst. The strong sensitivity of the emission parts of the P-Cygni lines of N V, Si IV, and C IV on the evolutionary stage of the O stars makes these spectra quite suitable as age tracers. Figure from Leitherer et al. 2001.

3 First generations of Stars – Ionization Efficiency of a Top-Heavy Initial Mass Function

Apart from the short evolutionary timescale of massive stars it is obviously the metallicity, and, connected to this, the steepness of the Initial Mass Function, which is responsible for the rarity of these objects in the present phase of the universe. It has to be the metallicity, because very recently strong evidence has been found that the primordial IMF has favored massive stars with masses $> 10^2 M_\odot$ (cf. Bromm et al. 1999). Thus, in the early universe, when only primordial elements were left over from the Big Bang, nature obviously preferred to form massive stars. This prediction is based on the missing metallicity which leads to a characteristic scale for the density and temperature of the primordial gas, which in turn leads to a characteristic Jeans mass of $M_J \sim 10^3 M_\odot$ (Larson 1998).

Finally, due to these physical conditions the Initial Mass Function becomes top-heavy and therefore deviates significantly from the standard Salpeter power-law (see, for instance, Bromm et al. 1999, 2001). Such an early population of very massive stars at very low metallicities (i.e., Population III stars) which have already been theoretically investigated by El Eid et al. (1983), recently turned out to be also relevant to cosmological issues, the most important being the cosmological question of when

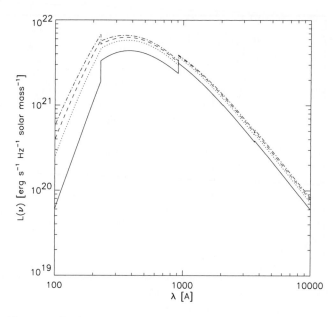

Figure 5: The normalized spectral energy distribution in the continuum of Population III stars – mass range 100–1000 $M_\odot$ at $Z = 0$. Note that the spectra attain an universal form for stellar masses $M > 300\ M_\odot$ Figure from Bromm et al. 2001.

and how the cosmic "dark ages" ended (Loeb 1998). We know that the dark ages ended because the absence of a Gunn-Peterson trough (Gunn and Peterson 1965) in the spectra of high-redshift quasars implies that the universe was reionized again at a redshift of $z > 5.8$ (Fan et al. 2000). For a long time, Carr et al. (1984) suspected that the first generations of stars have been relevant to control this process. Thus, Population III stars could contribute significantly to the ionization history of the intergalactic medium (IGM), but the contribution of the first generation of stars to the ionization history of the IGM depends crucially on their initial mass function. With regard to the first generations of stars the ionization efficiency of a top-heavy Initial Mass Function will, therefore, have to be investigated.

It is the enormous amount of UV and EUV radiation of these very massive stars which could easily change the status of the cold and dark universe at that time to become reionized again. This is indicated in Figure 5, which also shows that the total spectral luminosity depends solely on the total amount of mass, if the mass of the most massive stars exceeds 300 $M_\odot$ (cf. Bromm et al. 2001). Due to this top-heavy Initial Mass Function the total spectral energy distribution deviates significantly from that obtained with the standard Salpeter power-law, as is shown in Figure 6, and the flux obtained can contribute the decisive part to the unexplained deficit of ionizing photons required for the reionization of the universe (cf. Bromm et al. 2001).

However, in order to be able to make quantitative predictions about the influence of this extremely metal poor population of very massive stars on their galactic and intergalactic environment one primarily needs observations that can be compared to the predicted flux spectrum.

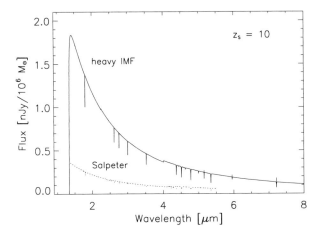

Figure 6: Predicted flux from a Population III star cluster at $z = 10$. A flat universe with $\Omega_\Lambda = 0.7$ is assumed. The cutoff below $\lambda_{obs} = 1216$ Å is due to complete Gunn-Peterson absorption, and the observable flux is larger by an order of magnitude for the case of the top-heavy IMF when compared to the case of the standard Salpeter power-law. Figure from Bromm et al. 2001.

Future observations with the Next Generation Space Telescope of distant stellar populations at high redshifts will principally give us the opportunity to deduce the primordial IMF from these comparisons. Kudritzki (2002) has recently shown that this is generally possible, by calculating state-of-the-art UV spectra for massive O stars in a metallicity range of $1 \ldots 10^{-4}$ $Z_\odot$. From an inspection of his spectra he concluded that significant line features are still detectable even at very low metallicities; thus, there will be diagnostic information available to deduce physical properties from starbursting regions at high redshifts that will eventually be observed with the Next Generation Space Telescope in the infrared spectral region.

As a second requirement we need to determine the physical properties of Population III stars during their evolution. A key issue in this regard is to obtain *realistic spectral energy distributions* calculated for metallicities different from zero for the most massive objects, since the assumption of a metallicity of $Z = 0$ is certainly only correct for the very first generation of Population III stars.

4 Theoretical Ionizing Fluxes of O Stars

Although less spectacular, we will now investigate the impact of massive stars on their environment in a more direct manner. Apart from the chemical enrichment of the ISM, the large amount of momentum and energy input into the ambient interstellar medium of these objects is primarily of importance. Especially the radiative energy input shortward of the Lyman edge, which ionizes and heats the Gaseous Nebulae surrounding massive Hot Stars, offers the possibility to analyze the influence of the EUV radiation of the photoionizing stars on the ionization structure of these excited H II regions.

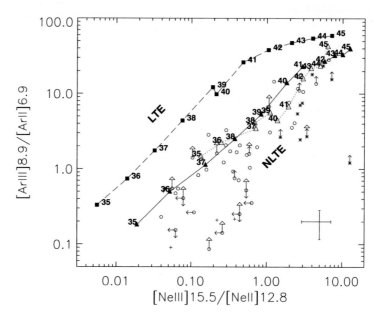

Figure 7: [Ar III] 8.99 μm / [Ar II] 6.99 μm versus [Ne III] 15.6 μm / [Ne II] 12.8 μm diagnostic diagram of observed and predicted nebular excitation of H II regions. Boldface numbers indicate models which are designated by their effective temperature in 10^3 K. Triangles are NLTE models and squares are LTE models; note that the NLTE models represent an impressive improvement in reproducing the [Ne III] emission. A representative error bar for the data is shown in the lower right corner. Figure from Giveon et al. 2002.

The primary objective of such investigations are studies of theoretical models of starburst regions, which for instance are used to determine the energy source in ultra-luminous infrared galaxies – ULIRGs – (cf. Lutz et al. 1996; Genzel et al. 1998). The interpretation of the corresponding extra-galactic observations obviously requires understanding the properties of the spectral energy distributions (SEDs) of massive stars and stellar clusters. Thus, the quality of the SEDs has to be probed in a first step by means of investigations of Galactic H II regions.

4.1 The Excitation of Galactic H II Regions

Giveon et al. (2002) recently presented a comparison of observed [Ne III] 15.6 μm / [Ne II] 12.8 μm and [Ar III] 8.99 μm / [Ar II] 6.99 μm excitation ratios, obtained for a sample of 112 Galactic H II regions and 37 nearby extragalactic H II regions in the LMC, SMC, and M33 observed with ISO-SWS, with the corresponding results of theoretical nebular models.

The authors have chosen infrared fine-structure emission lines for their investigation because these lines do not suffer much from dust extinction and the low energies of the associated levels make these lines quite insensitive to the nebular electron temperature. Moreover, the relative strengths of the fine-structure emission lines chosen are ideal for constraining the shape of the theoretical ionizing fluxes,

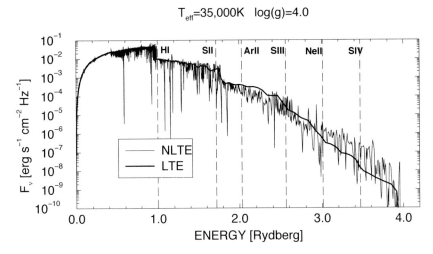

Figure 8: Comparison between the LTE (static atmosphere from Kurucz 1992) and NLTE (expanding atmosphere from Pauldrach et al. 2001) spectral energy distributions of a dwarf with an effective temperature of 35 000 K. The ionization edges of the relevant ions are indicated by vertical dashed lines. It is apparent that the difference in excitation produced by the LTE and NLTE atmospheres will be more pronounced in the Ne^{++}/Ne^{+} ratio than in the Ar^{++}/Ar^{+} ratio. Figure from Giveon et al. 2002.

since the [Ne III] 15.6 μm / [Ne II] 12.8 μm line ratio depends on photons emitted at $\geq$ 3 Rydbergs, while [Ar III] 8.99 μm / [Ar II] 6.99 μm is sensitive to the region $\geq$ 2 Rydbergs. Thus, these line ratios are extremely useful probes of the physical properties of H II regions and their associated ionizing sources especially with regard to the ionizing spectral energy distributions.

The complete set of observed emission line ratios of Ne and Ar is shown in Figure 7 together with the corresponding results of nebular model computations for which solar metallicity and a gas density of 800 cm^{-3} has been assumed, and which are based on LTE (static atmosphere from Kurucz, 1992) as well as NLTE (expanding atmosphere from Pauldrach et al., 2001) spectral energy distributions.

The diagnostic diagram clearly shows that the predicted nebular excitation increases with increasing effective temperature of the photoionizing star, and it also shows that the high excitation [Ne III] emission observed in H II regions is by far not reproduced by nebular calculations which make use of the ionizing fluxes of LTE models – the line ratios are under-predicted by factors larger than 10. This result is clearly an example of the well-known Ne III problem (cf. Baldwin et al. 1991; Rubin et al. 1991; Simpson et al. 1995).

It is quite evident, however, that the quality of the diagnostics depends primarily on the quality of the input, i.e., the spectral energy distributions. It is therefore a significant step forward that the ionizing fluxes of the NLTE models used represent an impressive improvement to the observed excitation correlation – note that the fit shown in Figure 7 may actually be even better, since a lot of the scatter is due to underestimated extinction corrections for the [Ar III] 8.99 μm line. The NLTE sequence also indicates that for most of the considered H II regions the effective temperatures of the exciting stars lie in the range of 35 000 to 45 000 K. Most importantly, the NLTE models can readily account for the presence of high excitation [Ne III] emission lines in nebular spectra. This result resolves the long-standing Ne III problem and supports the conclusion of Sellmaier et al. (1996), who found, for the first time, on the basis of a less comprehensive sample of H II regions and models that this problem has been the failure of LTE photoionization simulations primarily due to a significant under-prediction of Lyman photons above 40 eV. This issue is solved by making use of NLTE model atmospheres.

The reason for this improvement are the spectral shapes of the NLTE fluxes shortward of the Ar II and Ne II ionization thresholds which are obviously somewhat more realistic in the NLTE case. This is illustrated in Figure 8 where the spectral energy distributions of an LTE and an NLTE model are compared by example of a dwarf with an effective temperature of 35 000 K. As is shown, the SEDs harden in the NLTE case, meaning that the LTE model produces much less flux above the Ne^+ 40.96 eV threshold than the NLTE model. This is the behavior that is essentially represented in the excitation diagram.

4.2 Spectral Energy Distributions of Time-Evolving Stellar Clusters

The improvement obtained for the diagnostic diagram of the analysis of Galactic H II regions in Section 4.1 also has important implications for determining extragalactic abundances and population histories of starburst galaxies.

With regard to this, Thornley et al. (2000) carried out detailed starburst modelling of the [Ne III] 15.6 μm / [Ne II] 12.8 μm ratio of H II regions ionized by clusters of stars. As was shown above, the hottest stars in such models are responsible for producing large nebular [Ne III] 15.6 μm / [Ne II] 12.8 μm ratios. Hence, the low ratios actually observed led to the conclusion that the relative number of hot stars is small due to aging of the starburst systems. Thus, the solution of the Ne III problem has important consequences for the interpretation of these extragalactic fine-structure line observations, since the conclusion that due to the low [Ne III] 15.6 μm / [Ne II] 12.8 μm ratios obtained, the hottest stars as dominant contributors to the ionization of the starburst galaxies have to be removed, has obviously to be proven.

In order to tackle this challenge, Sternberg et al. (2002) computed spectral energy distributions of time-evolving stellar clusters, on the basis of a large grid of calculated NLTE spectral energy distributions of O and early B-type stars (being the major contributors to the Lyman and He I fluxes) in the hot, luminous part ($T_{eff} >$ 25 000 K) of the HR diagram. From this grid, models of stars following evolutionary tracks are suitably interpolated. The SEDs used rely on recent improvements of modelling expanding NLTE atmospheres of Hot Stars (Pauldrach et al. 2001). As an example of the individual models used, Figure 9 shows the calculated spectral energy distribution of a typical O star compared to the corresponding result of an NLTE model of Schaerer and de Koter (1997).

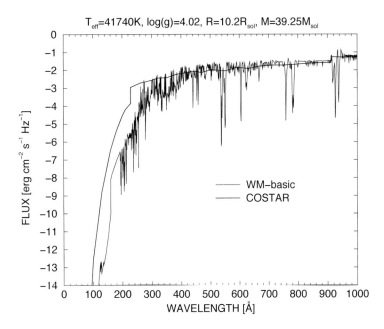

Figure 9: Calculated spectral energy distribution of a typical O star. The spectrum represents the result of the WM-basic model (Pauldrach et al. 2001), and the thick line represents the result of a corresponding COSTAR model (Schaerer and de Koter 1997).

The spectral energy distributions have been calculated for two modes of cluster evolution, a *continuous* and an *impulsive* one. The different spectral evolutions of these modes are shown in Figure 10 with regard to the spectral range of 0.8–2.5 Rydbergs.

In the continuous mode, the star-formation rate is assumed to be constant with time – on time scales which are longer than the lifetimes of massive stars – and the cluster is assumed to form stars at a rate of 1 $M_\odot$ per year. In the impulsive mode, all stars are formed "instantaneously", i.e., on time scales which are much shorter than the lifetimes of massive stars; the total mass of stars formed is $10^5\ M_\odot$, and the cluster is assumed to evolve passively thereafter. Furthermore, the computations are based on a Salpeter initial mass function, where for each mode two cases are assumed, one with an upper mass limit of $M_{\rm up} = 120\ M_\odot$ and the other one with an upper mass limit of $M_{\rm up} = 30\ M_\odot$.

Two striking effects are seen in Figure 10:

1. In the continuous mode, the cluster ionizing spectrum becomes softer – i.e., steeper – between 1 and 10 Myr. This is due to the increase of the relative number of late versus early-type OB stars during the evolution, since the less massive late-type stars have longer lifetimes. This effect is particularly noticeable for the cluster model with $M_{\rm up} = 120\ M_\odot$.

2. In the impulsive mode, the magnitude of the Lyman break increases drastically because the massive stars disappear with time.

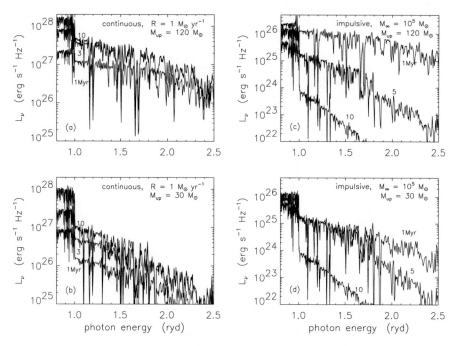

Figure 10: Spectral energy distributions of time-evolving stellar clusters. Two modes of cluster evolution are shown: *continuous* and *impulsive*. *(Left)*: Evolution for continuous star formation (lasting 10^{10} yr) for two extreme stellar compositions. In the upper panel evolving SEDs of a cluster with $M_{\rm up} = 120 \, M_\odot$, and in the lower panel with $M_{\rm up} = 30 \, M_\odot$ are shown. Note that at early phases the amount of both ionizing and non-ionizing photons increase due to the increasing number of new-formed stars. After a time period of ~ 5 Myr the number of ionizing photons remains roughly constant, because an equilibrium of stellar aging and stellar birth is achieved, whereas the number of non-ionizing photons remains proportional to age, because of accumulation effects of old stars which are the major contributors to this spectral range. *(Right)*: Evolution of an impulsive star formation burst (lasting 10^5 yr), for the same extreme stellar compositions as above. Note that the ionizing photons practically disappear at an age of ~ 10 Myr. Figure from Sternberg et al. 2002.

The evolution of the photon emission rates for photons above the Lyman ($Q_{\rm H}$) and the He I ($Q_{\rm He}$) ionization thresholds is shown in Figure 11 for both modes of the clusters. The key difference between the two modes leads to completely different shapes of the photon emission rates, which will be easy to distinguish in view of their photoionizing properties acting on their gaseous environments. It is also shown that increasing the upper IMF mass cut-off increases the photon emission rates, but does not change the shapes of the rates obtained during the evolution. The methods developed for determining extragalactic abundances and population histories from an analysis of H II regions are thus very promising.

From these investigations it is obvious that realistic spectral energy distributions of massive stars and stellar clusters are important for studies of their environments. The crucial question, however, whether the spectral energy distributions of massive

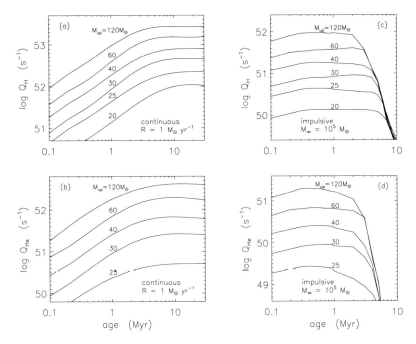

Figure 11: Evolution of the Lyman and He I photon emission rates for a range of values of the upper IMF mass cut-off. The continuous star formation mode is considered on the left hand side, and the impulsive star formation burst on the right. Figure from Sternberg et al. 2002.

stars are already realistic enough to be used for diagnostic issues of H II regions has not been answered yet, since on the basis of the results obtained for the diagnostic diagram in Section 4.1 it cannot be excluded that wrong fluxes could show the same improvements in the Ne^{++}/Ne^+ ratios just by chance.

An answer to this question requires an *ultimate test*!

This ultimate test is only provided by comparing the observed and synthetic UV spectra of the individual massive stars, based on the following reasons:

1. This test involves hundreds of spectral signatures of various ionization stages with different ionization thresholds which cover a large frequency range.

2. Almost all of the ionization thresholds lie within the spectral range shortward of the Lyman ionization threshold (cf. Figure 12); thus, *the ionization balances of all elements depend sensitively on the ionizing radiation throughout the entire wind*.

The ionization balance can be traced reliably through the strength and structure of the wind lines formed throughout the atmosphere. Hence, it is a natural and the only reliable step to test the quality of the ionizing fluxes by virtue of their direct product: *the UV spectra of O stars*.

But before we turn to this test we will continue our discussion with another astrophysically important stellar object.

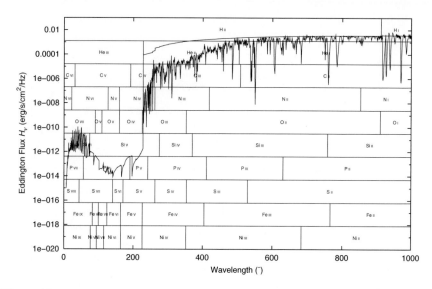

Figure 12: Spectral energy distribution shortward of the Lyman ionization threshold of an expanding NLTE model atmosphere for an O supergiant (Pauldrach et al. 2003). The small vertical bars indicate the ionization thresholds for all important ions; the ionization balance depends almost entirely on the ionizing flux, and this influence can be traced by the spectral lines in the observable part of the UV spectrum.

5 Supernovae of Type Ia as Distance Indicators

In order to complete the discussion of Hot Stars in the context of modern astronomy, we will concentrate now on the subject of Supernovae of Type Ia. With respect to diagnostic issues we will investigate the role of Supernovae of Type Ia as distance indicators. The context of this discussion regards, as a starting point, the current surprising result that distant SNe Ia at intermediate redshift appear fainter than standard candles in an empty Friedmann model of the universe.

Type Ia supernovae, which are the result of the thermonuclear explosion of a compact low mass star, are currently the best known distance indicators. Due to their large luminosities ($L_{\max} \sim 10^{43}$ erg/s) they reach far beyond the local supercluster (cf. Saha et al. 2001 and references therein). It is thus not surprising that Type Ia supernovae have become the most important cosmological distance indicator over the last years, and this is not only due to their extreme brightness, but primarily due to their maximum luminosity which can be normalized by their light curve shape, so that these objects can be regarded as *standard candles*. This has been shown by observations of local SNe Ia which define the linear expansion of the local universe extremely well (Riess et al. 1999), which in turn is convincing proof of the accuracy of the measured distances. On the other hand, observations of distant SNe Ia (up to redshifts of about 1) have yielded strong evidence that the expansion rate has been accelerated 6 Gyr ago (cf. Riess et al. 1998 and Perlmutter et al. 1999).

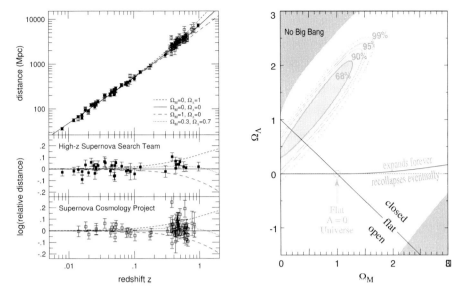

Figure 13: *Left*: Hubble diagram of Type Ia Supernovae. In the upper panel the distance modulus versus redshift is shown in the usual way, whereas the distance modulus has been normalized to an empty universe in the lower panels. Shown are the data of the High-z SN Search Team (filled squares; Riess et al. 1998) and the Supernova Cosmology Project (open squares; Perlmutter et al. 1999). In addition, lines of four cosmological models are drawn, where the full line corresponds to a cosmological model for an empty universe and the dotted line to a flat universe. Figure from Leibundgut 2001.
Right: Cosmological diagram showing the likelihood region as defined by SNe Ia in the Ω_Λ versus Ω_{Mass} plane. The results of 79 SNe Ia have been included – 27 local and 52 distant ones – from the sources above. Contours indicate 68 %, 90 %, 95 %, and 99 % probability, and the line for a flat universe is indicated. Note that the SN Ia luminosity distances clearly indicate an accelerated expansion of the universe. Figure from Perlmutter et al. 1999.

This surprising result becomes evident if we look at the Hubble diagram in the form of distance modulus versus redshift (cf. Figure 13). As is shown in the diagrams of the lower panels, which are normalized to a cosmological model for an empty universe, most SNe Ia at intermediate redshift are positioned at positive values of the normalized distance modulus; thus, the distant supernovae appear fainter than what would be expected in a empty universe. This means that deceleration from gravitational action of the matter content does not take place. Moreover, the SNe Ia appear even more distant indicating an accelerated expansion over the last 6 Gyr.

The implications of this are immediately recognized in the cosmological diagram on the right-hand side of Figure 13, where Ω_Λ versus Ω_{Mass} is plotted. The SN Ia results obviously exclude any world model with matter but without a a cosmological constant. Hence, the data show the need for an energy contribution by the vacuum. Thus, the SN luminosity distances indicate accelerated expansion of the universe!

However, this is not the only interpretation of the result obtained. There are other astrophysical explanations, such as obscuration by intergalactic dust or evolution of

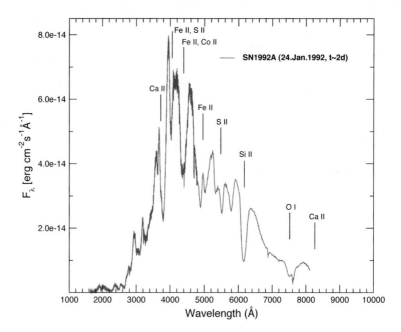

Figure 14: Observed HST spectrum of a standard Supernova of Type Ia at early phase – SN 1992A from 24 Jan. 1992. Figure from Kirshner et al. 1993.

supernovae of Type Ia (cf. Leibundgut 2001). The latter point refers primarily to the absolute peak luminosity of SNe Ia which might have changed between the local epoch and a redshift of 1.0, especially in the case of a decrease: then the fainter supernovae would not be a signature of larger distances, but rather of an evolution of these objects. Although present SN Ia models do not favor such an evolution, it cannot be excluded, because neither the explosions nor the radiation emerging from the atmosphere are understood in detail yet. Consequently, serious caveats for the cosmological interpretation of distant supernovae exist.

Thus, we are faced with the question: Are SNe Ia standard candles independent of age, or is there some evolution of the SN luminosity with age? Among other things spectroscopy is certainly a powerful tool to obtain an answer to this question by searching for spectral differences between local and distant SNe Ia.

Unfortunately, at present, we do not have a clear picture of the exact physical processes which take place in a SN Ia explosion and how the radiation released from it should be treated (for a recent review see Hillebrandt and Niemeyer 2000). In order to get an impression of the basic physics involved that affect the atmospheric structure of a SN Ia, a quick glance at an observed spectrum will help to point out the necessary steps for their further analysis.

Figure 14 shows a typical SN Ia spectrum at early epochs (< 2 weeks after maximum). The most striking features of this spectrum are the characteristic P Cygni line profiles which are quite similar to the signatures of O stars. But compared to the latter objects there are also important differences: the broad lines indicate that the velocities of the SN Ia ejecta are almost a factor of 10 larger (up to 30 000 km/s) and, as a second important point, SN Ia spectra contain no H and He lines. Instead, prominent absorption features of mainly intermediate-mass elements (Si II, O I, S II, Ca II, Mg II, ...) embedded in a non-thermal pseudo-continuum are observed.

To answer the question of whether SN Ia are standard candles in a cosmological sense, realistic models and synthetic spectra of Type Ia Supernovae are obviously required for the analysis of the observed spectra. As just shown, however, these models will necessarily need to be based on a similar serious approach for expanding atmospheres, characterized by non-equilibrium thermodynamics, as is the case for Hot Stars in general.

It will be shown in Section 7.1 that such spectra are in principle already available.

6 Concept for Consistent Models of Hot Star Atmospheres

In order to determine stellar abundances, parameters, and physical properties (and from these, obtain realistic spectral energy distributions) of Hot Stars via quantitative UV spectroscopy, a principal difficulty needs to be overcome: *the diagnostic tools and techniques must be provided.* This requires the construction of detailed atmospheric models.

As has been demonstrated in the previous sections, such a tool has not been made obsolete by general development in astrophysics. On the contrary, it is becoming more and more relevant to current astronomical research as spectral analysis of hot luminous stars is of growing astrophysical interest. Thus, continuing effort is expended to develop a standard code in order to provide the necessary diagnostic tool. In the following part of this section we describe the status of our continuing work to construct realistic models for expanding atmospheres.

Before we focus on the theory in its present stage, we should mention previous fundamental work which turned out to be essential in elaborating the theory. In this context I want to emphasize publications which refer to key aspects of theoretical activity. The starting point of the development of the radiation-driven wind theory is rooted in a paper by Milne (1926) more than 70 years ago. Milne was the first to realize that radiation could be coupled to ions and that this process subsequently may eject the ions from the stellar surface. The next fundamental step goes back to Sobolev (1957), who developed the basic ideas of radiative transfer in expanding atmospheres.

Radiation pressure as a driving mechanism for stellar outflow was rediscovered by Lucy and Solomon (1970) who developed the basis of the theory and the first attempt to its solution. The pioneering step in the formulation of the theory in a quasi self-consistent manner was performed by Castor, Abbott, and Klein (1975). Although these approaches were only qualitative (due to many simplifications), the theory was developed further owing to the promising results obtained by these authors. Regarding key aspects of the solution of the radiative transfer the work of Rybicki (1971) and Hummer and Rybicki (1985) has to be emphasized. With respect to the soft X-ray emission of O stars (detected by Seward et al. 1979 and Harnden et al. 1979), Cassinelli and Olson (1979) have investigated the possible influence of X-rays

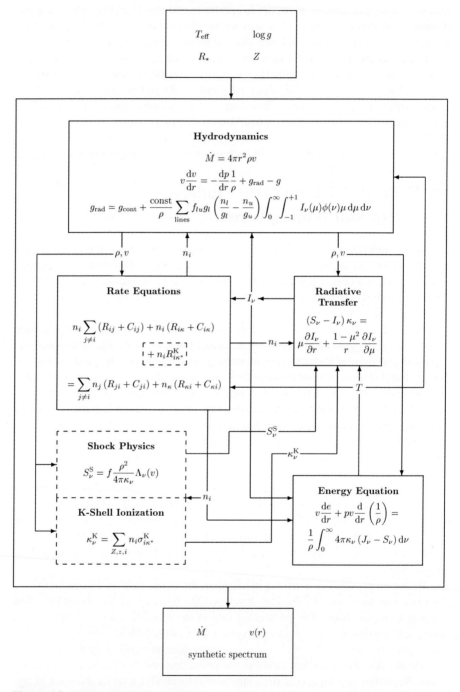

Figure 15: Schematic sketch of the basic equations and the non-linear system of integro-differential equations that form the basis of stationary models of hot star atmospheres – for details see text. Figure from Pauldrach et al. 1994b).

on the ionization structure. Finally, Lucy and White (1980) and Owocki, Castor, and Rybicki (1988) have distinguished themselves with basic theoretical investigations of time-dependent radiation hydrodynamics which describe the creation and development of shocks (for more details about the role of X-rays in the atmospheres of Hot Stars see the reviews of Pauldrach et al. 1994b and Kudritzki and Puls 2000).

6.1 The general method

The basis of our approach in constructing detailed atmospheric models for Hot Stars is the concept of *homogeneous, stationary, and spherically symmetric radiation driven winds*, where the expansion of the atmosphere is due to scattering and absorption of Doppler-shifted metal lines. Although these approximations seem to be quite restrictive, it has already been shown that the time-averaged mean of the observed UV spectral features can be described correctly by such a method (Pauldrach et al. 1994, 1994b).

The primary output of this kind of model calculation are the spectral energy distributions and synthetic spectra emitted by the atmospheres of hot stars. As these spectra consist of hundreds of strong and also weak wind-contaminated spectral lines which form the basis of a quantitative analysis, and as the spectral energy distribution from hot stars is also used as input for the analysis of emission line spectra which depend sensitively on the structure of the emergent stellar flux (cf. Section 4), a sophisticated and well-tested method is required to produce these data sets accurately.

However, developing such a method is not straightforward, since modelling the atmospheres of hot star involves the replication of a tightly interwoven mesh of physical processes: the equations of radiation hydrodynamics including the energy equation, the statistical equilibrium for all important ions with detailed atomic physics, and the radiative transfer equation at all transition frequencies have to be solved simultaneously. Figure 15 gives an overview of the physics to be treated.

The principal features are:

(1) *The stellar parameters* $T_{\rm eff}$ (effective temperature), $\log g$ (logarithm of photospheric gravitational acceleration), R_* (photospheric radius defined at a Rosseland optical depth of 2/3) and Z (abundances) have to be pre-specified.

(2) *The hydrodynamic equations* are solved (r is the radial coordinate, ϱ the mass density, v the velocity, p the gas pressure and $\dot{M}$ the mass loss rate). The crucial term is the radiative acceleration $g_{\rm rad}$ with minor contributions from continuous absorption and major contributions from scattering and line absorption. For each line the oscillator strengths f_{lu}, the statistical weights g_l, g_u, and the occupation numbers n_l, n_u of the lower and upper level enter the equations, together with the frequency and angle integral over the specific intensity I_ν and the line broadening function φ_ν accounting for the Doppler effect.

(3) *The occupation numbers* are determined by the *rate equations* containing collisional (C_{ij}) and radiative (R_{ij}) transition rates. Low-temperature dielectronic recombination is included and Auger ionization due to K-shell absorption (essential for C, N, O, Ne, Mg, Si, and S) of soft X-ray radiation arising from

shock-heated matter is taken into account using detailed atomic models for all important ions. Note that the hydrodynamical equations are coupled directly with the rate equations. The velocity field enters into the radiative rates while the density is important for the collisional rates and the equation of particle conservation. On the other hand, the occupation numbers are crucial for the hydrodynamics since the radiative line acceleration dominates the equation of motion.

(4) *The spherical transfer equation* which yields the radiation field at every depth point, including the thermalized layers where the diffusion approximation is applied, is solved for the total opacities (κ_ν) and source functions (S_ν) of all important ions. Hence, the influence of the spectral lines – the strong EUV *line blocking* – which affects the ionizing flux that determines the ionization and excitation of levels, is taken into account. Note that the radiation field is coupled with the hydrodynamics ($g_{\rm rad}$) and the rate equations (R_{ij}).

Moreover, the *shock source functions* (S_ν^S) produced by radiative cooling zones which originate from a simulation of shock heated matter, together with K-shell absorption (κ_ν^K), are also included in the radiative transfer. The shock source function is incorporated on the basis of an approximate calculation of the volume emission coefficient (Λ_ν) of the X-ray plasma in dependence of the velocity-dependent post-shock temperatures and the filling factor (f).

(5) *The temperature structure* is determined by the microscopic *energy equation* which, in principle, states that the luminosity must be conserved. *Line blanketing* effects which reflect the influence of line blocking on the temperature structure are taken into account.

(6) The iterative solution of the total system of equations yields the hydrodynamic structure of the wind (i. e., the *mass-loss rate* $\dot{M}$ and the *velocity structure* $v(r)$) together with *synthetic spectra* and ionizing fluxes.

Essential steps which form the basis of the theoretical framework developed are described in Pauldrach et al. 1986, Pauldrach 1987, Pauldrach and Herrero 1988, Puls and Pauldrach 1990, Pauldrach et al. 1994, Feldmeier et al. 1997, and Pauldrach et al. 1998.

The effect complicating the system the most is the overlap of the spectral lines. This effect is induced by the velocity field of the expanding atmosphere which shifts at different radii up to 1000 spectral lines of different ions into the line of sight at each observer's frequency. Since the behavior of most of the UV spectral lines additionally depends critically on a detailed and consistent description of the corresponding effects of *line blocking* and *line blanketing* (cf. Pauldrach 1987), special attention has to be given to the correct treatment of the Doppler-shifted line radiation transport of all metal lines in the entire sub- and supersonically expanding atmosphere, the corresponding coupling with the radiative rates in the rate equations, and the energy conservation. Another important point to emphasize concerns the atomic data, since it is quite obvious that the quality of the calculated spectra, the multilevel NLTE treatment of the metal ions (from C to Zn), and the adequate representation of line blocking and the radiative line acceleration depends crucially on the quality of the atomic models. Thus, the data have to be continuously improved whenever significant progress of atomic data modelling is made. Together with a revised inclusion of EUV and X-ray radiation produced by cooling zones which originate from the simulation of shock heated matter, these improvements have been implemented and described in a recent paper (Pauldrach et al. 2001). Very recently we have additionally focused on a remaining aspect regarding hydrodynamical calculations which are consistent with the radiation field obtained from the line-overlap computations, and we have incorporated this improvement into our procedure (see Pauldrach and Hoffmann 2003).

This solution method is in its present stage already regarded as a standard procedure towards a realistic description of stationary wind models. Thus, together with an easy-to-use interface and an installation wizard, the program package *WM-basic* has been made available to the community. The package can be downloaded from the URL given on the first page.

In the next section we will investigate thoroughly whether these kind of models already lead to realistic results.

7 Synthetic Spectra and Models of Hot Star Atmospheres

The first five sections of this review inevitably lead to the realization that line features of expanding atmospheres are one of the most important astronomical manifestations, since they can easily be identified even in spectra of medium resolution in individual objects like supernovae or in integrated spectra of starburst regions even at significant redshift. As a consequence, they can be used to provide important information about the chemical composition, the extragalactic distance scale, and several other properties of miscellaneous stellar populations. All that is required are *hydrodynamic NLTE model atmospheres* that incorporate the effects of spherical extension and expansion as realistic and consistent as possible. After almost two decades of work these model atmospheres now seem to be available, and in the following we will investigate thoroughly whether basic steps towards realistic synthetic spectra for Supernovae of Type Ia can already be presented, and whether UV diagnostic techniques applied to two of the most luminous O supergiants in the Galaxy lead to synthetic spectra which can be regarded as a measure of excellent quality.

7.1 Synthetic and Observed Spectra of Supernovae of Type Ia

Prominent features in the spectra of Type Ia supernovae in early phases are the characteristic absorption lines resulting from low ionized intermediate-mass elements, such as Si II, O I, Ca II, Mg II, cf. Figure 17. These absorption lines show characteristic line shapes due to Doppler-broadening resulting from the large velocity gradients. The formation of these lines results in a pseudo-continuum that is set up

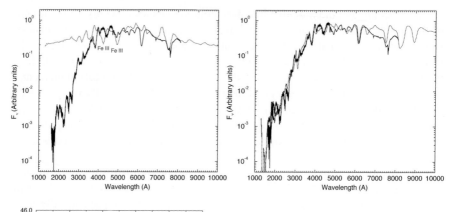

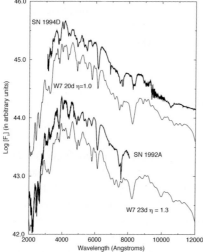

Figure 16: Synthetic NLTE spectra of a Supernova of Type Ia. *Top*: The dotted lines show the spectrum from a model without line blocking (left panel) and from one where line blocking is treated consistently (right panel), both compared to the observed spectrum of SN 1992A (solid line). (Figure from Pauldrach et al. 1996.) *Left*: The two curves in the lower part show a synthetic spectrum for a W7 model at 23 days compared to the observed spectrum of SN 1992A. (Figure from Nugent et al. 1997.) The observed spectrum is from Kirshner et al. 1993 (5 days after maximum light).

by the enormous number of these lines. Supernovae appear in this 'photospheric' epoch for about one month after the explosion and the spectra at this epoch contain useful information on the energetics of the explosion (luminosity, velocity- and density-structure) and on the nucleosynthesis in the intermediate and outer part of the progenitor star. Therefore, the primary objective in order to analyze the spectra is to construct consistent models which link the results of the nucleosynthesis and hydrodynamics obtained from state-of-the-art explosion models (cf. Reinecke et al. 1999) with the calculations of light curves and synthetic spectra of SN Ia. In a first step, however, this is done by using the standard hydrodynamical model W7 by Nomoto et al. 1984. Figure 16 exemplifies the status quo of the synthetic Spectra of Type Ia Supernovae at early phases obtained from this first step, compared to observations.

Obviously, the first comparison shown (top left panel of Figure 16) does not at all represent a realistic atmospheric model. The synthetic spectrum shown has been calculated for a model without line blocking, i.e., only continuum opacity has been considered in the NLTE calculation. Thus, the drastic effects of line blocking on the ionization and excitation and the emergent flux can be verified by comparing the

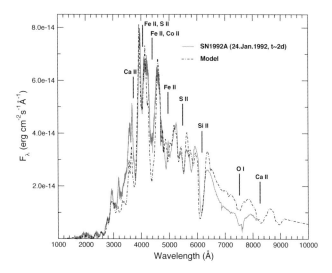

Figure 17: Comparison of the most recent synthetic NLTE spectrum at early phases and the observed spectrum of SN 1992A, illustrating that the method used is already on a quantitative level. The rest wavelengths of various metal lines are indicated by vertical bars. Figure from Sauer and Pauldrach 2002.

calculated flux with the observed flux, where the huge difference in the UV is due to the missing contribution of the lines to the opacity. It is an essential by-product of this result that the observed line features in the optical are also not reproduced: The two strongest features in the synthetic spectrum are due to Fe III lines, indicating that the ionization is too high because of the excessive UV flux. This behavior nicely illustrates that the ionization balance depends almost entirely on the ionizing UV flux and that this influence can be traced by the spectral lines in the observable part of the spectrum, as has been stated in Section 4.2.

The fact that the difference in the UV observed for this model is indeed due to line blocking becomes obvious if this important effect is properly taken into account, as in the case of the model shown in the upper right panel of Figure 16. The agreement of the resulting spectrum of this model with the observations is quite good, demonstrating that line blocking alone can diminish the discrepancy between the synthetic and the observed spectra. Both the flux level and most line features are now reasonably well reproduced throughout the spectrum (cf. Pauldrach et al. 1996). Additionally, in the lower left panel of Figure 16 a spectral fit to the same supernova observation by Nugent et al. (1997) is shown, which is based on a completely independent approach to the theory. The resulting spectrum is also quite reasonable.

Finally, Figure 17 shows a recently calculated synthetic spectrum that is based on the important improvements of the theory described in the previous section, compared to the observed "standard" supernova spectrum (SN 1992A). The synthetic spectrum reproduces the observed spectrum quite well, and the overall impression of this comparison is that the method used is already on a quantitative level, indicating that the basic physics is treated properly (cf. Sauer and Pauldrach 2002).

Although there is still need for improvements regarding, for instance, a proper treatment of the γ energy deposition in the outermost layers and the use of current hydrodynamic explosion models, these models form the basis for the primary objective of the subject, i.e., to search for spectral differences between local and distant SNe Ia. Concerning the facilities of spectral analysis we are thus close to the point where we can tackle the question of whether SNe Ia are standard candles in a cosmological sense.

7.2 Detailed Analyses of Massive O Stars

The objectives of a detailed comparison of synthetic and observed UV spectra of massive O stars are manifold. Primarily it has to be verified that the higher level of consistent description of the theoretical concept of line blocking and blanketing effects and the involved modifications to the models leads to changes in the line spectra with much better agreement to the observed spectra than the previous, less elaborated and less consistent models. Secondly it has to be shown that the stellar parameters, the wind parameters, and the abundances can be determined diagnostically via a comparison of observed and calculated high-resolution spectra covering the observable UV region. And finally, the quality of the spectral energy distributions has to be verified. The latter point, however, is a direct by-product of the other objectives if the complete observed high-resolution UV spectra are accurately reproduced by the synthetic ones, as illustrated in Section 4.2

In the following, the potential of the improved method will be demonstrated by an application to the O4 I(f) star ζ Puppis and the O9.5 Ia star α Cam.

7.2.1 UV Analysis of the Hot O Supergiant ζ Puppis

Referring to the beginning of the review, we will now start to examine carefully the old-fashioned UV spectrum of ζ Puppis. To put this into perspective, we will also briefly review the improvements of UV line fits following from gradual improvements of the methods used.

The first serious attempt to analyze the UV spectrum of this standard object goes back to Lamers and Morton in 1976. From a present point of view their model can not be regarded as a sophisticated one, because they *assumed* the complete model structure – the dynamical structure, the occupation numbers, and the temperature structure – and the line radiative transfer was treated in an approximative way. Nevertheless, we learned from this kind of spectrum synthesis that a solution to the problem is feasible, and that at least the dynamical parameters – the mass-loss rate ($\dot{M}$) and the terminal velocity (v_∞) – can, in principle, be determined from UV lines quite accurately. The UV line fits of Hamann in 1980 have been based on the same assumptions apart from a very detailed treatment of the radiative line transfer. As a result of this the calculated resonance lines of some ions were already quite well in agreement with the observed ones. But the real conclusion to be drawn from these comparisons is that a detailed analysis requires a more consistent treatment of expanding atmospheres.

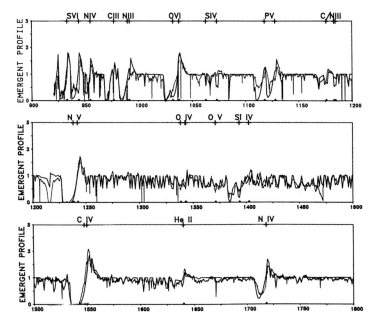

Figure 18: Calculated and observed UV spectrum of ζ Puppis. The observed spectrum shows the Copernicus and IUE high-resolution observations, and the calculated spectrum belongs to the final model of Pauldrach et al. 1994.

The situation improved considerably with the first attempts to find a consistent solution, meaning that the equations of hydrodynamics, non-equilibrium thermodynamics, and radiative transfer have been solved in a consistent way. Pauldrach (1987) introduced for the first time a full NLTE treatment of the metal lines driving the wind, and Puls (1987) investigated the important effect of multiple photon momentum transfer through line overlaps caused by the velocity-induced Doppler-shifts for applications in stellar wind dynamics. From this procedure, dynamical parameters, constraints on the stellar parameters, and, as has been verified by a comparison of a sample of calculated spectral lines with the observed spectrum, an ionization equilibrium and occupation numbers which were close to a correct description have already been obtained. But, among other approximations, the models still suffered from the neglect of radiation emitted from shock cooling zones and from a very approximate treatment of line blocking.

The further steps to reproduce most of the important observed individual line features in the UV required a lot of effort in atomic physics, in improved NLTE multilevel radiative transfer, and in spectrum synthesis techniques. An important step towards this objective was the paper by Pauldrach et al. (1994), where for the first time stellar parameters and abundances have been determined from individual UV line features. Figure 18 shows the corresponding synthetic spectrum that was already in overall agreement with the observations. Nevertheless, the treatment was still affected by a number of severe approximations regarding especially the neglect of the line blanketing effect, an approximate treatment of the line radiative transfer,

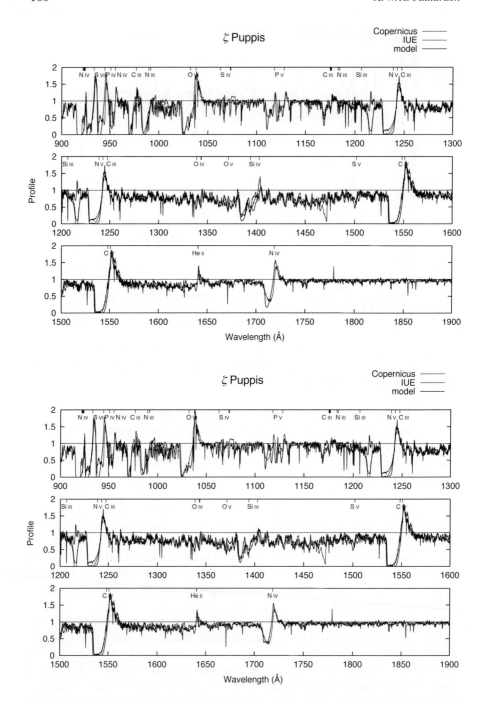

Figure 19: Calculated and observed UV spectrum for ζ Puppis. The observed spectrum shows the Copernicus and IUE high-resolution observations, and the calculated state-of-the-art spectra represent the final models of Pauldrach et al. 2003. For a discussion see text.

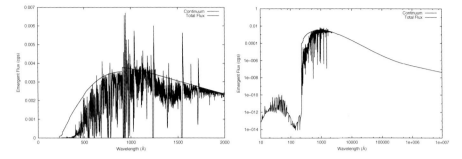

Figure 20: Calculated state-of-the-art spectral energy distribution of the final model of ζ Puppis of Pauldrach et al. 2003. *Left*: Linear scale for the flux. *Right*: Logarithmic scale for the flux.

and a simple treatment of emitted radiation from shock cooling zones. Thus, we conclude that a more consistent treatment of expanding atmospheres was still needed for a detailed analysis.

This more consistent treatment of expanding atmospheres that has now been applied and fixes the status quo is based on the improvements discussed in Section 6.1 (for a more comprehensive discussion see Pauldrach et al. 2001 and Pauldrach and Hoffmann 2003). As is shown in Figure 19, the calculated synthetic spectrum is quite well in agreement with the spectra observed by IUE and Copernicus. The small differences observed in the Si IV and the N IV lines just reflect a sensitive dependence on the parameters used to describe the shock distribution. This is verified by a comparison of the two panels of Figure 19, where just the shock-distribution has slightly been changed within the range of uncertainty of the corresponding parameters. Not only have the stellar and wind parameters of this object been confirmed by the model on which the synthetic spectrum is based, but also the abundances of C, N, O, P, Si, S, Fe, and Ni have been determined. We thus conclude that the present method of quantitative spectral UV analysis of Hot Stars leads to models which can be regarded as being realistic.

Consequently, we consider these kind of quantitative spectral UV analyses as the ultimate test for the accuracy and the quality of theoretical ionizing fluxes (cf. Figure 20), which can thus be used as spectral energy distributions for the analysis of H II regions.

7.2.2 UV Analysis of the Cool O Supergiant α Cam

In Figure 21 it can be seen that the method also works for cool O supergiants like α Cam. We recognize again that the synthetic spectrum is quite well in agreement with the IUE and Copernicus spectra (for a discussion see Pauldrach et al. 2001). In the uppermost panel of the spectrum, simulations of interstellar lines have additionally been merged with the synthetic spectrum in order to disentangle wind lines from interstellar lines (for a discussion of this point see Hoffmann 2002). In this case also, our new realistic models allowed to determine stellar parameters, wind parameters, and abundances from the UV spectra alone. (The operational procedure of the

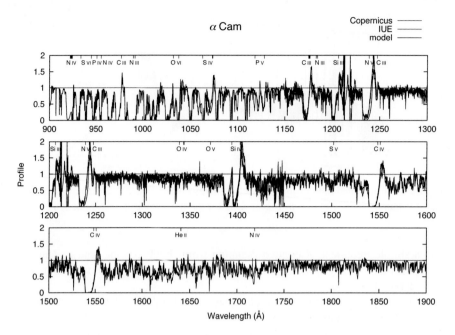

Figure 21: Comparison of the best model of Pauldrach et al. 2001 with spectra of α Cam observed by IUE and Copernicus, demonstrating the quality that can be achieved with the new model generation.

method begins with realistic estimates of R, T_{eff}, M, and a set of abundances. With these, the model atmosphere is solved and the velocity field, the mass loss rate $\dot{M}$, and the synthetic spectrum is calculated. The parameters are adjusted and the process is repeated until a good fit to all features in the observed UV spectrum is obtained. For a compact description of this procedure see Pauldrach et al. 2002). It turned out that the effective temperature can be determined to within a range of ± 1000 K and the abundances to at least within a factor of 2.

7.3 Wind properties of massive O stars

As a last point we want to illustrate the significance of the dynamical parameters of radiation-driven winds. The intrinsic significance is quite obvious: it is the consistent hydrodynamics which provides the link between the stellar parameters (T_{eff}, M, R) and the wind parameters (v_∞, $\dot{M}$). Thus, the appearance of the UV spectrum is determined by the interplay of the NLTE model and the hydrodynamics. (As was discussed in Section 6.1, the hydrodynamics is controlled by the line force, which is primarily determined by the occupation numbers, and the radiative transfer of the NLTE model, but the hydrodynamics in turn affects the NLTE model via the density and velocity structure.)

A tool for illustrating the significance of the dynamical parameters is offered by the so called *wind-momentum–luminosity relation*. This relation is based on two important facts: The first one is that, due to the driving mechanism of hot stars,

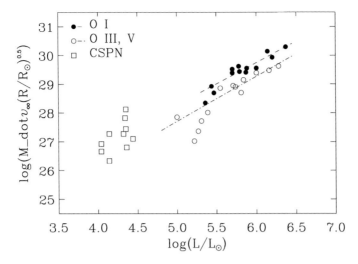

Figure 22: The wind-momentum–luminosity relation for massive O stars and Central Stars of Planetary Nebulae (CSPNs). Circles designate the O star analysis based on Hα profiles by Puls et al. 1996, and squares that of CSPNs by Kudritzki et al. 1997. Figure from Kudritzki and Puls 2000.

the mechanical momentum of the wind flow ($v_\infty \dot{M}$) is mostly a function of photon momentum (L/c) and is therefore related to the luminosity. The second one is that the expression $v_\infty \dot{M} R^{1/2}$ is an almost directly observable quantity. As the shapes of the spectral lines are characterized by the strength and the velocity of the outflow, the first term of this expression, the observed terminal velocity, can be measured directly from the width of the absorption part of the saturated UV resonance lines. Deducing the mass-loss rate from the line profiles is, however, more complex, since this requires the calculation of the ionization balance in advance. On the other hand, the advantage is that optical lines like Hα can be used for this purpose. Thus, the product of the last two terms of the expression $v_\infty \dot{M} R^{1/2}$ directly follows from a line fit of Hα (cf. Puls et al. 1996). The dynamical parameters obtained in this way are usually designated as *observed wind parameters*. Figure 22 shows that the wind-momentum–luminosity relation indeed exists for massive O stars, as they follow the linear relation predicted by the theory, in a first approximation and for fixed metallicities. With regard to the spread in wind momenta found by Kudritzki et al. (1997) for the Central Stars of Planetary Nebulae (lower part of Figure 22), Pauldrach et al. (2002, 2003) have given a solution which, however, is unlikely to be believed yet by the part of the community working on stellar evolution.

Moreover, it is important to note that as a by-product of this relation it can be used as an independent tool for measuring extragalactic distances up to Virgo and Fornax, since it is independent of the observationally unknown masses (for a recent and comprehensive review on this subject see Kudritzki and Puls 2000).

Figure 23 shows that the observed behavior of the wind momentum luminosity relation is represented quite well by our improved realistic models, particularly in the case of the two supergiants (ζ Puppis and α Cam) analyzed in Section 7.2. But,

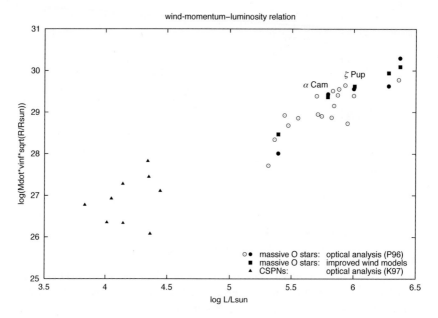

Figure 23: The wind-momentum–luminosity relation for massive O stars and CSPNs as in Figure 22. Also plotted as filled squares are the calculated wind momenta for five massive O stars; filled circles designate the corresponding observed values (cf. Pauldrach et al. 2002, 2003).

as already pointed out, this relation is independent of the stellar mass. Thus, in order to verify the statement of the previous section that the observed stellar and wind parameters are confirmed by the present models, we also have to investigate the relations of the individual dynamical parameters.

For this investigation we need to use as input for our models the same stellar parameters as have been used to obtain the observed wind parameters. On the basis of this requirement, Figure 24 (upper panel) shows that the observed and predicted values of the terminal velocities (v_∞) are in agreement to within 10%. Since v_∞ is proportional to the escape velocity ($v_{\rm esc}$)

$$v_\infty \propto v_{\rm esc} = \left(\frac{2GM}{R}(1-\Gamma)\right)^{1/2}$$

which strongly depends on the mass of the objects, *the mass is determined very accurately by the predicted values* (G is the gravitational constant and Γ is the ratio of radiative Thomson acceleration to gravitational acceleration). For the mass-loss rates we found agreement to within a factor of two, as shown in Figure 24 (lower panel). Due to the strong relation between the mass-loss rates and the luminosities ($\dot{M} \propto L$), *the luminosities can thus be precisely determined*.

Therefore, computing the wind dynamics consistently with the NLTE model permits not only the determination of the wind parameters from given stellar parameters,

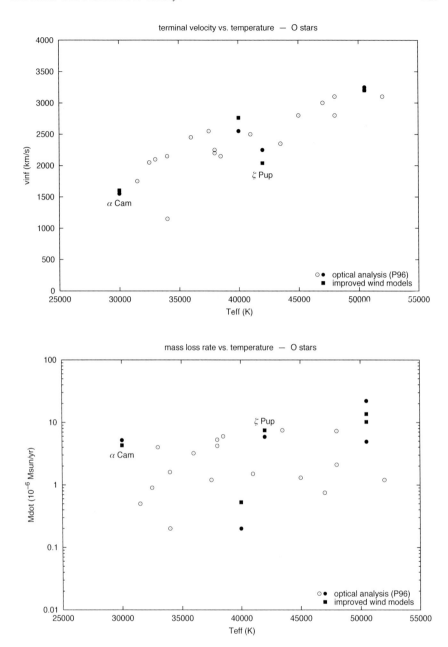

Figure 24: *Top*: Terminal velocities as a function of effective temperature for massive O-stars of the same sample as in Figure 23. *Bottom*: Mass-loss rates as a function of effective temperature for the same sample. The calculated values for the five massive O stars are designated by filled squares; filled circles designate the corresponding observed values (cf. Pauldrach and Hoffmann 2003 concerning ζ Puppis; the other objects have been treated by Hoffmann and Pauldrach 2002).

but conversely makes it possible *to obtain the stellar parameters from the observed UV spectrum alone*. As stated above, this means that, in principle, *the stellar parameters can immediately be read off by simply comparing an observed UV spectrum to a proper synthetic spectrum*. However, the whole procedure is not always an easy task, since in most cases a comprehensive grid of models and some experience will be required to do so. Although the idea itself is not new (cf. Pauldrach et al. 1988 and Kudritzki et al. 1992), only the new generation of models has reached the degree of sophistication that makes such a procedure practicable instead of purely academic. The corresponding conclusion is that *realistic models are characterized by at least a quantitative spectral UV analysis calculated together with a consistent dynamics*.

8 Conclusions and Outlook

The **need** for **detailed atmospheric models** of **Hot Stars** has been motivated in depth and a **diagnostic tool** with great astrophysical potential has been presented.

It has been shown that the models of the *new generation* are realistic. They are *realistic* with regard to a *quantitative spectral UV analysis* calculated, for the case of O stars, along with consistent dynamics, which, in principle, allows to read off the stellar parameters by comparing an observed UV spectrum to a suitable synthetic spectrum. The new generation of models has reached a degree of sophistication that makes such a procedure practicable.

The astronomical perspectives are enormous. Of course, there is still a large amount of hard and careful astronomical work that needs to be done. The diagnostic tool for expanding model atmospheres is in our hands, but this is (as usual) just the starting point; further elaboration, refinements, and modifications are required before the results are quantitatively completely reliable.

But this is not the handicap for the future. The present handicap are future observations. What is most urgently needed is the **Next Generation Space Telescope**.

With respect to the topics discussed in this review this telescope is needed for a variety of reasons:

To tackle the question whether **Supernovae Ia** are standard candles in a cosmological sense realistic models and synthetic spectra of Type Ia Supernovae require *spectral observations of objects at mid-redshift* and sufficient resolution. The context of this question is the current surprising result that distant SNe Ia at intermediate redshift appear fainter than standard candles in an empty Friedmann model. Consequently, the current SN-luminosity distances indicate an accelerated expansion of the universe.

In order to be able to make quantitative predictions about the influence of very massive, extremely metal-poor **Population III** stars on their galactic and intergalactic environment one primarily needs observations which can be compared to the predicted flux spectra that are already available for zero metallicity and can, in principle, be produced for metallicities different from zero. Observations with the Next Generation Space Telescope (NGST) of distant stellar populations at high redshifts will give us the opportunity to deduce the primordial IMF, thus allowing a quantitative investigation of the ionization efficiency of a Top-heavy IMF via *realistic spectral energy distributions* of these very massive stars.

NGST observations are also important for determining extragalactic abundances and population histories of **starburst galaxies** from an analysis of H II regions. This task also requires energy distributions of time-evolving stellar clusters, which are calculated on the basis of a grid of *spectral energy distributions* of O and early B-type stars. A crucial point with respect to this is whether the spectral energy distributions of massive stars are already realistic enough to be used for diagnostic issues of H II regions. It has been shown in this review that the *ultimate test* is provided by a comparison of observed and synthetic UV spectra of individual massive stars, since the ionization balance can be traced reliably through the strength and structure of the wind lines formed throughout the atmosphere.

Furthermore, NGST is important for directly exploiting the diagnostic perspectives of galaxies with pronounced current star formation. It has been demonstrated that massive stars dominate the UV wavelength range in **star-forming galaxies**, and that therefore the UV-spectral features of massive O stars can be used as tracers of age and chemical composition of starburst galaxies even at high redshift. This is in particular the case when the flux from these galaxies is amplified by gravitational lensing through foreground galaxy clusters; corresponding observations render the possibility of determining metallicities of starbursting galaxies in the early universe via *realistic UV spectra* (in the rest frame) of massive O stars.

As a final remark it is noted that the solution method of stationary models for expanding atmospheres is in its present stage already regarded as a standard procedure towards a realistic description. Thus, together with an easy-to-use interface and an installation wizard, the program package **WM-basic** has been made available to the community and can be downloaded from the author's home page.

Acknowledgments

I wish to thank my colleagues Tadziu Hoffmann and Tamara Repolust for proofreading the manuscript, and I am grateful to Amiel Sternberg and Claus Leitherer for providing me with figures from their publications. This research was supported by the Sonderforschungsbereich 375 of the Deutsche Forschungsgemeinschaft, and by the German-Israeli Foundation under grant I-551-186.07/97.

References

Baldwin, J.A., Ferland, G.J., Martin, P.G. et al. 1991, ApJ 374, 580

Bromm, V., Coppi, P.S., Larson, R.B. 1999, ApJ 527, L5

Bromm, V., Kudritzki, R.P., Loeb, A. 2001, ApJ 552, 464

Carr, B.J., Bond, J.R., Arnett, W.D. 1984, ApJ 277, 445

Cassinelli, J., Olson, G. 1979, ApJ 229, 304

Castor, J.I., Abbott, D.C., Klein, R. 1975, ApJ 195, 157

Conti, P.S., Leitherer, C., Vacca, W.D. 1996, ApJ 461, L87

El Eid, M.F., Fricke, K.J., Ober, W.W. 1983, A&A 119, 54

Fan, X. et al. 2000, AJ 120, 1167

Feldmeier, A., Kudritzki, R.-P., Palsa, R., Pauldrach, A. W. A., Puls, J. 1997, A&A 320, 899

Genzel, R. et al. 1998, ApJ 498, 579

Giveon, U., Sternberg, A., Lutz, D., Feuchtgruber, H., Pauldrach, A. W. A. 2002, ApJ 566, 880

Gunn, J.E., Peterson, B.A. 1965, ApJ 142, 1633

Hamann, W.R. 1980, A&A 84,342

Harnden, F.R., Branduardi, G., Elvis, M. et al. 1979, ApJ 234, L51

Hillebrandt, W., Niemeyer, J.C. 2000, ARA&A 38, 191

Hoffmann, T.L. 2002, Ph.D. thesis, LMU Munich

Hoffmann, T.L., Pauldrach, A.W.A. 2002, IAU Symposium 209, in press

Hummer, D.G., Rybicki, G.B. 1985, ApJ 293, 258

Kirshner, R.P., Jeffery, D.J., Leibundgut, B., et al. 1993, ApJ 415, 589

Kudritzki, R.P., Hummer, D.G., Pauldrach, A.W.A., et al. 1992, A&A 257, 655

Kudritzki, R.P., Méndez, R. H., Puls, J., McCarthy, J. K. 1997, in IAU Symp. 180, Planetary Nebulae, eds. H.J. Habing & H.J.G.L.M. Lamers, p. 64

Kudritzki, R.P., Puls, J. 2000, ARA&A 38, 613

Kudritzki, R.P. 2002, ApJ in press

Kunth, D., Mas-Hesse, J.M., Terlevich, R., et al. 1998, A&A 334, 11

Kurucz, R.L. 1992, Rev. Mex. Astron. Astrof. 23, 181

Lamers, H.J.G.L.M., Morton, D.C. 1976, ApJ Suppl. 32, 715

Larson, R.B. 1998, MNRAS 301, 569

Leibundgut, B. 2001, ARA&A 39, 67

Leitherer, C., Leão, J., Heckman, T.M., Lennon, D.J., Pettini, M., Robert, C. 2001, ApJ 550, 724

Lucy, L.B., Solomon, P. 1970, ApJ 159, 879

Lucy, L.B., White, R. 1980, ApJ 241, 300

Lutz, D. et al. 1996, A&A 315, 137

Loeb, A. 1998, in ASP Conf. Ser. 133, 73

Milne, E.A. 1926, MNRAS 86, 459

Morton, D.C., Underhill, A.B. 1977, ApJ Suppl. 33, 83

Nomoto, K., Thielemann, F.-K., Yokoi, K. 1984, ApJ 286, 644

Nugent, P., Baron, E., Branch, D., et al. 1997, ApJ 485, 812

Oey, M.S., Massey, P. 1995, ApJ 452, 210

Owocki, S., Castor, J., Rybicki, G. 1988, ApJ 335, 914

Pauldrach, A.W.A., Puls, J., Kudritzki, R.P. 1986, A&A 164, 86

Pauldrach, A.W.A. 1987, A&A 183, 295

Pauldrach, A.W.A., Herrero, A. 1988, A&A 199, 262

Pauldrach, A.W.A., Puls, J., Kudritzki R.-P., et al. 1988, A&A 207, 123

Pauldrach, A.W.A., Feldmeier, A., Puls, J., Kudritzki, R.-P. 1994b, in Space Sci. Rev. 66, 105

Pauldrach, A.W.A., Kudritzki, R.-P., Puls, J., et al. 1994, A&A 283, 525

Pauldrach, A.W.A., Duschinger, M., Mazzali, P.A., et al. 1996, A&A 312, 525

Pauldrach, A.W.A., Lennon, M., Hoffmann, T.L., et al. 1998, in: Proc. 2nd Boulder-Munich Workshop, PASPC 131, 258

Pauldrach, A.W.A., Hoffmann, T.L., Lennon, M. 2001, A&A 375, 161

Pauldrach, A.W.A., Hoffmann, T.L., Mendez, R.H. 2002, IAU Symposium 209, in press

Pauldrach, A.W.A., Hoffmann, T.L., Mendez, R.H. 2003, A&A, in press

Pauldrach, A.W.A., Hoffmann, T.L. 2003, A&A, in press

Perlmutter, S., Aldering, G., Goldhaber, G., et al. 1999, ApJ 517, 565

Pettini, M., Kellogg, M., Steidel, C.C., et al. 1998, ApJ 508, 539

Pettini, M., Steidel, C.C., Adelberger, K.L., Dickinson, M., Giavalisco, M. 2000, ApJ 528, 96

Puls, J. 1987, A&A 184, 227

Puls, J., Pauldrach, A.W.A. 1990, PASPC 7, 203

Puls, J., Kudritzki R.-P., Herrero A., Pauldrach, A.W.A., Haser, S.M., et al. 1996, A&A 305, 171

Reinecke, M., Hillebrandt, W., Niemeyer, J.C. 1999, A&A 374, 739

Riess, A.G., Filippenko, A.V., Challis, P., et al. 1998. Astron. J. 116, 1009

Riess, A.G., Kirshner, R.P., Schmidt, B.P., et al. 1999, Astron. J. 117, 707

Rybicki, G.B. 1971, JQSRT 11, 589

Rubin, R.H., Simpson, J.P., Haas, M.R., et al. 1991, PASP 103, 834

Saha, A., Sandage, A., Thim, F., Labhardt, L., Tammann, G.A., Christensen, L., Panagia, N., Macchetto, F. 2001, ApJ, 551, 973

Sauer, D., Pauldrach, A.W.A. 2002, Nucl. Astrophys., MPA/P13

Seward, F.D., Forman, W.R., Giacconi, R., et al. 1979, ApJ 234, L55

Schaerer, D., de Koter, A. 1997, A&A 322, 598

Sellmaier, F.H., Yamamoto, T., Pauldrach, A.W.A., Rubin, R.H. 1996, A&A, 305, L37

Simpson, J.P., Colgan, S.W.J., Rubin, R.H., et al. 1995, ApJ 444, 721

Sobolev, V. 1957, Sov. A&A J. 1, 678

Steidel, C.C., Giavalisco, M., Pettini, M., Dickinson, M., Adelberger, K.L. 1996, ApJ 462, L17

Sternberg, A., Hoffmann, T.L., Pauldrach A.W.A. 2002, ApJ, in press

Thornley, M.D., Förster-Schreiber, N.M., Lutz, D., et al. 2000, ApJ 539, 641

Walborn, N.R., Nichols-Bohlin, J., Panek, R.J. 1985, IUE Atlas of O-Type Spectra from 1200 to 1900Å (NASA RP-1155)

Gas and Dust Mass Loss of O-rich AGB-stars

Franz Kerschbaum[1], Hans Olofsson[2], Thomas Posch[1],
David González Delgado[2], Per Bergman[3], Harald Mutschke[4],
Cornelia Jäger[4], Johann Dorschner[4], Fredrik Schöier[5]

[1]Institut für Astronomie, Türkenschanzstraße 17, A-1180 Wien, Austria
[2]Stockholm Observatory, SCFAB, SE-10691 Stockholm, Sweden
[3]Onsala Space Observatory, SE-43992 Onsala, Sweden
[4]Astrophysikalisches Institut und Universitäts-Sternwarte Jena,
Schillergäßchen 2, D-07745 Jena, Germany
[5]Leiden Observatory, PO Box 9513, 2300 RA Leiden, The Nederlands

kerschbaum@astro.univie.ac.at

Abstract

This paper presents results from systematic surveys and subsequent modelling of circumstellar CO and SiO radio line emission of oxygen-rich semiregular and irregular variables of types SRa, SRb, and Lb, respectively. Results from high spatial resolution observations of a subsample of these objects with a more complex velocity structure in their CO profiles are presented and show clear evidence for deviations from the simple picture of spherically symmetric mass loss. The observations can be interpreted in terms of disks and/or axially symmetric outflows. Moreover, some insights into the thermal emission from dust particles formed around these late type variables are presented. An identification of metal oxides and amorphous silicates in low and intermediate mass loss rate objects has been possible.

1 Introduction

The asymptotic giant branch (AGB) is the crucial stage during the late evolution of stars with main sequence masses between 0.8 and 8 $M_\odot$. It is characterised by three important phenomena:

- First, the onset of repeated, explosive Helium burning in a shell (Helium shell flashes, thermal pulses) accompanied by deep convection, together responsible for the production and dredge-up of heavy elements like carbon and those of the s-process (e. g. Iben & Renzini 1983).

- Second, long-period pulsations combined with usually large variations of the surface radius and the formation of shock fronts in the stellar atmosphere (e. g. Willson 1988). Depending on the pulsational properties, these objects are classified as Miras, semiregular, or irregular variables.

- Finally, the development of a stellar wind with typical mass loss rates in the range 10^{-7}–10^{-5} $M_\odot$/yr producing a cool circumstellar envelope (CSE) where complex molecules and dust can form (see the review by Habing, 1996). Some objects, having the highest mass loss rates (OH/IR-stars, IR-Carbon stars), are totally obscured by their circumstellar material and only visible at infrared wavelengths.

AGB-stars are interesting objects because they are in one of the most dynamic stages during the life of the majority of all stars (including the sun). Because of their nucleosynthesis and heavy mass loss (in the form of gas and dust), they contribute significantly to the chemical evolution of the interstellar medium. In addition, because of their high luminosities (some 10^3 to 10^4 $L_\odot$), they belong to the most prominent objects in external galaxies – especially at infrared wavelengths.

1.1 Variable stars on the AGB

Most of the stars on the AGB are variable, mostly of the Mira or semiregular type, with reasonably well-defined periods. The pulsation period is an important quantity, empirically related to the luminosity (the so called P–L relation, e. g. Feast et al. 1989) and the mass loss rate (e. g. Whitelock 1990, Whitelock et al. 1994, Groenewegen 1994). This is true for both M- (often refered to as oxygen-rich) and C-type (often refered to as carbon-rich) variables. Luminosity and mass loss rate are both important parameters for our understanding of AGB evolution.

Light variations are observable at most wavelengths and occur on time scales in the range from weeks to years. In spite of its prominence, variability of AGB-stars has been almost exclusively studied in the visual range. Since the visual light changes are mainly due to opacity effects of a few temperature-sensitive molecules, the physically more relevant bolometric luminosity changes can only be investigated in the infrared, where most of the stellar radiation is emitted (e. g. Le Bertre 1988). However, near-infrared photometric time series covering many of the relevant object classes (Miras, semiregular (SRa/b; SRVs), and irregular (Lb; IRVs) variables of oxygen- and carbon-rich chemistry) are still rare – mainly because of the long time scales and the small number of observatories equipped with the required instrumentation.

Mira variables, OH/IR-stars and carbon stars are the most frequently studied AGB-objects, while the semiregular, or even more the irregular, variables have been almost neglected. This situation has improved during the last decade. In a series of papers (Kerschbaum & Hron 1992, Kerschbaum & Hron 1994, Kerschbaum 1995, Kerschbaum & Hron 1996 for semiregulars; Kerschbaum et al. 1996a, Kerschbaum 1999 for irregulars), systematic near-infrared photometric surveys, combined with visual and IRAS data from the literature, yielded a more detailed picture of the role of these variability groups within the evolution on the AGB.

The main outcome was: semiregulars of the type SRa are not a distinct class of variables but a mixture of 'intrinsic' Miras and SRbs. The O-rich SRbs consist of a 'blue' group with no indication of circumstellar material, $P < 150^d$ and $T_{\text{eff}} >$ 3200 K, and a 'red' group with temperatures and mass loss rates comparable to those of Miras, but with periods about a factor of two smaller.

The 'red' semiregulars (SRVs) have galactic scale heights and number densities similar to Miras. The 'blue' SRVs are about twice as numerous as the 'red' ones and only half as luminous. The C-rich SRVs generally have smaller periods and are more abundant than C-rich Miras. The scale heights for these two groups agree closely. Among all groups there are stars with much longer periods than expected from their temperatures and mass loss rates. On the basis of temperatures, luminosities, and number densities, the 'blue' SRVs appear to be objects on the early, non-thermally pulsing AGB. A possible interpretation for the O-rich 'red' SRVs is in terms of a phase preceding the Mira phase, with equal life times for both stages, and the assumption that they are likely to be overtone pulsators, while the Miras pulsate in the fundamental mode.

Most of the O-rich irregular variables (IRVs) can be divided into two groups showing near-infrared properties very similar to the 'blue' and the 'red' SRVs. The problem of variability classification from poorly sampled light curves is of special relevance to this group of objects, and one may even question whether the distinction between these variability types is artificial (Lebzelter et al. 1995).

1.2 Mass loss

Mass loss from the surface eventually becomes the sole factor that determines the time scale for the evolution on the AGB, at least during the final phases, and it is therefore important to identify the hitherto not well-known mechanism(s) behind the mass loss, and to investigate its (their) dependence on mass, metallicity, pulsational behaviour, etc ... It is possible that the low mass loss rate winds are purely pulsation-driven, while for the higher mass loss rate objects, radiation pressure on grains also plays an important role (Höfner et al. 1995).

Circumstellar CO radio line emission has proven to be a good measure of the gas mass loss rate of evolved stars. Extensive surveys have been performed, and presently more than 500 objects have been detected in CO lines (e. g., Knapp & Morris 1985; Zuckerman & Dyck 1986; Margulis et al. 1990; Nyman et al. 1992; Kastner et al. 1993; Olofsson et al. 1993; Bieging & Latter 1994; Groenewegen et al. 1996; Josselin et al. 1998; Knapp et al. 1998; Kerschbaum & Olofsson 1999). A significant fraction of this work was catalogued by Loup et al. (1993).

The majority of the surveys are dominated by high mass loss rate, evolved objects. The low end of the mass loss rate distribution, populated by objects located on the early AGB or at the beginning of the thermally pulsing AGB, has not been studied in detail. Consequently, as for the near IR photometry discussed above, we started a major survey of circumstellar CO emission towards cool, oxygen-rich SRVs and IRVs which is presented in the next section. A similar study for carbon-rich objects was initiated by Olofsson et al. (1993).

Thermal emission originating from small solid particles ("dust grains") is another observable, highlighting specific aspects of the mass loss of AGB-stars. For O-rich stars with CSEs optically thin at near- and mid-IR wavelengths, the shape of the 9–13 μm silicate solid state feature, as observed by the IRAS-LRS, was analysed by Sloan & Price (1995), Ivezić & Elitzur (1995), Hron et al. (1997), and Speck et al. (2000). Sloan & Price found no difference in the feature shapes between Miras and SRVs. Ivezić & Elitzur argued that the differences in the feature width are mostly caused by optical depth effects. Hron et al. used a more refined method to approximate the continuum radiation than Sloan & Price, and they found systematic differences between Miras and semiregulars. They also concluded that optical depth effects alone are not sufficient to explain these differences but that contributions from different grain materials must be considered. Ivezić & Elitzur (1996) and Winters et al. (1994), using quite different CSE models, have studied the effects of variability on the observed IR colours. While Ivezić & Elitzur only noted the effects due to the global change in optical depth, Winters et al. also found significant variations superposed on this global variability due to the shell-like structure of the dust CSE in their model.

A comparison of the shape and the variations in the dust features with existing models for dust formation in AGB-stars requires dedicated observing campaigns with an accuracy better than that provided by IRAS. Such data are now available through the Infrared Space Observatory (ISO), and also with new ground-based mid-infrared spectrometers. With ISO's broad wavelength coverage and its much higher spectral resolution compared to that of IRAS-LRS, a detailed mineralogical study of the dust CSEs around oxygen-rich AGB-variables is feasible. In addition to the classical 9.7 μm amorphous silicate feature, several other solid state bands were detected (see Sect. 3).

An important aspect of the mass loss process is its geometry. There is a drastic difference in morphology between the apparently largely spherical CSEs of AGB-stars, and the, almost exclusively, non-spherical, but axi-symmetric, descendants of the AGB-stars, the Planetary Nebulae (see Olofsson 1997 and references therein). There exists no viable explanation of this phenomenon, but it appears that the transition occurs during the very end of the AGB, or at the beginning of the post-AGB evolution. However, there are observational difficulties that severely limit our knowledge of the structure of AGB-CSEs. In particular, the mass loss geometry is determined fairly close to the star. We believe that the low mass loss rate objects can provide crucial information in this context.

After this introductory section, the following one, "Molecular line emission from the gaseous CSEs", presents results from systematic surveys and subsequent modelling of circumstellar CO and SiO emission of oxygen-rich semiregular and irregular variables of types SRa, SRb, and Lb.

Supplementing the gas mass loss results, the section "Composition of the dust CSEs" presents some insights into the thermal emission from solid dust particles which are formed around these late type variables. The resulting spectral energy distributions, the mineralogical composition, equilibrium temperatures, as well as the interrelation with pulsational properties are briefly discussed.

Finally, in the section "Mass loss geometry", results from high spatial resolution observations of objects with a more complex velocity structure in their CO radio profiles are presented.

2 Molecular line emission from the gaseous CSEs

2.1 Mass loss rates from CO radio lines

Two major surveys of circumstellar CO radio line emission of mainly low mass loss rate objects have been performed. Olofsson et al. (1993) observed a sample of optically bright carbon stars (mainly IRVs and SRVs), all carbon stars brighter than K = 2 mag in the General Catalogue of Cool Carbon Stars (Stephenson 1989). Kerschbaum & Olofsson (1999) observed a sample of M-type SRVs and IRVs. Most of them have IRAS 60 μm flux densities $S_{60} \geq 5$ Jy. No selection was made concerning the [12 μm] – [25 μm] colour, but objects bluer than about 0.36 were not detectable. As a consequence of this fact and the smaller number of IRVs, we tried to observe them down to IRAS-60 μm flux densities of about 2 Jy and only objects redder than [12 μm] – [25 μm]=0.36 were tried. Detailed radiative transfer modelling to derive mass loss rates was done by Schöier & Olofsson (2001) for the former sample, and by Olofsson et al. (2002) for the latter sample.

The (molecular hydrogen) mass loss rate distribution has a median value of 2.0×10^{-7} $M_\odot$ yr^{-1}, a minimum of 2.0×10^{-8} $M_\odot$ yr^{-1} and a maximum of 8×10^{-7} $M_\odot$ yr^{-1} for the M-stars. The mass loss rate distribution for the C-star sample is very comparable, suggesting that there is no dependence on chemistry for these types of objects. IRVs and SRVs with a mass loss rate in excess of 5×10^{-7} $M_\odot$ yr^{-1} must be very rare, and in this respect, the regularity and amplitude of the pulsation plays an important role. We find no significant difference between the IRVs and the SRVs in terms of their mass loss characteristics. Among the SRVs, the mass loss rate shows no dependence on the period. Thus, for these non-regular, low-amplitude pulsators it appears that the pulsational pattern plays no role for the mass loss efficiency. In fact, the mass loss rate characteristics of the M-star sample is very similar to that of a short-period M-Mira sample (Young 1995). The mass loss rates show no correlation at all with the stellar temperatures. It is notable that there is no example of the sharply double-peaked CO line profile, which is evidence of a large, detached CO shell, among the M-stars. About 10 % of the C-stars show this phenomenon which is possibly related to thermal pulses (Olofsson et al. 1990, 1996, 2000).

The gas expansion velocity distribution has a median of 7.0 km s^{-1}, a minimum of 2.2 km s^{-1} and a maximum of 14.4 km s^{-1} for the M-stars. The fraction of objects with low gas expansion velocities is high, about 30 % have velocities lower than 5 km s^{-1}. The gas expansion velocity distribution for the C-stars is clearly different. In particular, the fraction of low velocity sources is much lower. We find that the mass loss rate and the gas expansion velocity correlate well for both samples, even though for a given velocity (which is well determined), the mass loss rate may take a value within a range of a factor of five (the uncertainty in the mass loss rate estimate is lower than this within the adopted circumstellar model). In Fig. 1 we show the results for the M-star sample combined with those of a sample of M-Miras for which modelling of the circumstellar CO line emission has been done (González Delgado et al. 2003). The result is in line with theoretical predictions for an optically thin, dust-driven wind at low mass loss rates, that eventually turns into a gas expansion velocity of about 20 km/s independent of the mass loss rate.

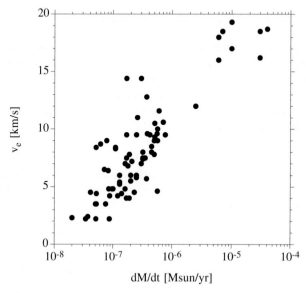

Figure 1: The circumstellar gas expansion velocity versus the mass loss rate for M-stars of type IRV, SRV, and Mira.

In addition to the very low gas expansion velocity sources, the M-star sample contains four sources with distinctly double-component CO line profiles: EP Aqr, RV Boo, X Her, and SV Psc (all SRVs). At present, the exact nature of these objects is unknown. They will be further discussed in Sect. 4.

Crude estimates of the dust properties, through the gas-grain collision heating term which is constrained by multi-line CO data, indicate that the two samples have similar gas-to-dust ratios and that these differ significantly from that of high mass loss rate C- and M-stars. This also means that the gas-CSEs due to low mass loss rates are cooler than expected from a simple extrapolation of the results for the well-studied high mass loss rate object IRC+10216.

2.2 An SiO radio line survey

A survey of 'thermal' SiO radio line emission (meaning emission from the ground vibrational state rotational lines which are normally not masering) of the M-stars detected by Kerschbaum & Olofsson (1999) was initiated a few years ago. SiO line emission was chosen because it is a useful probe of the formation and evolution of dust grains in CSEs, as well as of the CSE dynamics. A detailed study of the SiO radio line emission, including a complete analysis of circumstellar CO and SiO thermal line emission for a Mira comparison sample, is underway (González Delgado et al. 2003), and we present some of the major results here.

2.3 Observational results

A total of 60 M-type IRV/SRV stars were observed in circumstellar SiO radio line emission (i.e., about 85 % of the stars detected in circumstellar CO), and clear detections of SiO lines were obtained towards 36 sources, i.e., the detection rate was about 60 %. The detection rate appears lower in IRVs (44 %) than in SRVs (68 %). On average, the SiO(J = 2–1)/CO(J = 1–0) line intensity ratio is about 0.7, but for the lowest mass loss rate objects the SiO line can be substantially stronger than the CO line. This is illustrated in Fig. 2 where this ratio is plotted versus a simple mass loss rate measure. Can this be an indication of an SiO depletion, due to adsorption onto dust grains, which increases with the density of the CSE? This can only be answered after a thorough radiative transfer analysis has been performed (see below).

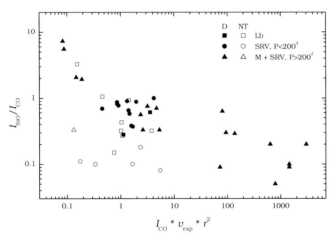

Figure 2: SiO(J = 2–1)/CO(J = 1–0) line intensity ratio versus a mass loss measure (open symbols denote non- or tentative detections)

The SiO and CO radio line profiles are clearly different from each other, although this conclusion is mainly based on the limited number of sources where the S/N-ratio of the data are high enough for both species. The SiO line profiles are significantly narrower in the sense that the main fraction of the emission comes from a velocity range significantly narrower (by about 10–20 %) than twice the expansion velocity determined from the CO data. On the other hand, the SiO line profiles have weak wings, such that the total velocity width of its emission is very similar to that of the CO emission. These "Gaussian"-like SiO line profiles were noted already by Bujarrabal et al. (1986), and they have been interpreted as being due to the influence of gas acceleration in the region which produces most of the SiO line emission.

2.4 SiO line modelling

Following the CO models presented in Schöier & Olofsson (2001) and Olofsson et al. (2002), González Delgado et al. (2003) have made a radiative transfer analysis of also the SiO radio line emission from the M-stars (IRV, SRV, and Miras). In some sense this is a more difficult enterprise. The SiO emission predominantly comes from a region closer to the star than does the CO line emission, and this is a region where we lack observational constraints. The SiO excitation is also normally far from thermal equilibrium with the gas kinetic temperature, and radiative excitation plays a major role. Finally, there exists no detailed chemical model for calculating the radial SiO abundance distribution. All this makes the SiO line modelling much more uncertain, and dependent on a number of assumptions.

González Delgado et al. adopted the assumption that the gas-phase SiO abundance stays high only very close to star, since further out, the SiO molecules are adsorbed onto the grains. Beyond this the abundance stays low until the molecules are eventually dissociated by the interstellar UV radiation. This size was estimated using both SiO multi-line modelling and interferometer data. Ingoring the contribution to the line emission from the high abundance region they derive (upper limits to) the (depleted) SiO abundance in the CSE.

The result of the line modelling is an SiO abundance distribution which for the IRV/SRV sample has a median value of 5×10^{-6}, a minimum of 1×10^{-6} and a maximum of 2×10^{-5}. This is clearly a factor of about ten lower than expected from theory (with solar abundances the maximum SiO fractional abundance is 7×10^{-5}, and detailed calculations on stellar atmosphere equilibrium chemistry give abundances of about 4×10^{-5}). The modelling of the Mira SiO line emission is more difficult, in particular for the high mass loss rates, and hence the results are more uncertain. For the high mass loss rate Miras ($\dot{M} > 5 \times 10^{-6}$ $M_\odot$ yr^{-1}), there is a distinct division into a low abundance group (on average 4×10^{-7}) and a high abundance group (on average 7×10^{-6}), while the low mass loss rate Miras have abundances similar to those of IRV/SRVs. González Delgado et al. see no reason why this division should be an effect of the modelling uncertainties, but cannot at this moment provide an astrophysical explanation for the division.

In Fig. 3 we show the abundances derived by González Delgado et al. as a function of the mass loss rate. The inclusion of the Miras shows that the trend of decreasing SiO abundance with increasing mass loss rate discernable for the IRV/SRV sample continues towards high mass loss rates. We interpret this as an effect of increased consumption of SiO into dust grains with increasing mass loss rate, i. e., the density of the envelope.

In order to limit the number of free parameters, González Delgado et al. (2003) assumed a gas expansion velocity which is constant with radius, i. e., there is no acceleration region. This made it impossible to fit in detail the SiO line profiles (described above). It is believed that the discrepancy is an effect of gas accelera-

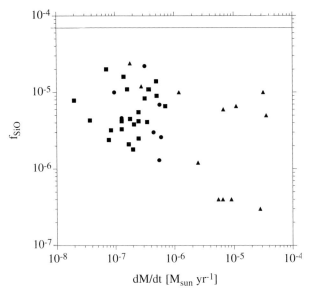

Figure 3: SiO abundance versus mass loss rate from modelling both the circumstellar CO and SiO lines (IRV:circle, SRV: square, Mira: triangle). The horizontal line indicates the maximum SiO fractional abundance expected from solar abundances.

tion in the region from which a significant fraction of the SiO line emission comes (see also Sahai & Bieging 1993). A high sensitivity multi-line study combined with interferometric observations are required to fully tackle this problem.

3 Composition of the dust CSEs

3.1 An evolutionary sequence for the dust emission of AGB-stars

It has already been mentioned in Sect. 1 that in general, SRVs and IRVs have mid-infrared spectra which are significantly different from those of Miras and OH/IR stars (see also Speck et al. 2000). It seems that this fact can be interpreted in terms of an *evolutionary sequence*, with the mass-loss rate as the key parameter. In this scenario, the 'youngest' AGB-stars are dominated by comparatively weak emission features at 13, 19.5, and 32 μm. These low mass loss rate AGB-stars ($\dot{M} < 3\times 10^{-7}$ $M_\odot$/yr) are mostly SRVs and IRVs. The candidate carriers of their characteristic dust features are *metal oxides* which will be discussed in detail in the next subsection. Most of the SRVs also show strong, broad *amorphous silicate* emission features, centered around 10 and 18 μm. These features, in turn, dominate the spectra of Mira stars with intermediate mass loss rates and 10 μm optical depths larger than 0.1. Mira and OH/IR-stars with very high mass loss rates ($\dot{M} > 10^{-5}$ $M_\odot$/yr), by contrast, show very pronounced *crystalline* silicate features (in addition to the amorphous silicate features that tend to appear in absorption in these stars). The end of the evolutionary sequence is marked by Planetary Nebulae. Their continuum emission peaks at wavelengths larger than 30 μm, so they do only weakly radiate at mid-IR wavelengths – apart from strong, very narrow emission lines which are due to forbidden atomic transitions and which are superimposed on the continuum radiation.

An important task in the interpretation of the mid-IR and FIR spectra of AGB-stars is to determine the interrelation between the *mineralogical composition of the dust* CSEs and other *physical and chemical parameters* characterizing the central star, like its C/O ratio, effective temperature, pulsational behaviour, mass loss geometry (see Sect. 4) and the already mentioned mass loss rate. Concerning the latter, it is still controversial whether the mass fraction of crystalline silicates in fact increases with the mass loss rate [as suggested by Suh (2002) and others], or whether the crystalline enstatite and forsterite grains only *radiate* more efficiently in case of high mass loss rates, but are present in maybe equal amounts in case of low mass loss rates (Kemper et al. 2001). For the other mentioned parameters, their relation to the mineralogy of circumstellar dust is still more complicated, and much work remains to be done in this field.

3.2 The metal oxide emission bands

The typical spectrum in the wavelength range between 8 and 35 μm of a semiregularly pulsating AGB-star, EP Aqr (variability type SRb, period 55 d), is shown in Fig. 4. Three prominent, sharp and three broad emission features have been highlighted by arrows. The positions of the three sharp features are 13.0, 19.5 and 31.8 μm. Their full widths at half maximum amount to 0.6, ~3, and 0.5 μm, respectively. The positions of the broad emission bands are 10, ~11, and 18 μm.

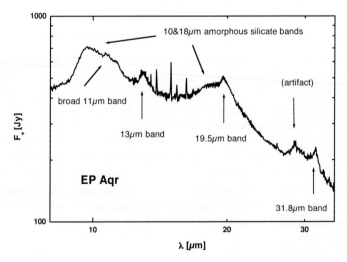

Figure 4: The mid-IR spectrum of an O-rich AGB-star with low mass loss rate (variability type SRb). The three sharp features which are indicated with arrows are metal oxide bands.

The carrier of the 11 μm feature can not yet be firmly established, whereas the remaining two broad bands are clearly attributable to some form of amorphous silicate (see below). It may very well be that amorphous Al_2O_3 produces the broad 11 μm emission in the spectra of many low mass loss rate AGB-stars. This was suggested, e. g., by Lorenz-Martins & Pompeia (2000). They also pointed out that Al_2O_3 can form rather close to the photospheres of the stars (i.e. at about 2 R_*) due to the small visual and near-IR opacity of alumina (see also Woitke et al. 2000). The same holds for metal oxides in general unless they are rich in transition metal ions like Fe, Cr, etc. Therefore, also $MgAl_2O_4$ (spinel) can form in the comparatively dense inner regions of dust CSEs. It seems that this is indeed the case for those stars in which we observe the 13.0 μm feature. In fact, the 13 μm feature was identified in IRAS and Kuiper Airborne Observatory spectra already, and it was originally attributed to (crystalline) α-Al_2O_3 (Glaccum 1995). However, as pointed out by Begemann et al. (1997), α-Al_2O_3, if present in CSEs (in the form of spherical, or nearly so, particles) should produce a second strong feature at 21 μm. On the basis of ISO spectra, it was concluded that such a 21 μm is not observable in O-rich CSEs. However, Posch et al. (1999, 2002b) and Fabian et al. (2001) found that, together with the 13 μm feature, a weak 16.8 μm and the above mentioned 31.8 μm feature *are* detectable. This is a strong argument for spinel being the carrier of these three bands (which have, as yet, only been seen in emission).

Another – in some cases striking – feature of AGB-stars with mass loss rates of a few 10^{-7} $M_\odot$/yr is the 19.5 μm emission band. This feature, too, was detected already in IRAS spectra (see Goebel et al. 1989, 1994), and recently Posch et al. (2002a) suggested that it is due to an iron rich magnesiowustite (Mg,Fe)O. Depending on the Fe:Mg ratio, magnesiowustite particles (if small compared to the wavelength and roughly spherical) can emit at wavelengths between 16.6 and 19.9 μm. The emissivity peak wavelength grows monotonically with x = [Fe]:[Mg]. For x = 9, an emissivity peak position of 19.6 μm results (see Fig. 5). Its position, but also its band profile, is very close to that of the observed 19.5 μm dust feature. This, together with the lack of known alternative carriers, makes it rather likely that $Mg_{0.1}Fe_{0.9}O$ particles produce the 19.5 μm emission band. If we assume other particle shapes (e. g. cubic shapes, since magnesiowustite belongs to the cubic crystal system), other peak wavelengths result: in this case, a smaller value of x would be more consistent with the present observational data. This point deserves further examination. Condensation experiments will provide constraints on the expected particle shapes and stoichiometries (i. e. x values) of circumstellar (Mg,Fe)O. Additional constraints from thermodynamical calculations are forthcoming (Ferrarotti & Gail, priv. comm.).

Further species of metal oxide dust which have been predicted to form in CSEs are calcium-aluminium oxides like $CaAl_{12}O_{19}$ (hibonite) and $CaAl_4O_7$ (grossite), as well as titanium oxides like TiO_2, Ti_2O_3, $CaTiO_3$ and $FeTiO_2$. Optical constants and opacities are now available for the former (Mutschke et al. 2002), from which some (not yet definitive) evidence of the presence of hibonite in CSEs can be derived (which is also supported by the investigation of presolar meteoritic grains). For the latter – the titanium oxides – optical constants in the mid-IR will be published soon (Posch et al., in prep.); however, spectroscopic evidence for their existence in CSEs is missing.

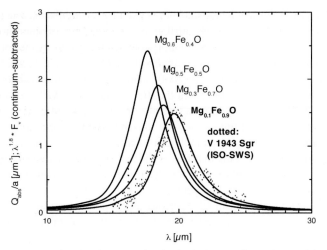

Figure 5: Emissivities Q_{abs}/a of sub-μm sized spherical magnsiowustite particles with different Fe:Mg ratios. Additionally, the 19.5 μm feature of the O-rich AGB-star V1943 Sgr (based on an ISO-SWS spectrum) has been overplotted.

3.3 The crystalline and amorphous silicate bands

The metal oxides whose features have been discussed above are certainly of crucial importance for the understanding of dust formation; however, the vast majority of oxygen-rich circumstellar minerals are still *silicates*. Two types of silicates must be distinguished: amorphous and crystalline. The 10 and 18 μm features, attributed to amorphous silicate, are several μm broad (Dorschner et al. 1995) and can be observed – in emission or self-absorption, depending on the optical depth – in a great variety of objects (from CSEs to interstellar clouds). The crystalline silicate features are narrow (FWHM values are normally smaller than 1 μm, see e.g., Molster et al. 1999) and are not so ubiquitous as the amorphous ones. However, there is a great manifold of them. Molster et al. (2002) classified them into seven complexes: a 7–13, 15–20, 22–25.5, 26.5–31.5, 31.5–37.5, 38–45.5, and a 50–72 μm complex, comprising, in total, almost 50 individual emissivity peaks. Presently, we are still rather far away from being able to precisely assign all the observed spectral features (or groups of features) to minerals with known optical constants. It has been possible, though, to gain some important insights in this field. To give just some representative examples (from Molster 2000, Suh 2002): first, it has been found that the carriers of most of the crystalline silicate bands are *Mg-rich* and *Fe-poor*; second, it has been recognized that crystalline silicate features are especially prominent in systems which are surrounded by a disc; and third, it became clear that the mass fraction of the silicates which are crystalline is limited to about 0.2 (note that it can be better constrained for optically thick dust shells, as pointed out by Kemper et al. 2001).

The carriers of the broad 10 and 18 μm bands have been traditionally assumed to be amorphous *magnesium-iron silicates* with the chemical formula $Mg_{2x}Fe_{2-2x}SiO_4$ and an x value not too far from 0.5. Optical constants of such (and other) glassy silicates for x = 0.4 and x = 0.5 have been published by Dorschner et al. (1995).

However, in view of the presence of almost pure crystalline *magnesium* silicates in O-rich CSEs, iron-free amorphous silicates such as Mg_2SiO_4 – potential precursors, through annealing processes, of the crystalline ones – also deserve interest. Such Fe-free amorphous silicates are also possible carriers of the 10 and 18 μm bands (see Jäger et al. 2003).

3.4 Equilibrium temperature of oxides and silicates in the dust shells of AGB-stars

At a given distance from the central star, the equilibrium temperature of a dust particle depends – in case of an optically thin dust shell – mainly on its absorption efficiency in the visual and near infrared range. This is due to the fact that most of the energy contained in the radiation field of an AGB star is concentrated in the wavelength region between 0.5 and 10 μm. In case of a thermodynamical equilibrium between the radiation field of the star (radius R_*, effective temperature T_{eff}) and the dust grains, the temperature T_d of the dust can be derived from the following energy balance:

$$\frac{1}{2}\left[1 - \sqrt{1 - \frac{R_*^2}{R^2}}\right]\int_0^\infty \pi a^2 Q_{abs}(\lambda) \pi B_\lambda(\lambda, T_{eff})\, d\lambda =$$
$$= \int_0^\infty 4\pi a^2 Q_{abs}(\lambda) \pi B_\lambda(\lambda, T_d)\, d\lambda$$

The first integral, together with the dilution factor of the radiation field, describes the energy *absorbed* by an individual dust grain with a radius a and an absorption efficiency factor $Q_{abs}(\lambda)$ in the stellar radiation field $B_\lambda(\lambda, T_{eff})$. The second integral describes the energy *emitted* by the same dust grain. $R(T_d)$ is the distance of the grain from the star. Note that the dilution factor used here – derived from the so-called Lucy-approximation – quickly converges against $R_*^2/R^2(T_d)$ for $R \gg R_*$, which means that the star can be considered as a point source for $R \gg R_*$.

Which equilibrium temperature profiles $T_d(R)$ do we get, in this model, for the oxides and silicates mentioned in the previous sections? Schematically, it can be said that oxides and silicates that do not contain transition metal ions (like Fe and Cr) show a very steep decrease of T_d as a function of R. By contrast, amorphous carbon, iron-containing oxides and silicates show a decrease of T_d with R that is roughly proportional to $1/\sqrt{R}$ – a behaviour which is expected for "grey" dust (dust with an emissivity that depends only weakly on the wavelength).

This fundamental difference is illustrated in Fig. 6 (see also Posch et al. 2002c). It should be noted, however, that such a clear distinction between the thermal behaviour of iron-containing dust grains and dust grains built from Mg, Al, Si etc. will only occur unless the latter contain transition metal ions in form of impurities. Such impurities can be iron or chromium particles which are not constituents of the crystal lattice but rather form cores, mantles, or inclusions. Volume fractions of 0.1 to 1 % of iron cores or mantles will strongly increase the opactity of an iron-free silicate such as $MgSiO_4$ in the 0.5–5 μm range. This will again lead to a heating of these

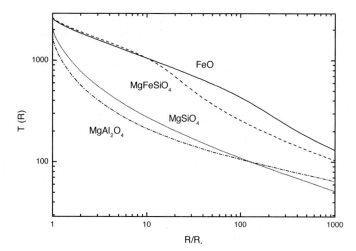

Figure 6: Equilibrium dust temperatures of iron-containing and iron-free oxides and silicates. The Fe-containing grains FeO and MgSiO$_4$ have significantly higher temperatures at a given distance from the central star. The stellar effective temperature T_{eff} was assumed to be 3000 K.

grains. This can be shown by calculating the opacity of core-mantle-particles, e.g. with Fe cores and MgSiO$_4$ mantles. An iron core with only 0.1 % volume fraction increases the opacity of an MgSiO$_4$ particle by a factor of 10 at 1 μm, leading already to a strong change in the particle's thermal behaviour (see Jäger et al. 2003).

4 Mass loss geometry

As outlined in the introduction, the circumstellar environment goes through a metamorphosis as the central star evolves from a red giant on the AGB to a white dwarf. The apparently overall spherical AGB-CSEs turn into, in most cases, highly axisymmetric Planetary Nebulae with additional complicated geometrical structures. There exists as yet no conclusive explanation for this phenomenon, but it appears that the transition occurs at the final stage of AGB-phase evolution or at the beginning of the post-AGB evolution. Most likely a change in the mass loss properties of the central star lies at the base. This means that only observations of very high angular resolution will provide the necessary information. Such knowledge, for instance, comes from observations of maser emission (e.g. Diamond et al. 1994), which is notoriously difficult to interpret, and interferometric (e.g. Lopez et al. 1997) and speckle imaging (e.g. Weigelt et al. 1998) in the near- to mid-IR range.

The low mass loss rate objects can provide crucial information here. In a few cases there exists reasonable evidence of non-spherical AGB-CSEs around such objects, e.g. the carbon star V Hya (SRa; Kahane et al. 1996) and the M-star X Her (SRb; Kahane & Jura 1996). In addition to spherical CSEs, the CO observations (performed with the IRAM 30 m telescope) reveal the presence of bipolar outflows in both cases. Presumably these outflows are identified because the spherical CSEs have low expansion velocities.

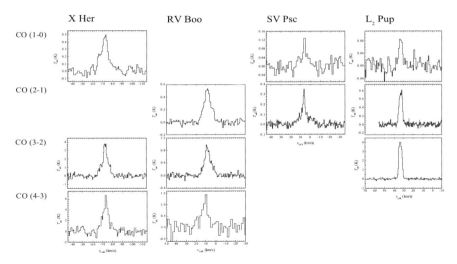

Figure 7: Various CO-transitions for four variables showing multi-component or very narrow expansion velocity profiles.

The vast majority of the low mass loss rate stars in the samples of Olofsson et al. (1993) and Kerschbaum & Olofsson (1999) shows simple CO radio line profiles (see above). However, among the M-stars about 10 to 20 % have double-component profiles or extremely low expanison velocities. These objects seem to have no other distinctive properties. Figure 7 gives some examples. Among these is X Her, which has been shown to have a complicated CSE. We therefore decided to observe interferometrically another of the double-component sources, RV Boo, and also to obtain additional data on X Her.

4.1 RV Bootis

A recent interferometric study in the CO($J = 1$–0 and 2–1) lines at the OVRO mm-array of the nearby O-rich SRV RV Boo clearly showed the presence of a non-spherical CSE (Bergman, Kerschbaum & Olofsson 2000; for single dish profiles see Fig. 7). Figure 8 illustrates the higher resolution CO(2–1) data in the form of a position-velocity diagram along the major axis of the extended 4×3 arcsecond brightness distribution. The structure suggests Keplerian rotation in a gas disk around a 1 to 3 solar masses AGB-star. Moreover, by comparison with a simple kinematical model shown in the lower half of the figure, there is indication of at least two circular gaps in the gas disk. This could, somewhat speculatively, be interpreted in terms of tidal effects of large bodies circling RV Boo.

To obtain even higher spatial resolution data, the VLA was used to observe the SiO($v = 0$, $J = 1$–0) line at a resolution of about 0.5 arcseconds. The emission is clearly detected, but the data are of too low S/N-ratio to allow a more detailed study. Additional runs at the VLA and complementary observations at infrared or visual wavelengths will shed light on this interesting object.

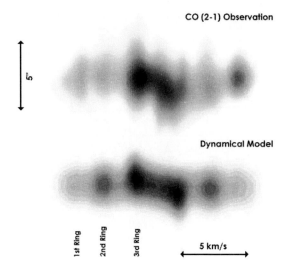

Figure 8: Position-velocity diagram along the major axis of the CO($J = 2$–1) brightness distribution around RV Boo (OVRO mm-array data).

4.2 X Herculis

X Her is another M-type SRV with a complex CO line profile (see Fig. 7). It is similar to RV Boo in most of its properties. We have obtained high spatial resolution interferometric CO emission observations of X Her using the OVRO mm-array. These new data support the findings of Kahane & Jura (1996) by showing a bipolar structure along the same orientation.

New VLA SiO($v = 0$, $J = 1$–0) line observations are shown in Fig. 9 in the form of a position-veleocity diagram along the minor axis of the CO distribution. The similarity of this diagram to that of RV Boo (Fig. 8) is striking. Once again, there is an indication of a rotating disk, in this case perpendicular to a bipolar outflow visible in the CO radio line data.

Since all observations of RV Boo and X Her suffer from bad S/N-ratios, additional observations are needed before we can firmly establish the circumstellar structure of these stars.

5 Summary

From our CO radio line survey (Kerschbaum & Olofsson 1999) and subsequent modelling (Olofsson et al. 2002) it turns out that O-rich IRV/SRVs have mass loss properties comparable to C-rich IRV/SRVs, and also short-period M-Miras; they seem quite independent of the chemistry and the pulsational periods. The derived mass loss rates range between 2.0×10^{-8} and 8.0×10^{-7} $M_\odot$/yr with a median of 2.0×10^{-7} $M_\odot$/yr. These objects show low gas expansion velocities of typically 7 km/s, while a few objects have extremely low values of only 2–4 km/s!

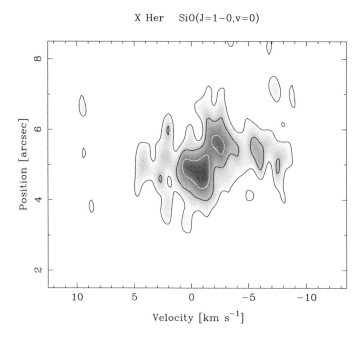

Figure 9: SiO($J = 1$–0) position-velocity diagram along the minor axis of the CO($J = 2$–1) brightness distribution towards X Her (VLA data; compare the similar diagram for RV Boo in Fig. 8).

Our new survey for thermal circumstellar SiO radio line emisson supplements the material presented above. A comparison of simple SiO/CO radio line intensity ratios with stellar observables like pulsational period, mass loss rate and visiblity of silicate dust emission indicates an anticorrelation of the mass loss rate with the relative SiO line strength. These trends are supported by the first results of a detailed modelling of the circumstellar SiO lines (González Delgado et al. 2003). The circumstellar SiO abundance is down by at least a factor of 10 compared to the expected, and there is possibly a trend in the sense that the abundance decreases with mass loss rate.

Although the vast majority of O-rich IRVs and SRVs shows simple (rectangular or parabolic) CO radio line profiles, a non-negligible fraction of about 5–10 % display a more complex velocity structure in their circumstellar gas. High spatial resolution observations of two objects, RV Boo and X Her, indicate the presence of disks and/or axially symmetric outflows.

There is increasing spectroscopic evidence that M-type SRVs and IRVs form a distinct group in terms of the composition of the dusty CSE. Their spectral energy distributions are dominated by emission features of oxide particles like magnesiowustite, spinel and amorphous Al_2O_3. Amorphous silicate features are also present in the spectra of SRVs and IRVs, but more prominent in the SEDs of M-Miras with intermediate mass loss rates. Stars with very high mass loss rates, on the other hand, are characterized by strong crystalline silicate emission bands.

References

Begemann, B., Dorschner, J., Henning, Th., et al. 1997, ApJ 476, 199

Bergman, P., Kerschbaum, F., Olofsson, H. 2000, A&A 353, 583

Bieging, J.H., Latter, W.B. 1994, ApJ 422, 765

Diamond, P.J., Kemball, A.J., Junor, W., Zensus, A., Benson, J., Dhawan, V. 1994, ApJL 430, L61

Dorschner, J., Begemann, B., Henning, Th., Jäger, C., Mutschke, H. 1995, A&A 300, 503

Fabian, D., Posch, Th., Mutschke, H., et al. 2001, A&A 373, 1125

Feast, M.W., Glass, I., Whitelock, P.A., Catchpole, R.M. 1989, MNRAS 241, 375

Glaccum, W. 1995, in: Airborne Astronomy Symposium on the Galactic Ecosystem, eds. M.R. Haas, J.A. Davidson, E.F. Erickson, ASP Conf. Ser. 73, 395

Goebel, J.H., Volk, K., Walker, H., et al. 1989, A&A 222, L5

Goebel, J.H., Bregman, J.D., Witteborn, F.C. 1994, ApJ 430, 317

Groenewegen, M.A.T. 1994, in: Circumstellar Matter, ed. G. Watt

Groenewegen, M.A.T., Baas, F., de Jong, T., Loup, C. 1996, A&A 306, 241

Habing, H.J. 1996, A&AR 7, 97

Hoefner, S., Feuchtinger, M.U., Dorfi, E.A. 1995, A&A 297, 815

Hron, J., Aringer, B., Kerschbaum, F. 1997, A&A 322, 280

Iben, I.Jr., Renzini, A. 1983, ARA&A 21, 271

Ivezić, Ž., Elitzur, M. 1995, ApJ 445, 415

Ivezić, Ž., Elitzur, M. 1996, MNRAS 279, 1011

Jäger, C., Dorschner, J., Posch, Th., Henning, Th. 2003, A&A, submitted

Josselin, E., Loup, C., Omont, A., Barnbaum, C., Nyman, L.A., Sevre, F. 1998, A&AS 129, 45

Kahane, C., Audinos, P., Barnbaum, C., Morris M. 1996, A&A 314, 871

Kahane, C., Jura, M. 1996, A&A 310, 952

Kastner, J.H., Forveille, T., Zuckerman, B., Omont, A. 1993, A&A 275, 163

Kemper, C., Waters, L.B.F.M., de Koter, A., Tielens, A.G.G.M. 2001, A&A 369, 132

Kerschbaum, F. 1995, A&AS 113, 441

Kerschbaum, F. 1999, A&A 351, 627

Kerschbaum, F., Hron, J. 1992, A&A 263, 97

Kerschbaum, F., Hron, J. 1994, A&AS 106, 397

Kerschbaum, F., Hron, J. 1996, A&A 308, 489

Kerschbaum, F., Lazaro, C., Habison, P. 1996a, A&AS 118, 397

Kerschbaum, F., Olofsson, H., Hron, J. 1996b, A&A 311, 273

Kerschbaum, F., Olofsson, H. 1998, A&A 336, 654

Kerschbaum, F., Olofsson, H. 1999, A&AS 138, 299

Knapp, G.R., Morris, M. 1985, ApJ 292, 640

Knapp, G.R., Young, K., Lee, E., Jorissen, A. 1998, ApJS 117, 209

Le Bertre, T. 1988, A&A 190, 79

Lebzelter, T., Kerschbaum, F., Hron, J. 1995, A&A 298, 159

Lopez, B., Danchi, W.C., Bester, M., et al. 1997, ApJ 488, 807

Lorenz-Martins, S., Pompeia, L. 2000, MNRAS 315, 856

Loup, C., Forveille, T., Omont, A., Paul, J.F. 1993, A&AS 99, 291

Margulis, M., van Blerkom, D.J., Snell, R.L., Kleinmann, S.G. 1990, ApJ 361, 673

Molster, F.J., Waters, L.B.F.M., Trams, N.R. 1999, A&A 350, 163

Molster, F.J. 2000, Thesis, University of Amsterdam

Molster, F.J., Waters, L.B.F.M., Tielens, A.G.G.M. 2002, A&A 382, 222

Mutschke, H., Posch, Th., Fabian, D., Dorschner, J. 2002, A&A 392, 1047

Nyman, L-Å, Booth, R.S., Carlström, U., et al. 1992, A&AS 93, 121

Olofsson, H., Bergman, P., Eriksson, K., Gustafsson, B. 1996, A&A 311, 587

Olofsson, H., Eriksson, K., Gustafsson, B., Carlström, U. 1993, ApJS 87, 267

Olofsson, H. 1997, Ap&SS 251, 31

Olofsson, H., González Delgado, D., Kerschbaum, F., et al., 2002, A&A 391, 1053

Posch, Th., Kerschbaum, F., Mutschke, H., et al. 1999, A&A 352, 609

Posch, Th., Kerschbaum, F., Mutschke, H., Dorschner, J., Jäger, C. 2002a, A&A 393, L7

Posch, Th., Kerschbaum, F., Mutschke, H., et al. 2002b, in: C. Gry et al. (eds.), Exploiting the ISO Data Archive, ESA SP-511

Posch, Th., Mutschke, H., Fabian, D. 2002c, Hvar Obs. Bull., accepted

Sahai, R., Bieging, J.H. 1995, AJ 105, 595

Schöier, F.L., Olofsson, H. 2001, A&A 368, 969

Sloan, G.C., Price, S.D. 1995, ApJ 451, 758

Speck, A.K., Barlow, M.J., Sylvester, R.J., Hofmeister, A.M. 2000, A&AS 146, 437

Stephenson, C.B., 1989, Publ. Warner & Swasey Obs., Vol.3, No.2

Suh, K.-W. 2002, MNRAS 332, 513

Sylvester, R.J., Kemper, F., Barlow, M.J., et al. 1999, A&A 352, 587

Weigelt, G., Balega, Y., Bloecker, T., Fleischer, A.J., Osterbart, R., Winters, J.M. 1998, A&A 333, L51

Whitelock, P. 1990, PASPC 11, 365

Whitelock, P., Menzies, J., Feast, M., et al. 1994, MNRAS 267, 711

Winters, J.M., Fleischer, A.J., Gauger, A., Sedlmayr, E. 1994, A&A 290, 623

Willson, L.A. 1988, in: Pulsation and Mass Loss in Stars, eds. R. Stalio, L.A. Willson, Kluwer Acad. Publ., p. 285

Woitke, P. 2000, in: R. Diehl, D. Hartmann (eds.), Astronomy with Radioactivities, p. 163

Young, K. 1995, ApJ 445, 872

Zuckerman, B., Dyck, H.M. 1986, ApJ 304, 394

Finding the Most Metal-poor Stars of the Galactic Halo with the Hamburg/ESO Objective-prism Survey

Norbert Christlieb

Hamburger Sternwarte
Gojenbergsweg 112
D-21029 Hamburg
Germany
E-Mail: nchristlieb@hs.uni-hamburg.de
URL: http://www.hs.uni-hamburg.de/DE/Ins/Per/Christlieb/

Abstract

I review the status of the search for extremely metal-poor halo stars with the Hamburg/ESO objective-prism survey (HES). 2 194 candidate metal-poor turn-off stars and 6 133 giants in the magnitude range $14 \lesssim B \lesssim 17.5$ have been selected from 329 (out of 380) HES fields, covering an effective area of 6 400 square degrees in the southern extragalactic sky. Moderate-resolution follow-up observations for 3 200 candidates have been obtained so far, and ~ 200 new stars with [Fe/H] < -3.0 have been found, which trebles the total number of such extremely low-metallicity stars identified by all previous surveys.

We use VLT-UT2/UVES, Keck/HIRES, Subaru/HDS, TNG/SARG, and Magellan/MIKE for high-resolution spectroscopy of HES metal-poor stars. I provide an overview of the scientific aims of these programs, and highlight several recent results.

1 Introduction

Extremely metal-poor (EMP) stars, defined here as stars with 1/1 000th of the metal content of the Sun (i.e., [Fe/H] < -3.0)[1], permit astronomers to study the earliest epochs of Galactic chemical evolution, since the atmospheres of these old, low-mass stars retain, to a large extent, detailed information on the chemical composition of the interstellar medium (ISM) at the time and place of their birth. These stars are hence the local equivalent of the high redshift Universe (for recent reviews on EMP stars see Cayrel 1996, Beers 1999, 2000; Preston 2000).

---
[1][A/B] = $\log_{10}(N_A/N_B) - \log_{10}(N_A/N_B)_\odot$, for elements A and B.

In the past decade, numerous studies have focused on abundance analysis of very metal-poor stars. In the early works of McWilliam et al. (1995b) and Ryan et al. (1996), it was found that many elements show a sudden change in the slope of their abundances relative to Fe, near [Fe/H] $= -2.5$. Other elements, including Ba and Sr, show a large scatter from star-to-star at very low [Fe/H], indicating that the gas clouds from which these stars formed were polluted by the products of only a few, if not single, supernovae (see, e.g., Shigeyama & Tsujimoto 1998; Tsujimoto et al. 2000). Thus, with individual EMP stars we can indirectly study the first generation of massive stars, which exploded in supernovae of type II (hereafter SN II). Furthermore, Karlsson & Gustafsson (2001) have shown that the inspection of patterns seen in abundance correlation diagrams of a large, homogeneously-analyzed sample of stars allows one to constrain theoretical yield calculations, and to perhaps even determine the mass function of the first generation of stars.

EMP stars also contribute to observational cosmology: The abundances of light elements (in particular Li), through models of Big Bang nucleosynthesis, put limits on the baryonic content of the Universe (e.g., Ryan et al. 1999), and individual age determinations of the oldest stars – either by nucleo-chronometry (e.g., Cowan et al. 1999; Cayrel et al. 2001), or by means of application of stellar evolutionary tracks (e.g., Fuhrmann 1998) – yield a lower limit for the age of the Galaxy, and hence the Universe.

However, it should be recognized that, owing to their rarity, the road to obtaining elemental abundances for EMP stars is long and arduous. The process involves three major observational steps (see Figure 1): (1) a wide-angle survey must be carried out, and candidates for metal-poor stars have to be selected; (2) moderate-resolution (~ 2 Å) spectroscopic follow-up observations of the candidates needs to be done in order to validate the genuinely metal-poor stars among them; and (3) high-resolution spectroscopy has to be obtained. Step (1) can be carried in many different ways (for a review, see Beers 2000), for example, with proper-motion surveys or photometric surveys. However, spectroscopic surveys are the most efficient way to find large numbers of EMP stars, and they have the advantage that the samples they yield are not kinematically biased, as opposed to proper-motion selected samples.

Until recently, the main source of EMP stars was the HK survey of Beers and colleagues (Beers et al. 1985; Beers 1999), which so far has yielded about 100 stars with [Fe/H] < -3.0. For most of them, high resolution ($R > 40\,000$) spectroscopy has now been obtained (see, e.g., Ryan et al. 1996; Norris et al. 2001). As a consequence, the abundance trends of many elements are now well-determined down to metalicities of [Fe/H] ~ -3.5.

Still larger samples of EMP stars are needed to identify the small subset of stars that are even more exotic then the EMP stars themselves, e.g., stars with strong enhancements of the r-process elements, suitable for nucleo-chronometric age dating (Cowan et al. 1997, 1999; Cayrel et al. 2001; Hill et al. 2002). The frequency of these stars among metal-poor giants with [Fe/H] < -2.5 has been found in the HK survey to be as low as $\sim 3\,\%$, a paucity confirmed by recent HES results (see Section 4.2 for details). HE 0107–5240 is another example of an extremely rare star. It is a giant with [Fe/H] $= -5.3$ (Christlieb et al. 2002a), which was found during follow-up observations of ~ 200 stars with [Fe/H] < -3.0 (see Section 4.4 below).

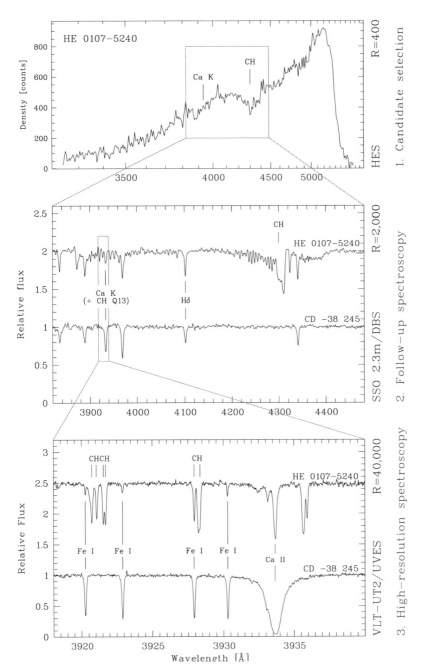

Figure 1: The three major observational steps towards obtaining elemental abundances of extremely metal-poor stars: (1) objective-prism spectra, yielding candidate metal-poor stars; (2) vetting of candidates by moderate-resolution follow-up spectroscopy; (3) high-resolution spectroscopy of confirmed candidates.

Furthermore, stars of the lowest metallicity place strong constraints on the nature of the Metallicity Distribution Function (MDF), which presumably arises from the ejected yields of the first generations of stars. For example, with the exception of the star at [Fe/H] $= -5.3$, it appears that the lowest abundance stars in the Galaxy reach no lower than [Fe/H] ~ -4.0, hence this is the level of early "pollution" of the ISM one might demand to be reached by the products of first generation stars. One also would like to test whether the MDF, as reflected by observations of halo stars in the Solar neighborhood, remains constant or varies with Galactocentric distance.

The Hamburg/ESO objective-prism survey (HES; Wisotzki et al. 2000) offers the opportunity to increase the number of EMP stars by at least a factor of five with respect to the HK survey, since the HES is more than 1.5 mag deeper, and covers regions of the sky not included in the HK survey (see Figure 2). For a detailed comparison of the HES with the HK survey, see Christlieb & Beers (2000).

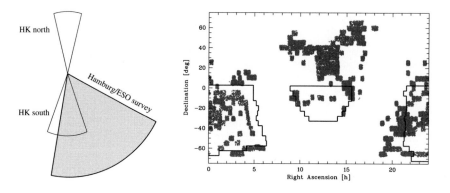

Figure 2: Comparison of the survey volumes of the HK and HE surveys (left panel); comparison of HES area (framed) with HK survey area (right panel). Grey areas denote HK survey plates.

The HES covers the total southern ($\delta < -2.5°$) extragalactic ($|b| \gtrsim 30°$) sky. The HES objective-prism plates were obtained with the ESO 1 m-Schmidt telescope, and scanned and reduced at Hamburger Sternwarte. The faintest HES metal-poor stars have $B \sim 17.5$ mag; stars this faint are observable at high spectral resolution using currently existing telescope/instrument combinations.

2 Selection of metal-poor candidates

Metal-poor turnoff stars are selected in the HES database of digital objective-prism spectra (for examples see Fig. 2) by automatic spectral classification (Christlieb et al. 2002b). Cooler stars ($B - V > 0.5$) are currently selected by applying an empirical cutoff line in $(B - V)$ versus KP space, where KP is the Ca II K line index defined by Beers et al. 1999).

The selection has so far been restricted to unsaturated point sources. For stars brighter than $B \sim 14$ mag, saturation effects (in particular in the red portions of the prism spectra) begin to occur. Investigations on methods for recovering at least a portion of these brighter candidates are underway (Frebel et al., in preparation). The importance of the bright sources comes from the relative ease with which high-resolution spectroscopy can be obtained. Extended sources are being rejected because spatial extension leads to smeared-out spectral lines in slitless spectra, and therefore many galaxies would otherwise enter the candidate sample as false positives.

A total of 2 194 candidate metal-poor turnoff stars have been selected from the 329 (out of 380) HES fields that are currently used for stellar work. These fields cover a nominal area of 8 225 deg^2 of the southern extragalactic sky. The covered area is reduced to an effective area of 6 400 deg^2 due to overlapping spectra (Wisotzki et al. 2000; Christlieb et al. 2001a). In addition to the turnoff stars, 6 133 metal-poor candidates with $0.5 < B - V < 1.2$ were selected, including 1 785 (i.e., $\sim 30\,\%$) candidates expected to have large over-abundances of carbon ([C/Fe] > 1.0). These stars are readily identifiable in the HES from their strong CH G bands (see Figure 4). A sample of 403 faint high latitude carbon (FHLC) stars has already been published (Christlieb et al. 2001a), and that sample contains many carbon-enhanced metal-poor (CEMP) stars as well, as recent follow-up observations have demonstrated.

As noted above, the HES is a very rich source for metal-poor giants. This is especially valuable because the present HK survey sample is dominated by hotter stars near the main-sequence turnoff, due to a temperature-related bias incurred by the initial visual selection of HK candidates. As a result, only roughly 20 % of the HK stars are metal-poor giants. However, efforts are under way to recover metal-poor HK survey giants, by means of Artificial Neural Network classifications of spectra extracted from digitized HK survey plates (Rhee 2000) in combination with JHK colors from the 2MASS survey (Skrutskie et al. 1997).

3 Follow-up observations

Moderate resolution (~ 2 Å) follow-up observations are needed to verify the metal-poor nature of the HES candidates. This is the bottleneck in every large survey for EMP stars, and the only way to cope with this is to establish large collaborations, involving many people, with access to many different telescopes. To provide the reader an impression of the extent of this effort, we list in Table 1 the telescopes that have been used in 2002 for follow-up of HES candidates, and the numbers of stars observed.

Follow-up spectra of $S/N > 20$ at Ca II K are required to determine [Fe/H] of the metal-poor candidates with an accuracy of 0.2–0.3 dex, using the technique of Beers et al. (1999). As can be seen from Table 1, on average ~ 20 stars per night can be observed at a 4 m-class telescope in single-slit mode, taking into account losses due to bad weather, moonlight, and technical downtime. Note that the candidate throughput at the 1 m-UK Schmidt is as high achieved as by telescopes with ~ 15 times more collecting area, due to the availability of the 150 fiber instrument 6dF

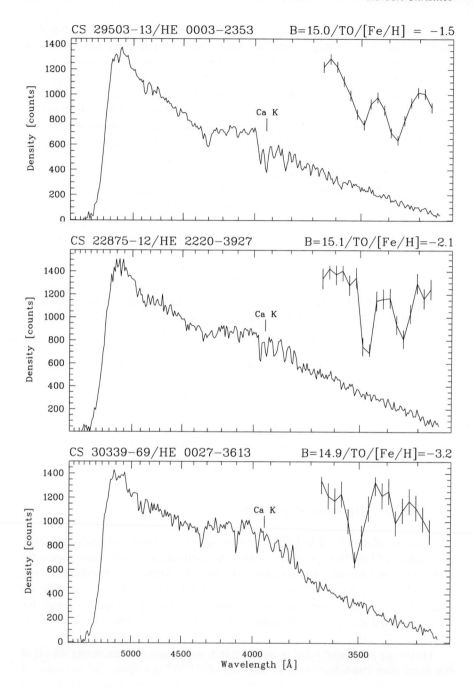

Figure 3: HES spectra of metal-poor stars found in the HK survey. Note that wavelength is decreasing from left to right. The sharp cutoff at ∼ 5400 Å is due to the IIIa-J emulsion sensitivity cutoff ("red edge").

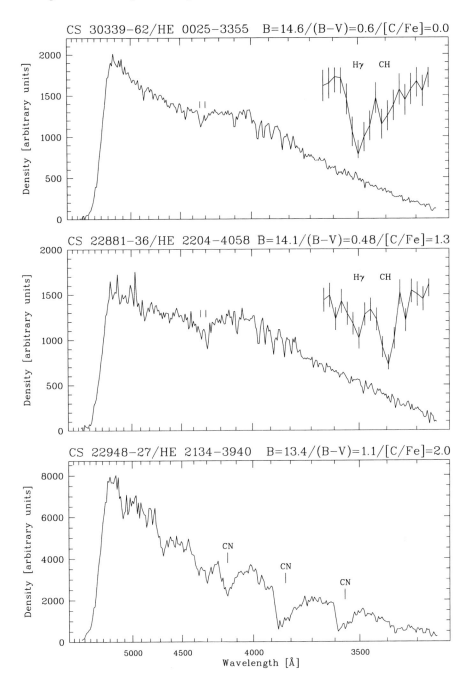

Figure 4: HES spectra of carbon-enhanced metal-poor stars from the HK survey.

Table 1: Follow-up observations of HES metal-poor candidates in 2002.

Telescope/Instr.	Nights	Stars	Stars/Night
AAT/RGOS	6	87	15
ESO 3.6m/EFOSC2	13	274	21
KPNO 4m/RCSP	5	109	22
Magellan I/B&C	33	713	22
Palomar 200"/DS	9	96	11
SSO 2.3 m/DBS	32	430	13
UK Schmidt/6dF	16	341	21
Sum/average	114	2050	18

(Watson 1998). Up to three fields with 10–60 metal-poor candidates, respectively, can be observed in one night. (The remaining fibers are filled with other interesting HES stars, e. g., field horizontal-branch stars, sdB stars, and white dwarfs.) However, the faintest HES candidates ($B \gtrsim 16.5$) cannot be observed with UK Schmidt/6dF, since the fiber diameter is $6.7''$ on the sky (chosen to suite the 6dF Galaxy Redshift Survey), and therefore observations of point sources are sky-background limited.

As of 1 February 2003, follow-up observations for 3 200 HES candidates have been obtained. Considering that an average pace of $\sim 1\,000$ candidates per year is the maximum one can hope to reach in the long run, and the fact that $\sim 1\,000$ additional, bright candidates are expected to be found in the HES, the follow-up will likely continue for another 3–5 years.

The "effective yield" of metal-poor stars in the HES, as compared to the HK survey, is presented in Table 2, and illustrated in Figure 5. The effective yield (EY) refers to the fraction of genuine stars below a given [Fe/H] in the observed candidate sample, as defined by Beers (2000). Based on a sample of 1 214 candidates, it appears that the HES is twice as efficient as the HK survey in finding stars with [Fe/H] < -3.0, when photoelectric BV photometry is used in the latter survey for pre-selection, and six times as efficient compared to the HK survey without usage of BV photometry. The higher yields of the HES can mainly be attributed to the fact that $B - V$ colors can be derived directly from the objective-prism spectra with an accuracy of better than 0.1 mag (Christlieb et al. 2001b), the higher quality of the HES spectra, and the employment of quantitative selection criteria, as opposed to visual selection in the HK survey.

It should be stressed that a bias towards higher EYs may be present in the HES candidate sample investigated so far, because the best (in terms of the weakness of the Ca II K lines visible in the HES spectra) and brightest candidates have been observed first. However, when the EYs within each of the candidate classes (as assigned during the visual verification of the HES candidates) are applied to the respective fraction of candidates within the total set of HES candidates, overall EYs of 4 % for stars with [Fe/H] < -3.0 result for turnoff stars and giants. If more restrictive selection criteria were to be used, the EYs could be increased to $\sim 10\,\%$, but in this case $\sim 1/4$ of the stars with [Fe/H] < -3.0 would not be identified.

Table 2: "Effective yields" of metal-poor stars (i.e., the fraction of genuine metal-poor stars below a given [Fe/H] in the observed candidate sample) in the HK and the HE surveys. N refers to the number of stars from which the statistic has been derived.

Survey	N	[Fe/H]		
		< -2.0	< -2.5	< -3.0
HK survey/no $B-V$	2614	11 %	4 %	1 %
HK survey/with $B-V$	2140	32 %	11 %	3 %
HES (turnoff stars)	571	59 %	21 %	6 %
HES (giants)	643	50 %	20 %	6 %

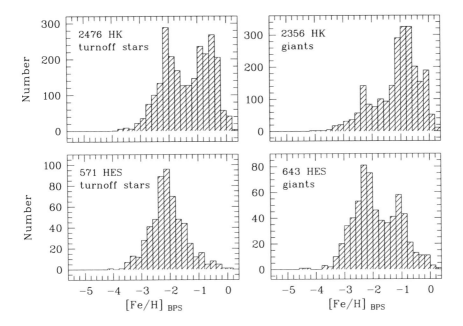

Figure 5: The observed Metallicity Distribution Functions of HES and HK turnoff stars (left panel) and giants (right panel)

Hence, a compromise is struck between the "purity" of the sample and the desire to include as many EMP stars as possible.

Using the numbers mentioned above, it can be estimated that at least 300 stars with [Fe/H] < -3.0 will eventually be found in the HES, provided that follow-up observations of *all* candidates are obtained (this estimate does not count the additional bright HES candidates, follow-up observations for which have only recently begun).

4 High-resolution spectroscopy projects

In this section, I provide an overview of ongoing high-resolution spectroscopic analyses of confirmed metal-poor stars (mainly) from the HES, and highlight some recent results.

4.1 The 0Z (zero Z) project

In a collaboration between Hamburger Sternwarte, Caltech, The Observatories of the Carnegie Institution of Washington (and formerly also Padua Observatory), we aim at obtaining high-resolution, high S/N spectroscopy of a large sample of the most metal-poor stars, with Keck/HIRES and Magellan/MIKE. First results have been published published in Cohen et al. (2002), Carretta et al. (2002), Lucatello et al. (2003), and Cohen et al. (2003).

Highlights include the discovery of HE 0024–2523. It is a peculiar main-sequence turnoff star, and a spectroscopic binary with a period of 3.14 days. Its abundance pattern is dominated by extreme enrichment of s-process elements, likely due to mass accretion from an already evolved companion which has now probably become a white dwarf. The relative lead abundance is [Pb/Fe] = +3.3 dex; the abundance ratio of Pb to Ba is even more extreme, [Pb/Ba] = +1.9 dex, among the highest ever determined in a metal-poor star. The unusually short period of this CH star suggests a past common envelope phase.

Stars like HE 0024–2523 provide the unique opportunity to study an extinct generation of extremely metal-poor AGB stars that left their fingerprints on their less massive companions. These stars place strong observational constraints on simulations of the structure, evolution, and nucleosynthesis of AGB stars, and allow us to study the s-process at very low metalicities. Recent calculations of Goriely & Siess (2001) predict that an efficient production of s-process elements takes place even in zero metallicity AGB stars, despite the absence of iron seeds, provided that protons are mixed into carbon-rich layers. Proton mixing results in the formation of ^{13}C, which is a strong neutron source due to the reaction ^{13}C$(\alpha, n)^{16}$O. The strong overabundances of Pb in HE 0024–2523 and other "Pb stars" (Aoki et al. 2000; VanEck et al. 2001; Aoki et al. 2002b) are in concert with these predictions.

Another exotic star found with Keck/HIRES observations is the EMP turnoff star HE 2148−1247. It appears that this star is enriched in both its r- and s-process elements, similar to CS 22898-027 (Preston & Sneden 2001; Aoki et al. 2002b), which is almost a twin of HE 2148−1247 with respect to stellar parameters as well as abundances, and a few other stars in the literature (Preston & Sneden 2001; Aoki et al. 2002b; Hill et al. 2000).

Cohen et al. (2003) propose a scenario for explaining the abundance pattern of HE 2148−1247. They suggest that the star is a member of a binary system (and in fact, the star shows significant radial variations) in which the former primary went through the AGB phase. In this stage, carbon and s-process elements were dumped onto the surface of HE 2148−1247 via mass transfer. The former primary then evolved into a white dwarf. Mass transfer then occurred into the *reverse* direction, i.e., from HE 2148−1247 to the white dwarf, which led to an accretion-induced collapse of the white dwarf into a neutron star. R-process nucleosynthesis then occurs in a neutrino-driven wind of the neutron star, and the nucleosynthesis products contaminate the surface layers of the star that we see today.

4.2 The HERES project – An ESO Large Programme

EMP stars with strong enhancements of r-process elements, like CS 22892-052 (McWilliam et al. 1995; Sneden et al. 1996; Cowan et al. 1997), and CS 31082-001 (Cayrel et al. 2001; Hill et al. 2002), are suitable for individual age determinations through the detection of the long-lived radioactive species thorium and/or uranium. Such old stars can therefore provide lower limits to the age of the Galaxy, and hence the Universe.

However, it appears that the pairs of elements used as chronometers need to be chosen very carefully. The results of Hill et al. (2002) suggest that comparison of the abundance of Th with that of the stable r-process element Eu does not yield reliable ages for all stars, and in particular an unrealistically small age for CS 31082-001. Th/Eu is roughly a factor of three higher in CS 31081-001 than observed in CS 22892-052, yielding a radioactive decay age for CS 31082-001 that is younger than the Sun! U/Th is expected to be a more accurate chronometer, due to the fact that ^{238}U, with a half-life as "short" as 4.5 Gyr (for comparison, ^{232}Th has a half-life of 14 Gyr), is involved. Furthermore, recent theoretical studies on r-process nucleosynthesis (Wanajo et al. 2002; Schatz et al. 2002) reveal that this chronometer is less sensitive to uncertainties of nuclear physics than any other chronometer that is amenable to observation in EMP stars.

Unfortunately, it is observationally very challenging to detect uranium in metal-poor stars; CS 31081-001 was the first one for which such a detection has been made. This is because the strongest spectral line in the optical, U II 3859.57 Å, is very weak (see Figure 9 of Hill et al. 2002), even if the enhancement of r-process elements is as high as [r/Fe] $= 1.7$ dex, as observed in CS 31081-001. Therefore, spectra of $S/N > 200$ per pixel and $R \geq 75\,000$ are required for derivation of accurate U abundances. Also, one has to contend with the difficulty that, if the carbon abundance is significantly enhanced with respect to iron, as is the case for CS 22892-052 ([C/Fe] $= 0.98$; McWilliam et al. 1995), the U line is blended with a CN line. As a result, measurement of an U abundance is impossible for CS 22892-052.

Quite apart from their obvious use in individual age determination, strongly r-process-enhanced metal-poor stars allow one to study r-process nucleosynthesis in general, and possibly to identify the site(s) of the r-process. While the abundance pattern of CS 22892-052 for the heavier stable neutron-capture elements ($Z \geq 56$), and the abundances of the elements in the range $56 \leq Z \leq 72$ in CS 31082-001 match a scaled solar r-process pattern very well, deviations are seen for some of the lighter neutron-capture elements in both stars, and for Th (as discussed above) and possibly Os in CS 31082-001. It has therefore been suggested (Sneden et al. 2000; Hill et al. 2002) that (at least) two r-processes must be at work.

Given these results, it is clear that a much larger sample of strongly r-process-enhanced metal-poor stars is required. The *Hamburg/ESO R-processed Enhanced* star Survey is aiming to increase the number of strongly r-process-enhanced stars to (conservatively) $\sim$ 5–10, and to study a major fraction of these by means of high-resolution, high signal-to-noise UVES spectra. The project is carried out in the framework of an ESO Large Programme (P. I. Christlieb), which was recently approved for an initial period of one year.

Stars with strong r-process enhancement can be readily recognized from their prominent Europium lines. For example, CS 22892-052 has an Eu II 4129.7 Å line as strong as 132 mÅ. This means that r-process-enhanced stars can be identified from "snapshot" high-resolution spectra, i.e., $R = 20\,000$ and $S/N = 30$. Such spectra can easily be obtained, e. g., with VLT-UT2/UVES, or with Subaru/HDS, in 20 min (including overheads) for a $B = 15$ mag star, even under unfavorable conditions (e. g., full moon, seeing $2''$, thin clouds). About half of our follow-up efforts presented above are seeking confirmation of giants with [Fe/H] ≤ -2.5 for this project.

To date, a total of 152 targets, 128 of them from the HES, have been observed in snapshot mode with UVES. Among the 98 stars with [Fe/H] < -2.5 dex, four with [r/Fe] $> +1.0$ dex have been found (three of them are HES stars). Among the 54 stars with [Fe/H] ≥ -2.5 dex, one with [r/Fe] $> +1.0$ dex has been found. Higher resolution and S/N spectra of most of these stars is currently being obtained with VLT/UVES, for attempts to detect uranium. An additional set of 117 stars is scheduled to be observed in snapshot mode in Period 72 (April–September 2003).

4.3 Carbon-enhanced metal-poor stars

One of the early results from the HK survey was the finding that the frequency of carbon-enhanced stars increases dramatically at the lowest values of [Fe/H], reaching $\sim$ 20–25 % at [Fe/H] < -2.5 (Norris et al. 1997a; Rossi et al. 1999). This result has been confirmed by follow-up observations of HES candidates.

It turns out that nature presents us with a "zoo" of carbon-enhanced metal-poor (CEMP) stars, with vastly differing properties. Above all, the origin of carbon in this class of stars remains unclear, despite extensive investigations by means of high-resolution spectroscopy (e. g., Norris et al. 1997a,b; Bonifacio et al. 1998; Hill et al. 2000; Aoki et al. 2002a; Norris et al. 2002; Lucatello et al. 2003). A larger sample of CEMP stars with abundance determinations, as well as long-term radial velocity monitoring, is needed to constrain the possible carbon enrichment scenarios realized in nature, and to separate known CEMP stars into the corresponding classes of objects.

We are currently increasing the number of CEMP stars in dedicated spectroscopic follow-up observations, mainly at the CTIO and KPNO 4 m telescopes. Confirmed CEMP stars from this and other efforts are studied at high resolution with Subaru/HDS and TNG/SARG. A program aiming at radial velocity monitoring of a large sample of CEMP stars has been proposed to be carried out with the Anglo-Australian Telescope, and will hopefully be continued for many years with this, and other 4m-class telescopes in the future.

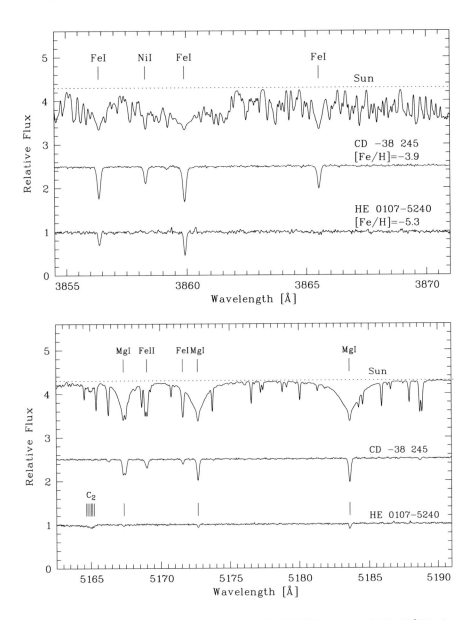

Figure 6: Spectrum of the Sun compared with the VLT/UVES spectrum of CD−38°245, the previously most metal-poor star, and with HE 0107−5240. The spectrum of CD−38°245 has been obtained with the same observational setup. The spectra are on the same scale and have been offset arbitrarily in the vertical direction. Note the very weak or absent Fe lines in the spectrum of HE 0107−5240, and the presence of the C_2 band at $\sim$ 5165 Å.

4.4 HE 0107–5240

Christlieb et al. (2002a) recently announced the discovery of HE 0107–5240, a giant with [Fe/H] = −5.3. This star provides constraints on low-mass star formation in the early Galaxy, and insights into the properties of the first generation of stars and supernovae. The abundance pattern of HE 0107–5240 is characterized by huge over-abundances of carbon and nitrogen with respect to iron ([C/Fe] = +4.0 dex; [N/Fe] = +2.3 dex), while the abundances of the other five elements of with presently detected spectral lines exhibit a behavior similar to what is seen in other EMP stars.

Umeda & Nomoto (2003) proposed *a posteriori* that the gas cloud from which HE 0107–5240 formed could have been enriched by a 25 $M_\odot$ Population III star exploding as a supernova of low explosion energy ($E_{\rm exp} = 3 \cdot 10^{50}$ erg). By assuming that the material produced during the SN event is homogeneously mixed over a wide range of the mass coordinate, and that a large fraction of the material falls back onto the compact remnant (the "mixing and fallback" mechanism), Umeda & Nomoto are able to reproduce the abundance pattern of HE 0107–5240 remarkably well (see their Figure 1). In particular, very high C/Fe and N/Fe ratios can be produced in this scenario (with the CNO elements produced in late stages of the evolution of the SN progenitor), offering an interesting possibility for explaining why a large fraction of the extremely metal-poor stars exhibit high enhancements of carbon (for a similar work on the abundance patterns of CS 29498-043 and CS 22949-037 see Tsujimoto & Shigeyama 2003).

The scenario of Umeda & Nomoto is also very attractive in that it does not require the existence of a binary companion for the CEMP stars. In fact, there are at least three unevolved CEMP stars known that do not show any radial velocity variations larger than 0.4 km s^{-1} over a period of 8 years (Preston & Sneden 2001), ruling out, at least for these three stars, the scenario in which a formerly more massive star went through its AGB phase and transfered dredged-up material onto the surface of the less massive companion, which we observe as a CEMP star today.

Umeda & Nomoto predict an O abundance of [O/Fe] = +2.8 dex for HE 0107–5240. A high-resolution, high S/N spectrum of HE 0107–5240, covering the near UV region down to the atmospheric cutoff, has now been obtained with VLT/UVES. Spectrum synthesis calculations suggest that the detection of UV-OH lines is possible if [O/Fe] > +0.3 dex. Therefore, the predictions of Umeda & Nomoto can soon be tested.

Acknowledgements

I am grateful to D. Reimers (P. I. of the HES) and L. Wisotzki (HES project scientist) for making my work on the HES possible. It is a pleasure to acknowledge the contributions of my numerous collaborators to HES metal-poor star research over the last years. Here I would like to especially thank those who have participated in the huge effort of vetting HES candidates: Paul Barklem, Tim Beers, Mike Bessell, Innocenza Busa, Judy Cohen, Doug Duncan, Bengt Edvardsson, Lisa Elliot, Birgit Fuhrmeister, Dionne James, Andreas Korn, Andy McWilliam, Solange Ramírez, Johannes Reetz,

Jaehyon Rhee, Silvia Rossi, Sean Ryan, Steve Shectman, Ian Thompson, and Franz-Josef Zickgraf. I thank Tim Beers for valuable comments, and proof-reading. This work is supported by Deutsche Forschungsgemeinschaft under grant Re 353/44-1, and was partly carried out during a Marie Curie Fellowship of the European Commission (contract number HPMF-CT-2002-01437) at Uppsala Astronomical Observatory, and a Linkage International Fellowship of the Australian Research Council (Project ID LX02114310) at Mt. Stromlo Observatory.

References

Aoki, W., Norris, J.E., Ryan, S.G., Beers, T.C., Ando, H. 2000, ApJ 536, L97

Aoki, W., Norris, J.E., Ryan, S.G., Beers, T.C., Ando, H. 2002a, ApJ 567, 1166

Aoki, W., Ryan, S.G., Norris, J.E., Beers, T.C., Ando, H., Tsangarides, S. 2002b, ApJ 580, 1149

Beers, T.C. 1999, in The Third Stromlo Symposium: The Galactic Halo, ed. B. Gibson, T. Axelrod, & M. Putman, ASP Conf. Ser. 165, 202–212

Beers, T.C. 2000, in The First Stars, Proceedings of the second MPA/ESO workshop, ed. A. Weiss, T. Abel, & V. Hill (Heidelberg: Springer), astro-ph/9911171

Beers, T.C., Preston, G.W., Shectman, S.A. 1985, AJ 90, 2089

Beers, T.C., Preston, G.W., Shectman, S.A. 1992, AJ 1987

Beers, T.C., Rossi, S., Norris, J.E., Ryan, S. G., Shefler, T. 1999, AJ 117, 981

Bonifacio, P., Molaro, P., Beers, T., Vladilo, G. 1998, A&A 332, 672

Carretta, E., Gratton, R., Cohen, J., Beers, T., Christlieb, N. 2002, AJ 124, 481

Cayrel, R. 1996, A&A Rev., 7, 217

Cayrel, R., Hill, V., Beers, T., Barbuy, B., Spite, M., Spite, F., Plez, B., Andersen, J., Bonifacio, P., Francois, P., Molaro, P., Nordström, B., Primas, F. 2001, Nature 409, 691

Christlieb, N., Beers, T. C. 2000, in Subaru HDS Workshop on stars and galaxies: Decipherment of cosmic history with spectroscopy, ed. M. Takada-Hidai & H. Ando (Tokyo: National Astronomical Observatory), 255–273, astro-ph/0001378

Christlieb, N., Green, P., Wisotzki, L., Reimers, D. 2001a, A&A 375, 366

Christlieb, N., Wisotzki, L., Reimers, D., Homeier, D., Koester, D., Heber, U. 2001b, A&A 366, 898

Christlieb, N., Bessell, M., Beers, T., Gustafsson, B., Korn, A., Barklem, P., Karlsson, T., Mizuno-Wiedner, M., Rossi, S. 2002a, Nature 419, 904

Christlieb, N., Wisotzki, L., Graßhoff, G. 2002b, A&A 391, 397

Cohen, J., Christlieb, N., Beers, T., Gratton, R., Carretta, E. 2002, AJ 124, 26

Cohen, J., Christlieb, N., Qian, Y., Wasserburg, G. 2003, ApJ, in press, astro-ph/0301460

Cowan, J.J., McWilliam, A., Sneden, C., Burris, D.L. 1997, ApJ 480, 246

Cowan, J.J., Pfeiffer, B., Kratz, K.-L., Thielemann, F.-K., Sneden, C., Burles, S., Tytler, D., Beers, T. 1999, ApJ 521, 194

Fuhrmann, K. 1998, A&A 338, 161

Goriely, S., Siess, L. 2001, A&A 378, L25

Hill, V., Barbuy, B., Spite, M., Spite, F., Plez, R.C.B., Beers, T., Nordström, B., Nissen, P. 2000, A&A 353, 557

Hill, V., Plez, B., Cayrel, R., Nordström, T.B.B., Andersen, J., Spite, M., Spite, F., Barbuy, B., Bonifacio, P., Depagne, E., François, P., Primas, F. 2002, A&A 387, 560

Karlsson, T., Gustafsson, B. 2001, A&A 379, 461

Lucatello, S., Gratton, R., Carretta, E., Cohen, J., Christlieb, N., Beers, T., Ramírez, S. 2003, AJ 125, 875

McWilliam, A., Preston, G., Sneden, C., Searle, L. 1995, AJ 109, 2757

Norris, J.E., Ryan, S.G., Beers, T.C. 1997a, ApJ 488, 350

Norris, J.E., Ryan, S.G., Beers, T.C. 1997b, ApJ 489, L169

Norris, J., Ryan, S., Beers, T. 2001, ApJ 561, 1034

Norris, J., Ryan, S., Beers, T., Aoki, W., Ando, H. 2002, ApJ 569, L107

Preston, G. 2000, PASP 112, 141

Preston, G., Sneden, C. 2001, AJ 122, 1545

Rhee, J. 2000, PhD thesis, Michigan State University

Rossi, S., Beers, T.C., Sneden, C. 1999, in The Third Stromlo Symposium: The Galactic Halo, ed. B. Gibson, T. Axelrod, & M. Putman, ASP Conf. Ser. 165, 264–268)

Ryan, S., Norris, J., Beers, T. 1996, ApJ 471, 254

Ryan, S.G., Norris, J.E., Beers, T.C. 1999, ApJ 523, 654

Schatz, H., Toenjes, R., Pfeiffer, B., Beers, T., Cowan, J., Hill, V., Kratz, K. 2002, ApJ 579, 626

Shigeyama, T., Tsujimoto, T. 1998, ApJ 507, L135

Skrutskie, M., Schneider, S., Stiening, R., Strom, S., Weinberg, M., Beichman, C., Chester, T., Cutri, R., Lonsdale, C., Elias, J., Elston, R., Capps, R., Carpenter, J., Huchra, J., Liebert, J., Monet, D., Price, S., Seitzer, P. 1997, in The Impact of Large Scale Near-IR Sky Surveys, ed. F. Garzón, N. Epchtein, A. Omont, W. Burton, & P. Persi (Dordrecht: Kluwer), 25–32

Sneden, C., McWilliam, A., Preston, G.W., Cowan, J.J., Burris, D.L., Amorsky, B.J. 1996, ApJ 467, 819

Sneden, C., Cowan, J., Ivans, I., Fuller, G., Burles, S., Beers, T., Lawler, J. 2000, ApJ 533, L139

Tsujimoto, T., Shigeyama, T., Yoshii, Y. 2000, ApJ 531, L33

Tsujimoto, T., Shigeyama, T. 2003, ApJ 584, L87

Umeda, H., Nomoto, K. 2003, Nature, in press, astro-ph/0301315

Van Eck, S., Goriely, S., Jorissen, A., Plez, B. 2001, Nature 412, 793

Wanajo, S., Itoh, N., Ishimaru, Y., Nozawa, S., Beers, T. 2002, ApJ 577, 853

Watson, F. 1998, AAO Newsletter, 11

Wisotzki, L., Christlieb, N., Bade, N., Beckmann, V., Köhler, T., Vanelle, C., Reimers, D. 2000, A&A 358, 77

A Tale of Bars and Starbursts
Dense Gas in the Central Regions of Galaxies

Susanne Hüttemeister

Astronomisches Institut
Ruhr-Universität Bochum
44870 Bochum, Germany
huette@astro.rub.de, www.astro.rub.de/huette/

Abstract

The dense, i. e. molecular component of the interstellar medium, the site of star formation, is often strongly concentrated in the central region of galaxies, feeding nuclear starburst activity. We first examine how the gas is funneled toward the central region by means of gas flows in a bar. Within the bar, the gas is subjected to tidal forces and characteristic shocks, which lead to changes in its properties. These can be traced with diagnostic line ratios, which are examined for a sample of gas rich barred galaxies. Not all gas-filled bars turn out to be equal, but a diffuse, unbound gas component plays a role in many cases. This component is shown to offset standard determinations of molecular masses, and is also important in the starburst centers of galaxies. These central regions are examined more closely in the second part of this contribution. The fraction of gas residing in dense cores is shown to differ to a surprising degree in starburst nuclei. Specific molecular tracers, sensitive to certain types of environment like UV-rich PDR regions are beginning to prove their usefulness as probes into the heart of deeply obscured starburst or composite starburst/AGN nuclei, thus allowing insight into the evolutionary stages of the starburst process.

1 The tale's first part: Bars

In the first part of this contribution, the influence of bars on galaxy evolution, achieved chiefly through gas flowing along the bar and accumulating in the center, will be briefly summarized. We will also describe the general principles of using molecular line ratios as diagnostics of global gas properties and encounter examples of bar environments studied in several molecular transitions.

1.1 Bars and galaxy evolution

Bars are regarded as an important transport mechanism of material toward the central regions of galaxies, fuelling nuclear starbursts and driving galaxy evolution through the concentration of mass close to the nucleus (e. g. Sakamoto et al. 1999, Combes et al. 1990 and references therein). Bars are a very frequent ingredient in the structure of galaxies: $\sim 1/3$ of all spiral galaxies in the local universe are strongly barred, corresponding to Hubble types SB. Another third of all local spirals, including the Milky Way, show a weak bar, which is often confined to the inner few kpc of the galaxy or less and may be identified more clearly in the Near Infrared than at optical wavelengths. With more detailed observations of a larger sample of galaxies, the number of galaxies where a bar in the nuclear region could be found has risen steadily over the last years.

Also, many cases of nuclear bars beeing nested within larger bars, often as dynamically decoupled structures, have been identified (e. g., Laine et al. 2002).

Bars are very easily triggered in interactions between galaxies, but are also able to grow in galactic disks as spontaneous m = 2 bar instabilities in density wave theory (e. g. Athanassoula 1984). Still, observations and numerical simulations have suggested for a long time that bars are more frequent in interacting galaxies. They form readily, e. g. through temporary resonances, when two gas rich disk galaxies interact, both in the individual galaxies before an actual merging event takes place and after the two disks have merged (see simulations by Barnes 1992). It has been argued that bars can self-destruct when the central mass concentration becomes large (e. g. Friedli and Benz 1993, Hasan et al. 1993, Norman et al. 1996, Das et al. 2003). On the other hand, in a number of simulations bars in disks seem to be very stable structures that, once formed, may persist for billions of years (Athanassoula 2002).

Despite this controversy on bar lifetimes, there is wide agreement about their role as a driving force of galaxy evolution: They efficiently transfer angular momentum outward and allow gas to flow into the central region along characteristic orbits. The gas flows and orbits typical for bars are summarized in Fig. 1. Orbits elongated along the major bar axis (x_1-orbits) sustain the bar. The oriention of the orbits changes by $90°$ at the Inner Lindblad Resonance (ILR), where

$$\Omega_P = \Omega - \kappa/2$$

with Ω_P: bar pattern speed, Ω: angular velocity of circular rotation and κ: epicyclic frequency. The resulting x_2-orbits are slightly elongated perpendicular to the bar. Thus, if the ILR is found at a large distance from the center, which may be possible if the central mass concentration is high, and x_2-orbits dominate, the bar self-destructs. Since the orbits change orientation again at corotation (CR, $\Omega_P = \Omega$), no self-consistent bar can extent further out into the disk. There may, however, be bars that end well before corotation, due to disk scale lengths that are smaller than the corotation radius (Combes and Elmegreen 1993).

In hydrodynamic (e. g. Athanassoula 1992) models of the gas flow, the close relation between density maxima, characteristic bar shocks and dust lanes observed in many strongly barred galaxies is apparent. Due to the viscosity of the gas, the change in orbit orientation becomes more gradual, leading to the curved dust lanes observed. Orbital energy is dissipated both in the shock regions and due to tidal forces acting on the clouds approaching the center.

To concentrate mass in the central region and thus affect galaxy evolution, obviously the gas flow inward through the bar has to be active. However, long bars that

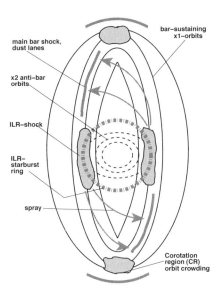

Figure 1: Gas flows and characteristic orbits in a two-dimensionsional, i.e. thin, bar. The two major orbit types are indicated along with regions of gas accumulation close to the ILR and the CR and the location of the bar shocks.

are filled with dense gas that can be traced in CO, the molecule most commonly used to map the distribution of the molecular phase of the Interstellar Medium, are rare and thus likely transient.

Among the few specimens known are NGC 1530 (Reynaud and Downes 1998), M100 (Sempere and García-Burillo 1997), UGC 2855 (Hüttemeister et al. 1998), NGC 7479 (Hüttemeister et al. 2000), NGC 2903 and NGC 3627 (Regan et al. 1999). These bars differ significantly in their properties, i.e. velocity field, linewidth and derived shock structure. This is somewhat surprising, since the gas should in all cases be responding to a strong bar potential in a similar way. The dynamics of this response should then have consequences for the properties of the gas in the bar and the nucleus. For example, a diffuse component unbound from clouds should be produced (e.g. Das and Jog 1995). The evolutionary state of the bar and the degree of central mass concentration may be instrumental in regulating the conditions of the gas in the bar (see models by Athanassoula 1992). More specifically, the gas is funnelled from outer x_1-orbits to inner x_2-orbits as the bar evolves in time (see e.g. the simulations by Friedli and Benz 1993).

1.2 Diagnosing dense gas properties: Molecular lines and their ratios

To derive the physical conditions, most importantly the density of molecular hydrogen $n(H_2)$ and the kinetic temperature T_{kin}, a diagnostic tool is needed.

Often, the lowest rotational transition of the ^{12}CO molecule ($J = 1\text{–}0$ at 115 GHz) is the only molecular line that is observed in an external galaxy. If observations at high resolution are available, the kinematics of the dense gas can be derived from these observations, but almost nothing is revealed about the physical conditions of the gas.

The ^{12}CO molecule is the most abundant tracer of molecular gas, with a typical ^{12}CO/H$_2$ abundance ratio of $\sim 10^{-4}$. H$_2$ itself cannot be observed directly in cool molecular clouds, since as a symmetric molecule it has no permanent dipole moment. It does have vibrational transitions, but those are excited at very high temperatures (for molecular clouds, several 100 K) only. Thus, the ^{12}CO(1–0) rotational transition serves as an ubiquitous, all purpose tracer of molecular gas, easily excited at H$_2$ volume densities of a few 100 cm^{-3}. It is usually assumed to be optically thick, a reasonable assumption in most cases, but not always entirely valid, as will become apparent later.

The investigation of the physical properties of the molecular gas requires more information than ^{12}CO(1–0) alone can provide. If several CO lines have been observed (arising from higher J levels of ^{12}CO as well as from the rarer isotopomers ^{13}CO or even C^{18}O), it becomes possible to solve a simplified version of the radiative transfer problem, and obtain estimates of n(H$_2$) and T_{kin}.

A classical way to simplify the global problem of radiative transfer, which generally requires detailed knowledge of the velocity structure of the entire molecular cloud, is provided by the *Large Velocity Gradient* (LVG) approximation. Here, it is assumed that the motion within the cloud is dominated by a large, systematic velocity gradient. Once photons have left the region where they originated, they can freely escape from the cloud. Thus, a global problem becomes a local one that can be solved easily. The LVG approximation has first been developed by Sobolev (1960) for expanding stellar atmospheres. Therefore, it is sometimes also called the Sobolev approximation. A model calculation using this approximation iteratively solves a system of rate equations for a given H$_2$ density, relative molecular abundance, velocity gradient, continuum background temperature and kinetic gas temperature (see Henkel 1980, de Jong et al. 1975, Scoville and Solomon 1974 for details of the modelling process). The results of an LVG calculation usually agree with those of another variant of localizing the problem, by using *Maximum Escape Probability* models.

Because of its simplicity, the LVG approximation is widely used. However, we have to keep in mind that – while it is certainly superior to an analytic or Local Thermal Equilibrium (LTE) treatment – the assumption of a large velocity gradient is not likely to be realized in a real cloud where turbulence always plays a role. In an extragalactic environment, where we have an ensemble of clouds or clumps within the telescope beam, we also have an unknown, clumpy cloud structure with relative movement of the subcomponents, and possibly systematic rotation. In this sense, the LVG approximation is 'unphysical'. However, without the knowledge of the 3-D velocity structure of the cloud complex, other approximations may be just as unrealistic. Also, it has been shown (e.g. White 1975, Linke and Goldsmith 1980) that LVG calculations and the opposite extreme, microturbulence, agree in the resulting molecular densities to a factor of three or (often) better.

Another, potentially more serious, limitation of the LVG approximation is the fact that it models just one gas component, i.e. one set of gas density and temperature, while a cloud will usually have density and temperature structure. Still, since

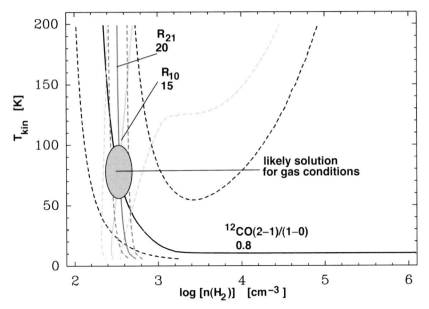

Figure 2: An example for a non-LTE radiative transfer model. Dark gray lines represent $^{12}CO/^{13}CO(2-1) = \mathcal{R}_{21}$ line intensity ratios, light gray lines give $^{12}CO/^{13}CO(1-0) = \mathcal{R}_{10}$ ratios, and black lines show $^{12}CO(2-1)/^{12}CO(1-0)$ line ratios. Dashed lines represent 1σ errors, calculated from observed line intensities. In this example, the observations yield $\mathcal{R}_{21} = 15$, $\mathcal{R}_{10} = 20$ and $^{12}CO(2-1)/^{12}CO(1-0) = 0.8$. The shaded area shows the region where the modelled line ratios come closest – ideally, they would intersect in one point – and thus indicates the most likely solution for the molecular hydrogen density and the kinetic temperature of the dominant gas component.

the method is transparent in its assumptions, and yields useful insights when the properties of the dominant gas component are to be estimated, it is very useful in analysing gas properties in external galaxies, especially if only a few lines have been measured, which is almost always the case.

Fig. 2 shows an example of an LVG model calculation, based on observational data. The lines represent observed 'diagnostic' line ratios, i. e. the line intensity ratios of $^{12}CO/^{13}CO(1-0) = \mathcal{R}_{10}$, $^{12}CO/^{13}CO(2-1) = \mathcal{R}_{21}$ and $^{12}CO(2-1)/^{12}CO(1-0)$. If the lines intersect in one point, this gives the most likely solution for gas density and kinetic temperature. Often, the lines do not quite intersect, but the most likely solution can still be found, e. g. by applying a χ^2-test. Line ratios are used instead of line intensities since, assuming that the lines originate in the same volume of gas, the beam filling factor, typically a small multiplicative and unknown quantity in extragalactic observations, does not affect the results. As long as low J CO transitions are considered, the assumption of the emission arising from the same volume is reasonable. This is, of course, no longer the case when high density tracing molecules like HCN, HNC or CN are also taken into account.

The obtain a solution, clearly at least three lines have to be observed. For practical reasons (line intensity, transparency of the Earth's athmosphere), these are typically ^{12}CO(1–0), ^{12}CO(2–1) and ^{13}CO(1–0). Clearly, more information is gained if more lines can be included, e. g. ^{13}CO(2–1), ^{12}CO(3–2) or C^{18}O(1–0). If four or more line have been observed, it becomes possible to go beyond one-component models to more complex scenarios involving several gas components. It is then also possible to not assume the relative isotopic abundances and velocity gradients, but solve for these parameters, too, in a multi-dimensional iterative process (see Weiß et al. 2001 for an example for the central region of the starburst galaxy M82).

Studying the models and relating them to know environments, it becomes clear that the the value of a diagnostic line ratio, e. g. $\mathcal{R}_{10}$, alone can already be an indication of the nature of the dominating gas component. Low values of $\mathcal{R}_{10} \approx$ 4–6 are typical for spiral arms in galactic disks (Polk et al. 1988). In the starburst centers of galaxies, $\mathcal{R}_{10}$ is elevated to typical values of 10–15 (Aalto et al. 1995). Even higher values of > 15 are found in a few centers of luminous merging galaxies (Casoli et al. 1992, Aalto et al. 1991). Physically, high values of $\mathcal{R}_{10}$ go along with low global $n(H_2)$ densities of $\leq 10^3$ cm^{-3}. Alternatively, the ^{13}CO isotopomer, that, due its smaller abundance, is less able to shield itself against destructive UV radiation, may be selectively dissociated and thus physically removed from the gas, if the effective optical depth is low. This may be accomplished by a low (column) density are a high degree of turbulent motion, which may be expected in the environment of a starburst nucleus or a bar shock.

1.3 Examples of gas-filled bars

We have examined $\mathcal{R}_{10}$ as the most easily accessible diagnostic line ratio in a small sample of three galaxies with gas-filled bars. For two galaxies, high-resolution interferometric observations in ^{12}CO(1–0) are also available. Despite being of the same Hubble type, SBc, the dominant physical properties of the gas in the bar of the three galaxies differ significantly.

NGC 7479 (Fig. 3) is an SBc starburst galaxy with a LINER (or possibly Seyfert 2) nucleus. The asymmetric morphology of the galaxy characterised by a much more pronounced southern spiral arm has been related to a minor merger event (Laine and Heller 1999). The gas-filled bar has a length of $\sim$ 10 kpc and has been mapped several times in ^{12}CO(1–0) using interferometers (Quillen et al. 1995, Laine et al. 1999, and, most sensitively, Hüttemeister et al. 2000). The typical kinematic signature of a bar (shock), namely S-shaped isovelocity contours, is easily recognised in NGC 7479. In addition, a kinematic high-velocity component is found in the central region. This can be interpreted as a region where x_2-orbits dominate close to an ILR. Alternatively, it may also be due to a tilted rotating inner disk fed directly by mass infall along the bar.

Our study of $\mathcal{R}_{10}$ reveals strong variations within the bar and the starburst center of the galaxy: Within the central 1 kpc, $\mathcal{R}_{10}$ varies between 10 and 30, the global value for the central condensation is 22, i. e. somewhat higher than for typical starburst nuclei. Within the bar, we find variations between 4 and 40 within a few arcseconds (or a few 100 pc; 1" $\sim$ 155 pc at an assumed distance of 32 Mpc). It is striking that the low, disk-like values for $\mathcal{R}_{10}$ occur well away from the peak intensities in ^{12}CO, which in turn are well correlated with the dust lanes indicating the main bar shock.

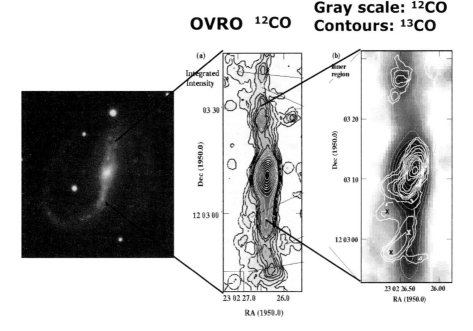

Figure 3: The gas-filled bar of NGC 7479. The optical image in the left panel was taken with HoLiCam at Hoher List Observatory in the V-band (courtesy M. Altmann). The middle panel shows the bar in the ^{12}CO(1–0) transition, the right panel depicts the inner part of the bar in ^{12}CO(1–0) (grayscale and thin gray contours) and ^{13}CO(1–0) (white contours, adapted from Hüttemeister et al. 2000, based on OVRO data).

Using the result from the models that high values of $\mathcal{R}_{10}$ indicate low gas densities, we can infer that the shock region is dominated by low density molecular gas that is probably 'diffuse' in the sense of not being bound to clouds. This gas is expected to have an optical depth of (only) τ 1–2 in the ^{12}CO(1–0) transition, a property that has consequences for the way this transition traces the total gas mass (see section 1.4). It is unlikely to be involved in current star formation activity.

Star formation is more likely encountered in the region of low $\mathcal{R}_{10}$. Here, downstream from the bar shock, the conditions are more quiescent and thus favorable to star formation. The gas is in the form of bound clouds.

The high global central value of $\mathcal{R}_{10}$ indicates that even in the center, where active star formation is known to go on, and thus the dense gas fraction is expected to be much higher than in the bar, diffuse gas may play a role, e.g. as an intercloud medium. This is aspect we will return to in Part 2. Here, we just point out that diffuse gas may be important in starburst nuclei, too. However, the influence of changing kinetic temperatures must not be neglected: In a high $T_{\rm kin}$ environment, the higher J levels are more strongly populated, resulting in relatively stronger emission from especially the ^{13}CO(2–1) transition. Indeed, the central $\mathcal{R}_{21}$ (measured with a single dish telescope a low resolution) is 13, significantly lower than $\mathcal{R}_{10}$.

UGC 2855 is another SBc spiral, ~ 20 Mpc distant and tidally interacting with UGC 2866, at a projected distance of 60 kpc. Its center is characterised by a weak starburst, and it has a gas-filled bar of ~ 8 kpc length. The high resolution ^{12}CO map we took at OVRO (Hüttemeister et al. 1999) shows solid body rotation all along the bar, with no indication of a bar shock and only slight perturbations, indicative of streaming motions along the bar. In the center, a high velocity feature, related to either a clumpy nuclear disk or an ILR ring, is found.

$\mathcal{R}_{10}$ in the center and along the bar is almost constant at 5–10, again indicating quiescent conditions. $\mathcal{R}_{21}$ in the center has been determined as ~ 14, also only slightly elevated. Thus, the conditions encountered in this galaxy are very different from what is found in NGC 7479. The diagnostic $\mathcal{R}_{10}$ line ratio can serve as an indicator of a different evolutionary stage of the bar in UGC 2855: The bar shock has not yet developed, possibly because the central mass concentration is still low. Models by Athanassoula (1992) indeed indicate that bars with a low central mass concentration may lack a distinctive shock signature. Thus, UGC 2855 may be a galaxy in a pre-starburst state, with an evolutionary very young active bar.

Our final example, *NGC 4123* (Fig. 4), is also an SBc galaxy and member of a wide pair. We only have single dish data at low resolution ($\sim 45''$ in the (1–0) lines, $\sim 22''$ in the (2–1) transitions) for this galaxy, which hosts a central starburst of moderate strength. $\mathcal{R}_{10}$ in the center (which in this case includes the inner part of the bar) is high at 26, again indicating an important role of the diffuse gas component in this region. $\mathcal{R}_{21}$ is determined for a smaller region, dominated by the starburst center. It is 14, suggesting that in the starburst center the kinetic temperature may be elevated, while diffuse gas still plays a role.

Remarkably, at the bar end, $\mathcal{R}_{10}$ and $\mathcal{R}_{21}$ drop to ~ 4–5. This is a strong argument against one possible explanation of changing $\mathcal{R}$, namely an inflow of gas that is underabundant in ^{13}CO. This scenario is very unlikely in a bar environment, which is expected to be well mixed, in any case, but the example of NGC 4123 clearly shows that the physical properties of the gas (possibly combined with selective photodissociation) dominate the changes in $\mathcal{R}$.

1.4 Diffuse gas and mass determinations

We have seen in the examples presented above that very high values of the diagnostic line ratio $\mathcal{R}_{10}$ that have so far only been found in a few extreme starburst mergers are also characteristic for a number of gas-filled bars. While for starburst mergers the possibility of abundance differences as an explanation of the faint ^{13}CO emission (as proposed by Casoli et al. 1992) cannot be excluded (even though we think it unlikely and suggest temperature effects as the main reason), in a bar environment it is clear that the reason for changing and high values of $\mathcal{R}_{10}$ is excitation, most importantly the relative amount of diffuse, unbound gas of low density.

Differences in kinetic temperature can also influence $\mathcal{R}$. To distinguish density and temperature effects, both $\mathcal{R}_{10}$ and $\mathcal{R}_{21}$ have to be known. If diffuse gas dominates, both values are expected to be high. If high T_{kin} contributes significantly, $\mathcal{R}_{21}$ should be lower than $\mathcal{R}_{10}$. This is an effect that we have found in the starburst

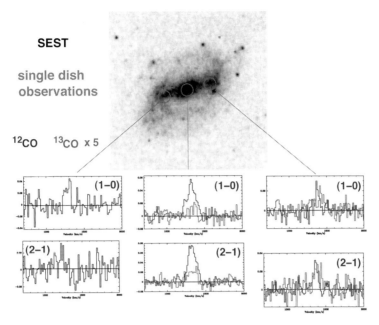

Figure 4: The gas-filled bar of NGC 4123. The optical image in the upper panel is from the Digital Sky Survey. Only single-dish, i. e. low resolution, data exist for this galaxy. The lower panels show overlays of the ^{12}CO(1–0) and ^{12}CO(2–1) lines (shown in black) and the corresponding ^{13}CO transitions (grey lines, multiplied by 5 for clarity), taken with the SEST 15 m telescope at La Silla.

nucleus of NGC 7479 and also some luminous mergers. Within a bar, especially close to the bar shock, the existence of a widespread component of diffuse, unbound, probably warm molecular gas is the most likely explanation of high $\mathcal{R}_{10}$.

Of course, the molecular medium within a bar that is 10 kpc long is complex – describing it as just one (diffuse) phase does not do it justice. Certainly, a denser second phase exists, and with high quality observations like the ones we present for NGC 7479, we can identify and locate it in relation to the bar shock. However, globally the diffuse gas is likely to dominate in many bars with well-developed shocks.

As pointed out briefly above, the ^{12}CO(1–0) transition in the diffuse and turbulent gas component is expected to have an optical depth of $\tau \approx 1$–2. This means that strong emission results, but the assumption of high optical depth, often taken for granted for ^{12}CO transitions, cannot be made. The most important consequence of this concerns the estimate of molecular gas masses.

Very often, the column density of molecular gas is simply taken to be proportional to the integrated intensity of the ^{12}CO(1–0) transition: $I(\mathrm{CO}) \cdot X_{\mathrm{CO}} = \mathcal{N}(\mathrm{H}_2)$. The 'standard' conversion factor X_{CO} is usually assumed to be $X_{\mathrm{CO}}(\mathrm{SCF}) = 1.8$–$2.3 \cdot 10^{20}$ cm^{-2}(K km s$^{-1})^{-1}$ (Strong et al. 1988, Dame et al. 2001). While this conversion factor may indeed be valid in galactic disks, it is *significantly too high* as soon as a diffuse gas component becomes important. Then, the *molecular mass is overestimated* by up to an order of magnitude.

The assumptions made when deriving the standard conversion factor break down when the emission is dominated by diffuse gas: ^{12}CO does not have a high optical depth anymore, and the gas is not virialized and bound in clouds. Both conditions need to be fulfilled if the direct proportionality between I(CO) and $\mathcal{N}$(H$_2$) is to hold.

Environments that give rise to a diffuse gas component are both bars and galactic (starburst) nuclei: This component becomes important whenever shocks, high turbulence or high pressure start to dominate. Consequently, many recent investigations have found that the standard conversion factor overestimates the gas mass in central regions of galaxies significantly, by factors between 3 and 10. This seems to be true in galaxy centers of almost all types, from our own galactic center (Dahmen et al. 1998) to galaxies hosting a strong starburst, as indicated by luminous or even ultraluminous Far Infrared emission (e. g. Yao et al. 2003, Downes and Solomon 1998). Since CO emission is usually strongest in galaxy centers, and most extragalactic CO surveys refer to these regions, a large percentage of CO-based molecular gas masses published in the past are highly suspect.

Bars, especially bar shocks, may be regions where diffuse gas dominates low J ^{12}CO emission even more clearly than in starburst centers. It is possible that there are regions along bars where dense cores are lacking entirely.

However, there are also unsolved questions related to both the gas within the bar and the fate of the material reaching the center. Some bars show significant star formation, while others do not. This may be related to the bar strength and the exact way shocks act on the gas, inhibiting or facilitating star formation. In any case, it is undisputed that a temporary or long-lived bar transports significant amounts (2–4 M$_\odot$ in the case of NGC 7479) of largely diffuse gas to at least its ILR region, typically $\sim$ 1 kpc away from the center. The details of the degree and location of the compression the gas is subjected to once it reaches the central region and the strength of the starburst that may result are less well understood, though starburst rings close to the ILR seem to be common features.

Independent of the details, though, we can be state that the gas that has been funneled into the central region by the bar provides the raw material a nuclear starburst feeds on. Thus, we will now examine the properties of starbursts of different strength, to gain insight into the fate of the dense gas once it has reached the central region of a galaxy.

2 The tale's second part: Starbursts

Starbursts are defined as objects having star formation rates that are so high that the activity can be sustained for a very limited time only, typically a few 10^7 years, before the fuel has been consumed. The star formation rate and thus strength of the starburst is often related to the Far Infrared (FIR) luminosity of the objects, since the starburst regions are deeply embedded in clouds of dense gas and dust, resulting in the optical and UV radiation of the newly born stars being reradiated at FIR wavelengths. Moderate starbursts like the nearby M 82 or the three barred galaxies we presented in the section 1.3 have FIR luminosities of a few 10^{10} L$_\odot$. A starburst

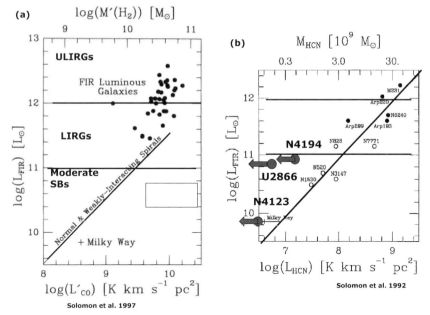

Figure 5: The correlation of molecular gas and FIR luminosity. Panel (a) depicts the relation between ^{12}CO luminosity or inferred molecular gas mass (probably somewhat overestimated using standard conversion factors) for normal spiral galaxies, moderate, luminous and ultraluminous starbursts (adapted from Solomon et al. 1997). Panel (b) shows the relation between $L_{\rm FIR}$ and the luminosity (or mass) of the all purpose high density tracer HCN (adapted from Solomon et al. 1992). Three galaxies defying the trend have been added to the plot.

galaxy is classified as a Luminous IR galaxy (LIRG) if its FIR luminosity exceeds 10^{11} L$_\odot$. The most extreme cases are Ultraluminous IR galaxies (ULIRGs), with $L_{\rm FIR} \geq 10^{12}$L$_\odot$. These objects, rivalling the brightness of quasars, are almost always the products of galaxy mergers, which are characterized by very efficient gas funneling (probably by means of temporary bars) into the central region.

2.1 Molecular gas, star formation and FIR luminosities

It is an undisputed paradigm that star formation takes place in the dense cores of molecular clouds. This naturally implies that starburst galaxies have a larger amount of molecular gas, and possibly also a larger fraction of *dense* molecular gas, i.e. gas with densities exceeding $n({\rm H}_2) \sim 10^{4-5}$ cm^{-3}.

The correlation between FIR luminosity and CO luminosity is indeed well-known and reasonably tight for normal spirals and moderate starburst galaxies (e.g. Solomon and Sage 1988, see also Fig. 5 (a)). However, the $L_{\rm FIR}$-$L_{\rm CO}$ relation changes for LIRGs and ULIRGs (Solomon et al. 1997, confirmed by Yao et al. 2003 for a large sample of IR-luminous galaxies, Fig. 5 (a)). (U)LIRGs are overluminous in FIR compared to the CO luminosity and (even more so, taken the changing conversion factor into account) the molecular mass. Still, the amount of molecular gas in the center of a LIRG or ULIRG is still very high at up to 10^{10} M$_\odot$.

The systematic change of the L_{FIR}/L_{CO} ratio by about an order of magnitude (albeit with a scatter that also reaches about a factor of 10) suggests differences in the structure of the ISM between normal galactic nuclei, moderate starbursts and (U)LIRGs (see below). Luminous starbursts should also have a higher star formation rate (SFR) per unit molecular mass. This may be explained if the fraction of dense ($n(H_2) \geq 10^{4-5}$ cm^{-3}) gas is increased in luminous starbursts, maybe due to the high pressure environment and radiative shock compression of the diffuse molecular medium when it enters the extreme environment of the starburst nucleus (Jog and Das 1992).

2.2 The fraction of dense gas

This dense component of the molecular ISM is no longer well traced by CO, once ^{12}CO emission from gas of lower density but higher filling factor may easily overwhelm the emission originating in the dense core component. Thus, other molecular tracers that are only (collisionally) excited at a gas density of $\sim 10^{4-5}$ cm^{-3} are needed. The HCN molecule has become an 'all-purpose' tracer of this component, since its $J = 1$–0 line is reasonably strong and very accessible for mm telescopes. Typically, the CO/HCN intensity ratio is ~ 10 in a starburst nucleus.

Studies of HCN emission from starburst galaxies do indeed show a generally good correlation between L_{FIR} and L_{HCN} (Solomon et al. 1992, Fig. 5 (b)). This correlation has been claimed to be tighter than the one between L_{FIR} and L_{CO}, and it holds for starbursts of all strengths, thus confirming the notion that the dense gas fraction in a (U)LIRG is higher than in a starburst of low or moderate intensity or a normal spiral galaxy. The most simple picture of molecular gas in starbursts thus sees the dense gas fraction, as traced by HCN, as the dominating factor: It grows with FIR luminosity, while the star formation efficiency per unit of *dense* molecular mass remains approximately constant.

In this case, the L_{CO}/L_{HCN} is expected to depend only on L_{FIR}, falling as the FIR luminosity rises. Indeed, a trend in this direction is apparent from the data (e. g. Aalto et al. 2002, Aalto et al. 1995). However, there is a lot of scatter in this relation – a LIRG with, e. g. $L_{FIR} \sim 3 \cdot 10^{11}$ L$_\odot$ may have a L_{CO}/L_{HCN} ratio of 4 (NGC 34) or 16 (NGC 3256). In general, the range of L_{CO}/L_{HCN} is between ~ 2 and 40. Thus, there are more paramters describing starbursts than just the dense gas fraction, taken to be directly related to star formation rate and FIR luminosity.

In the regime of *moderate* starburst activity (L_{FIR} between 10^{10} L$_\odot$ and 10^{11} L$_\odot$), we have found a few galaxies that seem to defy the trend of an elevated dense gas fraction going along with a starburst entirely. In three galaxies (NGC 4123, one of our SBc sample objects, UGC 2866, the starburst companion of UGC 2855, and NGC 4194, a starburst merger of moderate FIR luminosity) we could not detect any HCN emission. These objects are plotted in Fig. 5. Especially NGC 4194 and UGC 2866 are very significantly underluminous in HCN. The two objects are unarguably starbursts – NGC 4194, also known as the Medusa, has a star formation rate of 40 M$_\odot$ yr^{-1} (Storchi-Bergmann et al. 1994), and UGC 2866 is one of only 20 galaxies rated as H II galaxies in the IRAS catalogue.

Of course, we do not claim to have found galaxies with active star formation that does not take place in dense molecular cores. We do suggest, however, that in these cases the star formation rate per unit mass of gas residing in dense cores may be quite different from what is found in the bulk of the starburst nuclei which follow the general trend. The gas may spend a shorter time in the dense phase, being converted into stars more efficiently. This may result in a SFR that is comparable to other starbursts, while the dense gas fraction is lower. After all, to see a large amount of gas at densities of $\sim 10^5$ cm^{-3}, the gas has to be stable at these densities for a considerable amount of time – it cannot instantanously collapse to form stars. Indeed, if we take the ratio $L_{\rm FIR}/L_{\rm CO}$ to be a measure of overall star formation effciency (SFE), as is commonly done, the SFEs in NGC 4194 and the most nearby ULIRG, Arp 220, are very similar.

We do not have a large enough sample of starburst galaxies lacking HCN emission yet to derive statistically sound statements about their typical luminosities and starburst properties. However, we do expect these objects not to be ULIRGs, but to have a more fragmented ISM. We suggest that the starburst in these objects takes place in a low(er) pressure environment, possibly connected to a shallower gravitational potential.

We have obtained interferometric ^{12}CO(1–0) maps of both UGC 2866 (Hüttemeister et al. 1999) and NGC 4194 (Aalto and Hüttemeister 2000). These studies show that the starbursts in both galaxies have something in common: Compared to bursts in ULIRGs, they encompass very extended regions. The molecular gas distributions in HCN-bright Ultraluminous IR Galaxies (ULIRGs) are very compact (less than 1 kpc, often only 500 pc in size). In contrast, both UGC 2866 and NGC 4194 show CO distributions, corresponding to starburst regions, that are several kpc in size.

In a scenario explaining a lower dense gas fraction combined with a high SFR by starburst extent, the decisive parameter is indeed the *pressure* in the starburst region. In the low pressure environment of an extended starburst, gas is not maintained at high densities but rather passes through a high density phase quickly on its way to becoming stars. The star formation out of dense cores may be very efficient since it may take place in an environment of reduced shear (solid-body like rotation, possibly again related to the presence of a barlike structure), where gravitational instabilities easily dominate and collapse into stars results. In NGC 4194, a starburst merger for which we favour a minor merger scenario, and too some degree also in UGC 2866, this dynamical signature is indeed found.

In contrast, a very compact starburst in a deep potential well results in high average gas densities. Densities high enough to excite HCN are required to just stabilize clouds against tidal shear. Lower density material will exist as diffuse, unbound, but still molecular intercloud gas – this is the component that dominates the ^{12}CO and offsets the molecular mass determination even in ULIRGs. Still, the dynamical and radiative environment in such a compact central molecular concentration may not be particularly favorable for star formation. The large FIR luminosity characterising ULIRGs of course points to significant star formation – but part of the FIR luminosity may also originate from an embedded AGN or densely packed stars from in aged starburst that heat the ISM. Large amounts of HCN-emitting dense gas may not at present be involved in active star formation, yielding the very high $L({\rm HCN})/L({\rm CO})$ ratios often observed in ULIRGs.

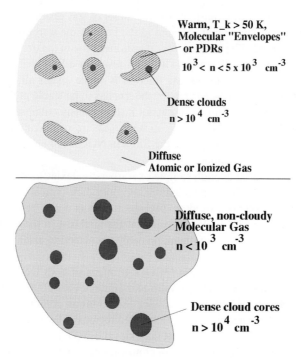

Figure 6: Scenarios for the structure of the ISM in extended, low pressure and possibly HCN-faint starbursts like NGC 4194 ('fried eggs', top) and the compact, very high pressure gas distribution typical for a ULIRG ('raisin roll', bottom).

Fig. 6 shows cartoons of the two cases. In the 'fried egg' scenario we propose for extended, low pressure starbursts with a lower dense gas fraction, a few dense cloud cores, which form stars efficiently, are surrounded by warm molecular envelopes of density $10^3 \leq n(H_2) \leq 5 \cdot 10^3$ cm^{-3} and $T_{kin} > 50$ K. The intercloud medium is neutral (H I) or ionized. Still, the ^{12}CO emission may be largely dominated by the low density, warm envelopes, which take the place of the diffuse, unbound gas in bars and ULIRGs. This is shown by a high global ratio $\mathcal{R}_{10}$ of 20 (with considerable variations) in NGC 4190, and a $\mathcal{R}_{21}$ which also seems to be high.

In a ULIRG, the global structure of the ISM in the starburst region may be envisioned as a 'raisin roll' (bottom part of Fig. 6). The intercloud medium is molecular, unbound and non-cloudy. Starburst cores which need to have densities exceeding 10^4 cm^{-3} just to be tidally stable, are suspended in this diffuse molecular 'matrix'. Thus, a ULIRG can have a high dense gas fraction, as traced by HCN, while the diffuse gas still plays a very important role.

2.3 The dense cores – approaching chemistry

We have seen in the preceeding section that the fraction of dense gas and the efficiency the gas in these dense cores is converted into stars may differ in a sample of starburst nuclei. The next question to address is whether at least the cores themselves are similar.

As we had to go beyond ^{12}CO to learn more about the properties of the molecular gas component as a whole, we now have to examine high density tracers other than HCN to assess the similarities and differences of the dense component which is the actual site of star formation. The multimolecule chemical studies that are needed are still in the beginning stage. Few interferometric, high resolution studies of LIRGs or ULIRGs in high density tracers other than HCN exist, since these observations are still at (or beyond) the sensitivity limit of present day instruments. There are, however, a few sensitive studies carried out with large single dish telescopes, studies that, of course, lack resolution, but can give first insights at least into whether significant differences are found, and possible interpretations.

We have carried out one such study of the high density tracers HNC and CN in a sample of 14 (U)LIRGs, all easily detected in HCN, using the SEST 15 m telescope at La Silla and the Onsala 20 m telescope (Aalto et al. 2002). HNC and CN are chosen since they are expected to trace dense gas with very different dominant properties: HNC, the isomer of HCN, should be transferred into HCN at high temperatures. Thus, the HCN/HNC intensity ratio should grow with rising $T_{\rm kin}$ (and also $n(H_2)$, Schilke et al. 1992). HNC therefore is often taken as a tracer of quiescent, cool dense gas. In sharp contrast, the radical CN traces ionisation fronts from H II regions and, in general, regions strongly affected by UV radiation, i. e. Photon Dominated Regions (PDRs). Thus, we might expect that strong HNC and CN emission are mutually exclusive, and that CN emission is enhanced in the environment of a strongly star forming ULIRG and, possibly even more so, in sources with contributions from AGNs, since CN is expected to thrive in an environment dominated by X-ray chemistry.

A survey of HNC in nearby spiral galaxies (Hüttemeister et al. 1995) already cast doubt onto the notion that HNC unambigously traces quiescent conditions. This is confirmed by our survey of LIRGs, where HNC is detected in 9 out of 11 objects. The HCN/HNC ratio varies between 1 and > 4, with no (anti)correlation to $L_{\rm FIR}$ or the presence of an AGN. HCN/HNC is high in ULIRGs like Arp 220 (1.6), and HNC is very bright in some sources with an AGN component (Mrk 231, HCN/HNC ~ 1, NGC 7130 (1.2) and NGC 7479 (1.2), see Fig. 7). In ULIRGs like Arp 220 or Mrk 231, there is no evidence for a significant cool, dense gas component. Thus, while HNC might emerge from the cold centers of cores in moderate starbursts (e. g. NGC 1808, which has HCN/HNC ~ 2), the 'standard' scenario is very unlikely in the galaxies which are brightest in HNC.

We suggest that bright HNC emission may emerge from a moderately dense $(n(H_2) \geq 10^5$ cm$^{-3})$ medium where the chemistry is dominated by ion-neutral instead of neutral-neutral reactions. Then, the temperature dependence of the HCN production is strongly suppressed. Other possibilities involve optical depth effects and the non-collisional excitation of both HNC and HCN by mid-infrared pumping, which requires optically thick mid-IR sources. These may indeed exist in the centers of the most luminous ULIRGs. Thus, it is possible to explain HNC emission from ULIRGs with a number of scenarios that may be distinguished as more detailed observations, also including higher J transitions, become available.

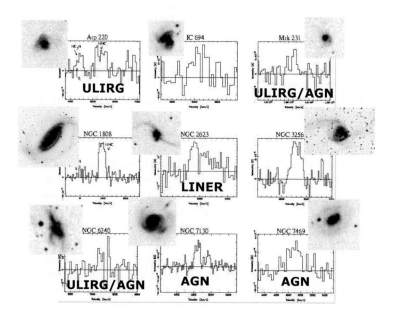

Figure 7: Results of a sensitive single dish survey of HNC in LIRGs and ULIRGs (data from Aalto et al. 2002). The images of the galaxies are taken from the DSS.

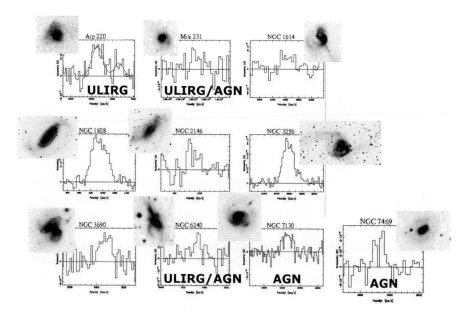

Figure 8: Results of a sensitive single dish survey of CN in LIRGs and ULIRGs (data from Aalto et al. 2002). As in Fig. 7, the images of the galaxies are from the DSS.

We were, however, quite surprised by our results from the CN survey (Fig. 8). CN emission is detected in 10 out of 14 galaxies, but, contrary to initial expectations, it is faint or undetactable in a number of bright sources with AGN contributions. In fact, 3 of our 4 non-detections host an AGN (Mrk 273, NGC 34 and NGC 2623). On the other hand, CN emission is as bright or brighter than HCN emission in 3 galaxies, all of which are (pure) starbursts.

Thus, it is becoming clear that not all dense gas fractions are equal. The relative intensities of dense gas tracers vary significantly in ways the details of which are ill understood so far.

In the case of CN, from our existing low resolution observations it is impossible to tell wether the CN deficiency in bright sources occurs in gas associated with the AGN or in the gas involved in the starburst activity further away from the nucleus. It is, however, possible to formulate scenarios that can be tested observationally:

1. The CN faintness occurs near the AGN: in this scenario, the bulk of the dense gas is situated near the AGN and not in a starburst ring or disk. In this case, the models of the chemical impact of an AGN on its surrounding gas need some modification – and a very useful AGN-starburst diagnostic tool might have been found. To disentangle this, the distribution of the CN emission in a starburst/AGN composite source at high resolution is required.

2. The lack of CN is related to the evolutionary stage of the starburst: in this scenario the dense gas is mostly associated with the starburst. It is known that CN is very bright in some starburst galaxies, like NGC 3690, NGC 1614 or Arp 220. Possibly the CN luminous objects are in a particularly UV intense phase of their development. If the starburst is in an early phase, the filling factor of UV radiation may be low compared to the amount of warm dense gas. It is of course also possible that the CN faint galaxies are not going through a starburst at all, and that the dense gas and dust is being heated by another mechanism (dynamical heating, heating by an existing compact stellar population). Imaging CN in galaxies where there are several regions of starburst activity of different age will help disentangle the starburst evolution scenario.

3 Outlook

We are on the verge of being able to trace both starburst evolution and the starburst-AGN connection that is found in an increasing number of composite sources by means of extinction-free molecular tracers that allow direct insights into the hearts of the active centers of infrared luminous galaxies.

It is possible to make significant progress on this as well as on the detailed properties and processes of gas in bars and the role of the diffuse gas phase with existing instruments, but we are pushing their limits in both sensitivity and resolution. Thus, the next generation of interferometers, but also large single dish mm telescopes like the LMT will be very important for further insights. CARMA, the combined BIMA/OVRO array will offer new opportunities in the northern hemisphere, but the ALMA array will mean a quantum jump in both sensitivity and resolution.

Of course, progress in modelling has to go along with observational progress, so that the emerging complex picture of starburst/AGN evolution as traced by molecular emission can be properly interpreted. Finally, a multiwavelength approach, also drawing on data from forthcoming instruments like SOFIA and Herschel, will complement information gained directly from the molecular component in galactic nuclei.

Acknowledgments

Special thanks are due to my friend and colleague Susanne Aalto (Onsala). Most of this work was done in close collaboration and in long in discussions with her. I am also grateful to Steve Curran, Mousumi Das, Antonis Polatidis, and William F. Wall who all contributed to parts of the work presented here.

References

Aalto. S., Black, J.H., Johansson, L.E.B., Booth, R.S. 1991, A&A 249, 323

Aalto, S., Booth, R.S., Black, J.H., Johansson, L.E.B. 1995, A&A 300, 369

Aalto, S., Hüttemeister, S. 2000, A&A 362, 42

Aalto, S., Polatidis, A.G., Hüttemeister, S., Curran, S.J. 2002, A&A 381, 783

Athanassoula, E. 1984, Phys. Rep. 114, 321

Athanassoula, E. 1992, MNRAS 259, 345

Athanassoula, E. 2002, ApJ 569, L83

Barnes, J.E. 1992, ApJ 393, 484

Casoli, F., Dupraz, C., Combes, F. 1992, A&A 264, 55

Combes, F., Elmegreen, B.G. 1993, A&A 271, 391

Combes, F., Dupraz, C., Gerin, M. 1990, In: Wielen, R. (ed.) Dynamics and Interactions of Galaxies. Springer, Berlin, p. 305

Dahmen, G., Hüttemeister, S., Wilson, T.L., Mauersberger, R. 1998, A&A 331, 959

Dame, T.M., Hartmann, D., Thaddeus, P. 2001, ApJ 547, 792

Das, M., Jog, C.J. 1995, ApJ 451, 167

Das, M., Teuben, P.J. Vogel, S.N, et al. 2003, ApJ 582, 190

de Jong, T., Chu, S.-I., Delgarnp, A. 1975, ApJ 199, 69

Downes, D., Solomon, P.M. 1998, ApJ 507, 615

Friedli, D., Benz, W. 1993, A&A, 268, 65

Hasan, H., Pfenniger, D., Norman, C. 1993, ApJ, 409, 91

Henkel, C. 1980, PhD thesis, Bonn

Hüttemeister, S., Henkel, C., Mauersberger, R., et al. 1995, A&A 295, 571

Hüttemeister, S., Aalto, S., Wall, W.F. 1998, A&A 346, 45

Hüttemeister, S., Aalto, Das, M., S., Wall, W.F. 2000, A&A 363, 93

Jog, C.J., Das, M. 1992, ApJ 400, 476

Laine, S., Heller, C.H. 1999, MNRAS 308, 557

Laine, S., Kenney, J.P.D., yun, M.S., Gottesman, S.T. 1999, ApJ 511, 709

Laine, S., Shloman, I., Knapen, J.H., Peletier, F. 2002, ApJ 567, 97

Linke, R.A., Goldsmith, P.F. 1980, ApJ 235, 437

Norman, C., Sellwood, J. A., Hasan, H. 1996, ApJ, 462, 114

Polk, K.S., Knapp, G.R., Stark, A.A., Wilson, R.W. 1988, ApJ 332, 432

Quillen, A.C., Frogel, J.A., Kenney, J.P.D., et al. 1995, ApJ 441, 549

Regan, M.W., Sheth, K., Vogel, S.N. 1999, ApJ 526, 97

Reynaud, D., Downes, D. 1998, A&A 337, 671

Sakamoto, K., Okumura, S.K., Ishizuki, S., Scoville, N.Z. 1999, ApJ 525, 691

Schilke, P., Walmsley, C,M. Pineau de Forets, G., et al. 1992, A&A 256, 595

Scoville, N.Z., Solomon, P.M. 1974, ApJ 187, L67

Sempere, M.J., Garcia-Burillo, S. 1997, A&A 325, 769

Sobolev, V.V. 1960, In: Moving Envelopes of Stars, Harvard University Press, Cambridge

Solomon, P.M., Sage, L.J. 1988, ApJ 334, 613 Solomon, P.M., Downes, D., Radford, S.J.E. 1992, ApJ 387, L55

Solomon, P.M., Downes, D., Radford, S.J.E., Barrett, J.W. 1997, ApJ 478, 144

Storchi-Bergmann, T., Calzetti, D., Kinney, A. 1994, ApJ 429, 572

Strong, A.W., Bloemen, J.B.G.M., Dame, T.M., et al. 1988, A&A 207, 15

Weiß, A., Neininger, N., Hüttemeister, S., Klein, U. 2001, A&A 345, 571

White, R.E. 1975, ApJ 211, 744

Yao, L., Seaquist, E.R., Kuno, N., Dunne, L. 2003, AJ in press (astro-ph/0301511)

Tip-AGB Mass-Loss on the Galactic Scale

Klaus-Peter Schröder

Astronomy Centre, University of Sussex (CPES)
Falmer, Brighton, BN1 9QJ, UK
kps@star.cpes.susx.ac.uk

Abstract

For a mature galaxy, dust-driven "superwinds" ($\dot{M} > 10^{-5}$ $M_\odot$/yr) of moderate-mass tip-AGB giants provide an important mechanism of stellar mass injection back into the ISM. $M_{\rm Bol}$ versus (in particular) J–K diagrams, extracted from contemporary IR-photometry surveys (DENIS, ISOGAL), reveal the most relevant tip-AGB stars as the J–K index depends strongly on the mass-loss. At the same time, realistic hydrodynamical computer-simulations of dust-driven winds in a cool, pulsating atmosphere are now available to match observed properties and quantify the mass-loss of individual tip-AGB objects.

The straight-forward object-by-object approach to derive the collective tip-AGB mass-loss, however, is likely to remain incomplete because the most significant contributors are, inevitably, the most obscured objects. Instead, we here present an alternative method, by computing a matching synthetic (complete) tip-AGB stellar sample: Synthetic stars of a range of masses and ages are created randomly and distributed on a grid of evolution tracks, using the most appropriate SFR and IMF. For the evolution of the C-rich majority of tip-AGB objects, we consider mass-loss by a relation derived from the Berlin wind models. The respective, time-averaged IR model colours are found to be a function of mass-loss and luminosity.

A synthetic ($M_{\rm Bol}$, J–K) diagram, modelled on the solar neighbourhood and based on 1.4 million present-day stars brighter than $M_V = 4.0$, reveals that about 50 % of the collective mass-loss (or 2.6×10^{-4} $M_\odot$/yr) is provided by just 10 extreme (C-rich) tip-AGB objects with J–K > 6. At the distance of the LMC, these objects would be well below the J-threshold of the DENIS survey. Another 10 giants, with J–K between 4.0 and 6.0 and mass-loss rates between 10^{-6} and 10^{-5} $M_\odot$/yr, provide a further 0.7×10^{-4} $M_\odot$/yr.

1 Introduction

For a young galaxy, the cycle of star formation, which consumes part of the ISM and leads to star death with mass injection back into the ISM, is mainly closed by the hot winds of WR stars and by spectacular SN explosions. With a mature galaxy, however, this dominance of massive stars is broken by the collective mass-loss of the vast number of evolved moderate mass stars (i.e., 1.3 $M_\odot$ < M_i < 8 $M_\odot$).

After several billion years, these stars had the time to develop into red giants which suffer mass-loss by a "cool wind" and, finally, end their AGB evolution by a brief "superwind" (Renzini 1981). This phase lasts only about 30 000 years, but produces a significant mass-loss with rates in excess of 10^{-5} $M_\odot$/yr. Such tip-AGB giants become strongly enshrouded in their dust-forming circumstellar envelopes, and IR observations are required to detect them with sufficient sensitivity.

The importance of the tip-AGB for the galactic mass-recycling process and the chemical evolution has long been fully recognized and is reflected in the large observational and theoretical progress made in the past two decades (see the substantial review by Habing 1996). Still, it is not possible to account for the dust-enshrouded tip-AGB stars of the solar neighbourhood in an absolute way by observations alone, see e. g. van Loon (2000). Rather, any mass-loss assessment requires the involvement of a radiative transfer code and depends on the model assumptions made (e. g., van Loon et al. 1999). Worse, the distances (and, therefore, the luminosities) are mostly mere estimates since there are no objects of this kind near enough to yield precise Hipparcos parallax measurements. Rather, the superwind-phase is so short-lived that a very large volume is required to find a reasonable number of these extreme tip-AGB stars. This has motivated the recent deep photometric IR survey DENIS, which includes large numbers of tip-AGB stars at the known distances of the Magellanic clouds (see Cioni et al. 2001). This gives very large samples which are complete within their space volume and IR luminosity thresholds. However, at the very tip-AGB, some of the most dust-enshrouded objects may remain undetectable in, at least, the J band and can be missed in an object-by-object assessment of the collective tip-AGB mass-loss.

Here, a complementary approach to derive the collective tip-AGB mass-loss is presented. We compute synthetic stellar samples of tip-AGB stars, which are based on the results of computer models and which are free of observational bias. Several steps are required in this process: (1) The mass-loss rates and IR colours for individual tip-AGB stars are derived from time-averages of hydrodynamic computer models for the dust-driven, pulsating winds (see below). (2) The mass-loss characteristics are derived from a large set of such models and then (3) combined with actual stellar models obtained from a fast and well calibrated evolution code (Schröder et al. 1999). (4) A grid of such evolution tracks with mass-loss is computed, reaching from the ZAMS (zero age main sequence) to the very tip-AGB and superwind phase. (5) Synthetic stars are randomly created and then placed on the grid according to their age and mass. The distribution in mass and time considers a matching choice of the initial mass function (IMF) and a star formation rate SFR(t).

For the first step, mass-loss rates are based on a larger number of computer models of a pulsating, dust-driven cool wind. These wind models, which compute the hydrodynamics, the dust-formation chemistry (in detail), radiative transport and wind acceleration in a selfconsistent way, have been developed over the past 10 years by the Sedlmayr group. Their complex treatment of dust-driven, carbon-rich winds has been described by a number of papers, e. g.: Fleischer et al. (1992), Sedlmayr (1994), Sedlmayr and Winters (1997), Winters et al. (2000).

From the combination of the model mass-loss with an evolution code (Schröder et al. 1999), the following picture emerges: Only stars with an initial mass M_i of more than 1.3 $M_\odot$, just after turning into carbon stars, become bright enough on the tip-AGB to exceed their Eddington-type, critical luminosity L_c for a stable, dust-driven wind. Once L_* exceeds L_c, nearing the tip-AGB, an increasingly strong, dust-driven superwind depletes the envelope mass of the giant as effective temperature $T_{\rm eff}$ and gravity g decrease further, until the core becomes nearly exposed. The timescale of the superwind, which arises in a consistent interaction with stellar evolution, as well as the total mass lost in the process, agree very well with the requirements of the formation of planetary nebulae, i.e., between 0.5 and more then 1 $M_\odot$ are lost within just 30 000 years. In addition, brief but large changes in the highly temperature-sensitive mass-loss (see below) are induced by thermal pulses – consistent with the pronounced density structure observed in CS shells on large scales (for pioneering observations of CS shell structure, we like to mention C. H. Townes and collaborators, e. g., Sutton et al. 1977).

The scope of this paper is now the step from individual to collective tip-AGB stellar mass-loss, thereby connecting stellar with galactic astrophysics, by means of synthetic stellar samples. A first attempt of this kind, carried out for the IMF and SFR of the solar neighbourhood, was undertaken by Schröder & Sedlmayr (2001), entirely based on a grid of stellar evolution tracks with mass-loss according to the Berlin wind models. Meanwhile, we have improved the mass-loss relation, and we also derive a relation between mass-loss and IR-colours. This enables us to present our synthetic tip-AGB stars in the very useful (J–K, $M_{\rm Bol}$) diagram.

2 The mass-loss and IR-colours of computer models for dust-driven, C-rich winds

2.1 The mass-loss relation for C-rich tip-AGB objects

Our description of the mass loss during the tip-AGB evolution phases is now based on a substantially updated set of self-consistent, dynamical wind models for dust-forming, carbon-rich, pulsating atmospheres (Table 1), which are all chosen to operate well above the critical, Eddinton-like luminosity L_c for the dust-driven winds as found by Schröder et al. (1999). Wachter et al. (2002) describe in detail, how we derived the respective mass-loss relation, and only a short summary shall be given here.

To obtain a physically relevant representation of the theoretical mass-loss rates given by the large set of computed wind models, we applied the multidimensional maximum-likelihood method (as described by Arndt et al. 1997). The input parameters for each wind model are completely independent of each other. Hence, for each model the data set to be described consists of a maximum of 6 independent variables (stellar mass M, effective temperature $T_{\rm eff}$, luminosity L, chemical abundance ratio $\epsilon = \epsilon_C/\epsilon_O$, pulsational period P and piston velocity-amplitude Δv), and one dependent variable (the time-averaged mass-loss $\dot{M}$). Before giving a mass-loss approximation described only by the three fundamental stellar parameters M, L and $T_{\rm eff}$, any possible dependence of the mass-loss on ϵ_C/ϵ_O, P or Δv was considered (while Arndt et al. (1997) averaged over the mass-loss rates from models which differed in those input parameters).

Table 1: Listing of the selected wind models used here, including IR and mass-loss properties.

$M_\star/M_\odot$	T_{eff}/K	$L/L_\odot$	ϵ_C/ϵ_O	P/d	$\langle \dot{M} \rangle$	J–K	H–K	K–L	L–M	K–[12]	W-No.
0.63	3000	8000	1.30	820	3.0e-5	8.92	3.80	4.22	1.16	7.58	w155
0.70	3000	12000	1.30	1100	6.2e-5	10.55	4.49	5.04	1.43	9.15	w156
0.70	3500	12000	1.30	650	1.0e-5	3.89	1.72	2.08	0.52	3.87	w174
0.80	2200	15000	1.30	300	1.1e-4	13.66	5.89	6.57	2.00	12.11	w38
0.80	2400	7500	1.50	104	1.4e-5	7.42	3.23	3.61	1.04	6.62	w125
0.80	2550	7500	1.50	104	6.0e-6	4.51	2.00	2.36	0.62	4.31	w130
0.80	2600	5000	1.30	300	7.4e-6	5.89	2.57	2.96	0.83	5.52	w49
0.80	2600	5000	1.30	350	6.4e-6	7.06	3.05	3.47	0.99	6.40	w110
0.80	2600	5000	1.30	400	1.3e-5	7.51	3.23	3.64	1.03	6.73	w44
0.80	2600	5000	1.30	500	1.3e-5	7.94	3.41	3.85	1.12	7.14	w111
0.80	2600	5000	1.30	600	1.6e-5	8.93	3.84	4.29	1.25	7.89	w112
0.80	2600	6000	1.30	400	1.6e-5	8.00	3.45	3.90	1.14	7.25	w51
0.80	2600	7000	1.30	450	1.9e-5	7.96	3.43	3.89	1.13	7.24	w60
0.80	2600	7500	1.30	300	1.0e-5	5.77	2.53	2.93	0.82	5.44	w113
0.80	2600	7500	1.30	450	2.5e-5	8.17	3.53	4.02	1.17	7.55	w48
0.80	2600	7500	1.30	600	5.1e-5	11.78	5.06	5.66	1.68	10.41	w114
0.80	2600	7500	1.30	800	3.6e-5	10.02	4.30	4.79	1.39	8.79	w141
0.80	2600	10000	1.30	640	5.0e-5	10.27	4.43	4.95	1.43	9.05	w63
0.80	2600	12000	1.30	800	7.0e-5	10.58	4.57	5.13	1.50	9.48	w61
0.80	2600	15000	1.30	1000	9.9e-5	11.81	5.12	5.72	1.71	10.57	w62
0.80	2700	5000	1.30	300	3.9e-6	4.37	1.91	2.27	0.59	4.26	w167
0.80	2700	5000	1.30	350	5.6e-6	5.56	2.41	2.79	0.75	5.12	w168
0.80	3000	7500	1.50	400	9.0e-6	5.84	2.52	2.89	0.77	5.33	w124
0.80	3000	7500	1.80	450	8.3e-6	6.00	2.59	2.93	0.78	5.35	w32
0.80	3000	7500	1.80	650	1.0e-5	6.32	2.73	3.11	0.85	5.78	w31
0.80	3000	15000	1.50	300	1.7e-5	5.28	2.29	2.71	0.72	5.21	w30
0.80	3000	15000	1.50	650	3.0e-5	7.72	3.30	3.76	1.04	6.98	w28
0.80	3000	15000	1.50	800	4.1e-5	9.04	3.85	4.34	1.22	7.97	w29
0.84	3000	20000	1.30	1200	7.9e-5	9.80	4.18	4.69	1.32	8.53	w157
0.84	3500	20000	1.30	880	2.0e-5	4.69	2.06	2.46	0.66	4.60	w177
0.84	3700	20000	1.30	710	1.1e-5	3.05	1.39	1.79	0.46	3.45	w178
0.94	3000	25000	1.30	1300	8.8e-5	9.50	4.05	4.55	1.28	8.28	w158
0.94	3500	25000	1.30	1000	2.3e-5	4.78	2.09	2.48	0.67	4.68	w180
0.94	3700	25000	1.30	810	1.2e-5	2.87	1.31	1.68	0.42	3.23	w181
0.94	3900	25000	1.30	1300	2.5e-5	3.99	1.75	2.15	0.58	4.14	w183
1.00	2400	12000	1.30	300	2.6e-5	6.96	3.04	3.44	0.99	6.43	w122
1.00	2400	12000	1.30	500	5.9e-5	10.89	4.71	5.23	1.55	9.66	w144
1.00	2400	12000	1.30	600	7.7e-5	11.59	5.02	5.57	1.67	10.33	w120
1.00	2400	12000	1.30	800	7.9e-5	11.80	5.11	5.66	1.71	10.49	w145
1.00	2600	10000	1.30	640	4.3e-5	9.59	4.12	4.64	1.36	8.62	w64
1.00	2600	10000	1.80	650	2.6e-5	9.35	4.02	4.48	1.31	8.25	w13
1.00	2800	7000	1.30	400	5.4e-6	4.24	1.87	2.22	0.58	4.14	w78
1.00	2800	8000	1.30	400	5.1e-6	4.22	1.86	2.22	0.58	4.09	w97
1.00	2800	10000	1.30	640	1.8e-5	6.54	2.80	3.22	0.88	5.97	w69
1.00	2900	10000	1.25	578	9.6e-6	4.42	1.94	2.30	0.60	4.34	w72
1.00	2900	10000	1.30	578	1.3e-5	5.43	2.35	2.75	0.75	5.21	w71
1.20	2600	7000	1.30	400	6.1e-6	5.74	2.50	2.88	0.79	5.24	w53
1.20	2800	10000	1.40	400	9.8e-6	4.94	2.16	2.50	0.66	4.65	w126
1.20	2800	10000	1.50	400	1.6e-5	6.93	2.99	3.39	0.95	6.38	w108
1.20	2800	10000	1.80	400	1.3e-5	6.46	2.79	3.16	0.88	5.93	w109

We found that the dependence on ϵ was too subtle to be of any significance. The choice of piston amplitude Δv, however, does make a difference for the mass-loss characteristics and the modulation of the IR brightness. From a comparison with observed objects, we were able to constrain our model selection to $\Delta v = 5.0$ km/s. Finally, there is a dependence on the pulsational period P which we substituted for an additional dependence on the luminosity L, by applying the observed period-luminosity relation of Groenewegen & Whitelock (1996). The resulting mass-loss relation, as used here, is:

$$\log \dot{M}_{\rm fit} = -4.52 - 6.81 \cdot \log(T_{\rm eff}/2600\,{\rm K}) + 2.47 \cdot \log(L/10^4\,{\rm L}_\odot) - 1.95 \cdot \log(M/{\rm M}_\odot)$$

where $\dot{M}_{\rm fit}$ is given in units of ${\rm M}_\odot$/yr.

This relation shows a very good correlation (with a coefficient of 0.965) with the set of actual, time-averaged model mass-loss rates listed in Table 1. The typical deviation of $\dot{M}_{\rm fit}$ from any of the actual rates represented by it, is $\pm 20\,\%$, see Figs. 1a to 1c. The resulting strong dependence of $\dot{M}$ on $T_{\rm eff}$ of such a dust-driven wind is the macroscopic consequence of the temperature-sensitive microscopic physics and chemistry of the dust-formation processes considered by the wind model.

2.2 The relation between IR-colour and mass-loss on the tip-AGB

Direct, quantitative relations between the mass-loss rate and the observed IR colours are most desirable, since they can be applied to large observed samples to assess individual and collective tip-AGB mass-loss. For a small sample of about 40 well observed sources in the solar neighborhood, tight (semi-)empirical relations between the individual mass loss rate and near infrared color indices (i.e., J–K) have indeed been found by Le Bertre (1997) for carbon stars, and by Le Bertre & Winters (1998) for oxygen-rich Mira variables. These relations have been derived by a phase-dependent spectral modelling and, in principle, allow to determine the mass-loss rate from a suitable near infrared color index alone, independent of the distance to the source. Of course, these relations depend on the assumptions made in the spectral modelling (e.g., the dust-to-gas ratio is assumed to be constant), and they should be confirmed on a larger sample of well observed sources (i.e. a near infrared monitoring over several pulsation periods is required for this approach). In addition, the (semi-)empirical relations are restricted to J–K < 6.5.

Our computer-model-based relation between time-averaged IR-colours and mass-loss (and luminosity, see below) is fully complementary in both the J–K range and in its approach. It also leads to a new link between theory and the wealth of IR photometric data available today (e.g., Omont et al. 1999, Loup et al. 1999): the synthetic (J–K, $M_{\rm Bol}$) diagram, which compares directly with observed tip-AGB stellar IR data. Our relation is based on the same selection of wind computations as above and as Wachter et al. 2002. Again, the maximum-likelihood procedure of Arndt et al. (1997) was used. However, now the time-averaged IR-colours (J–K, H–K, K–L, and L–M) take, one by one, the rôle of the dependent variable which here is the quantity to be represented. Consequently, mass-loss is now one of the independent variables.

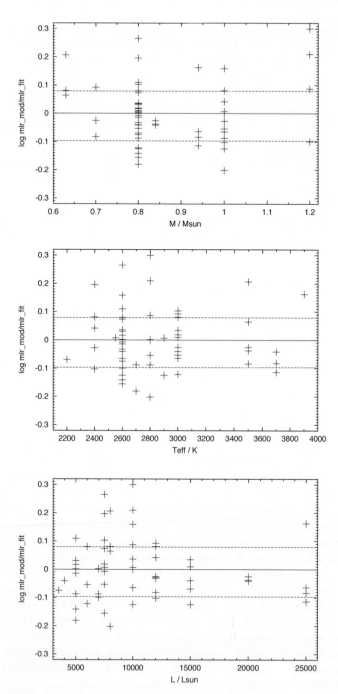

Figure 1: Deviations of the mass-loss rate given by the fit formula from the model mass-loss rates represented by it. The dashed lines mark a deviation of 20%. a) Logarithmic mass-loss ratio ($\log \dot{M}_{\mathrm{mod}}/\dot{M}_{\mathrm{fit}}$) over stellar mass $M/M_\odot$, b) ... over effective temperature T_{eff}, c) ... and over luminosity $L/L_\odot$.

In order to obtain a reliable, multi-parameter representation with the limited number of models available (50, see Table 1), we had to restrict the number of relevant variables as much as possible. The primary parameter is, of course, the mass-loss $\dot{M}$. In addition, inspection of pairs of models, which had all but one parameter matching each other, revealed a consistent dependence on the luminosity which clearly dominates any statistical variation of the time-averaged quantities. The same method did not show any conclusive dependence on period, effective temperature, nor on the C/O ratio. If there is any, it would be too subtle to be resolved by the present set of wind models. That does not mean, however, that there is no relation between the IR properties and those parameters. Rather, any such dependence is already accounted for indirectly by the primary term – of course, as shown above, the mass-loss rate itself depends on, e. g., the effective temperature.

With stellar properties given in solar units, the time-averaged IR colour of the pulsating wind models can be expressed in terms of the time-averaged mass-loss rate by the following relation:

$$J - K = 7.05 + C_{J-K} \cdot (\log \dot{M} + 4.66) - 7.4 \cdot (\log L_* - 4.0)$$

For $0.5 \cdot 10^{-5} < \dot{M} < 2.2 \cdot 10^{-5}$, we find $C_{J-K} = 6.39$. For larger mass-loss rates, the slope increases to $C_{J-K} = 9.25$.

Another sensitive and popular indicator of mass-loss is the K–[12] index, in case the IRAS 12 μm band flux is available. For this, we find

$$K - [12] = 6.48 + C_{K-[12]} \cdot (\log \dot{M} + 4.66) - 5.85 \cdot (\log L_* - 4.0)$$

with $C_{K-[12]} = 5.27$ for $0.5 \cdot 10^{-5} < \dot{M} < 2.2 \cdot 10^{-5}$, changing to 7.67 for larger mass-loss rates.

Since these relations strictly apply to strong mass-loss ($\dot{M} > 0.5 \cdot 10^{-5}$), they are valid only for large IR colours: J–K and K–[12] > 4 (depending on the stellar luminosity) – i. e., for extreme tip-AGB objects with $L_* > L_c$ and $T_{\text{eff}} < 3500$ K.

Fig. 2 is a 3D display of the individual, time-averaged J–K model colours over both $\log \dot{M}$ and $\log L_*$, together with the plane of the above given best-fit representation (seen edge-on as a line). The intrinsically variable conditions in the wind models are reflected in a stochastic spreading of the individual values. However, both formulae of aboce give a very good representation: The mean deviation is about 6 %, and all correlation coefficients are about 0.96.

This kind of relation means that individual superwind mass-loss rates can be determined from photometric data alone – provided, of course, the distance is known: $\log \dot{M} = -8.20 + 0.156 \cdot (J-K) - 0.463 \cdot M_{\text{Bol}}$. This relation is valid over the J–K range covered by our wind models: $4 < J-K < 11$.

With large data sets of stars observed by IR photometry now becoming available, we want to address the exact *collective* tip-AGB mass-loss of a stellar population. It can be derived from a *complete* sample of giants by finding a best matching synthetic sample with respect to the (J–K, M_{Bol}) diagram. To illustrate this approach, we devote the next section to computing such a synthetic sample which should match the solar neighbourhood stellar population.

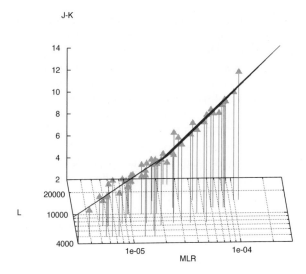

Figure 2: Individual, time-averaged wind model J–K IR-colours in a 3D plot, over both mass-loss rate $\dot{M}$ (MLR) and stellar luminosity (L). In order to obtain a true account of any model-to-model variation and of the dependence with increasing mass-loss, the distribution is shown along the plane of the best-fit representation (seen edge-on as a line) which gives smaller J–K values for larger luminosities.

3 Synthetic giants and a (J–K, M_{Bol}) diagram for the solar neighbourhood tip-AGB

Among the common IR colour indices, J–K is one of the most sensitive mass-loss indicators. Consequently, we have chosen to plot our representative synthetic sample of tip-AGB stars in a (J–K, M_{Bol}) diagram. In this visualization of our synthetic stars, the dust-enshrouded objects stand out well to the right side, and a direct comparison with observation can easily be made.

Our individual evolution tracks, computed with the well-tested Eggleton code (see Pols et al. 1998 and references therein), consider the above mass-loss relation on the AGB as soon as the critical luminosity L_c for the dust-driven winds has been exceeded – provided that the respective star can be considered as carbon-rich. To account for the moderate mass-loss of AGB stars with luminosities still below L_c and with warmer photospheres, we use an empirically calibrated mass-loss relation, mainly based on the Reimers relation (for details, see Schröder et al. 1999). This earlier RGB and AGB mass-loss, as well as the empirically adjusted overshoot and mixing parameters for the main-sequence evolution (Schröder et al. 1997) determine the mass of the H-rich stellar envelope of the later tip-AGB star. The gradual onset of the superwind is mostly driven by the decreasing T_{eff} of the tip-AGB giant in combination with the characteristic, strong temperature-sensitivity of the dust-driven wind. The brief but strong changes in the mass-loss rate are a reaction to the thermal pulses of the star. Finally, with noticeably decreasing mass, decreasing gravity and

T_{eff} lead to a self-acceleration of the mass-loss, ending the tip-AGB phase of the star with a nearly exposed stellar core.

The individual evolution tracks, which are the same as those used by Wachter et al. (2002), now form the basis for the generation of the synthetic tip-AGB giant stars. Table 2 gives an overview of the basic properties of every second of all those evolution models which have a C-rich superwind phase, and Figs. 3a, 3b show examples of the resulting evolution of the mass-loss on the tip-AGB (see Wachter et al. 2002 for more).

For initial masses $M_i > 3 M_\odot$, hot-bottom-burning sets in on the tip-AGB (Blöcker 1999), the stars cannot turn C-rich, and dust-driven superwinds are based on an oxygen-rich chemistry, for which we have no synthetic mass-loss description yet. Our grid of evolution tracks extends to $M_i = 16\ M_\odot$ (see Schröder & Sedlmayr 2001), but these must lack the short, final (O-rich) superwind phase. Hence, a small number of related objects remains unaccounted for by the synthetic stellar sample presented here (see our discussion).

The procedure to randomly distribute synthetic stars in time and mass on a fine-meshed grid of our evolution models is the same as the one used and described by Schröder & Sedlmayr (2001). In the same paper, we derived the IMF of the solar neighbourhood by finding a matching synthetic stellar sample to a complete, observed sample, based on the Hipparcos catalogue entries for $d < 50$ pc distance. The synthetic sample presented here has been generated with the same IMF, $dN/d\log M_i \propto M_i^{-1.7}$ for $M_i < 1.8\ M_\odot$, and $\propto M_i^{-1.9}$, else.

In order to obtain sufficient numbers of tip-AGB giants, we use a sample which is 1000 times larger than that of the real stars within 50 pc distance: $1.4 \cdot 10^6$ present-day stars more luminous than $M_V = 4.0$ were generated by an apparent SFR of $1.14 \cdot 10^{-3} \cdot e^{-(t_{*9}/6.3)}$ stars with $0.9 < M_i/M_\odot < 12$ per year, with t_{*9} as the star's age in Gyrs ($0 < t_{*9} < 12$). On this basis, there is a total of 5067 red giants (B − V > 1.4) with a mass-loss ranging between $\approx 10^{-9}$ and $10^{-4}\ M_\odot$/yr (see Fig. 4).

In order to apply the appropriate description of IR properties to each synthetic star, we distinguish between four cases:

(1) Stars with very little mass-loss, where the J–K index is defined by the "naked" photosphere of the giant: For this case, which applies to most objects which are still ascending on the AGB (or RGB), we adopted the J–K colours of model photospheres from Bessel et al. (1989), represented by the expression J–K $= 0.928 - 0.00428 \cdot (T_{\text{eff}}/K - 4000)$.

(2) Stars with mass-loss which are still O-rich, as they have yet to advance further on the tip-AGB, or as some of them are very luminous, massive stars with hot-bottom burning that destroys the dredged-up carbon: The IR properties of these O-rich stars are best described by the (semi-)empirical relation derived from O-rich winds by Le Bertre & Winters (1998): J–K $= 0.65 - 2.5/(\log \dot{M} + 4.25)$

(3) Tip-AGB stars which have just turned C-rich, with a moderate, dust-driven mass-loss: Since the observed lower luminosity limit for C-rich stars lies just a little below the critical luminosity for the dust-driven, C-rich winds (see Schröder et al. 1999, and references therein), we adopted $\log L_* > \log L_c - 0.1$ for this group. For the J–K index of these stars we use a semi-empirical relation which is a best match

Table 2: Every 2nd evolution model used to generate the synthetic, C-rich giants with a superwind. Listed are: initial stellar mass M_i, mass lost on the RGB, on the AGB (except superwind), by the superwind in the final 30 000 years, and the final stellar mass M_f, all in units of $M_\odot$.

M_i	$\int \dot{M}_{RGB}$	$\int \dot{M}_{AGB}$	$\int \dot{M}_{SW}$	M_f	
1.00	0.24	0.20	—	0.55	
1.10	0.12	0.38	0.01	0.56	[1]
1.20	0.09	0.47	0.03	0.58	[1]
1.30	0.08	0.30	0.28	0.60	
1.40	0.07	0.37	0.31	0.61	
1.50	0.06	0.39	0.38	0.62	
1.60	0.05	0.41	0.46	0.63	[2]
1.70	0.04	0.42	0.55	0.63	
1.80	0.03	0.45	0.62	0.64	
1.90	0.02	0.50	0.68	0.65	[3]
2.05	0.001	0.58	0.79	0.66	
2.25	—	0.63	0.92	0.68	
2.50	—	0.67	1.11	0.70	
2.80	—	0.74	1.32	0.72	

[1]) only brief superwind burst(s)
[2]) onset of core overshooting on MS at $M_i \approx 1.6\ M_\odot$
[3]) RGB evolution ends with He flash for $M_i \leq 1.95\ M_\odot$

of the C-rich mass-loss rates given by Le Bertre (1997), specifically in the range of $2.5 <$ J–K < 4: J–K $= -0.10 - 6.0/(\log \dot{M} + 4.00)$. By a random spread of $\pm 0.5^{\mathrm{mag}}$ we account for the observed variability.

(4) The objects which exceed the critical luminosity L_c of our models to drive a stable superwind: Based on the results of Schröder et al. (1999, Fig. 1), we approximated $L_c(M_*, T_{\mathrm{eff}})$ by the relation $\log L_c/M_* = 3.80 + 4 \cdot (\log T_{\mathrm{eff}} - 3.45)$. For synthetic stars of this group, J–K is given by the relation derived in the previous section, with J–K > 4.

Among the so characterized synthetic giants, there is a total of 20 tip-AGB stars in group (4) with a mass-loss rate $\dot{M}$ well in excess of $10^{-6}\ M_\odot$/yr (see Fig. 5 and a detailed listing in Table 3). This includes 10 dust-enshrouded objects with J–K > 6.0 which would be difficult to observe. All of these objects are in their superwind phase (i. e. their final 30 000 yrs on the tip-AGB, with $\dot{M} > 10^{-5}\ M_\odot$/yr). They are easily identified in Fig. 6, the (J–K, M_{Bol}) diagram to visualize the IR properties of our synthetic sample.

The collective mass-loss rate of the whole synthetic sample presented here is $5.0 \cdot 10^{-4}\ M_\odot$/yr. To this, the just 20 stars with J–K > 4.0 contribute $3.3 \cdot 10^{-4}\ M_\odot$/yr. The 10 superwind objects alone contribute $2.6 \cdot 10^{-4}\ M_\odot$/yr, which is about half of the collective stellar mass-loss rate! This theoretical prediction is consistent with the relative numbers derived from an observed sample of galactic mass-losing AGB stars by Le Bertre et al. (2001).

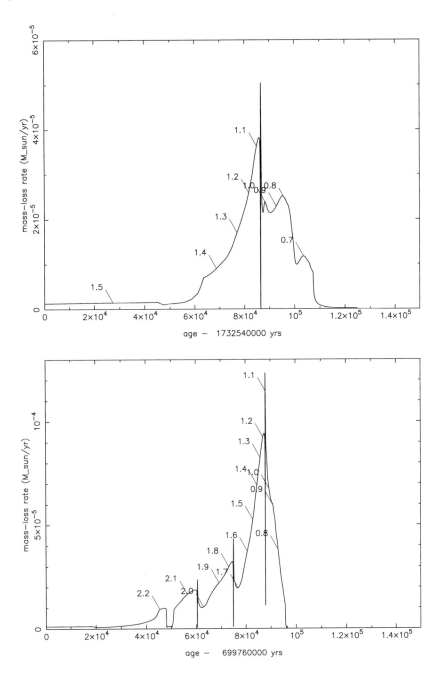

Figure 3: Tip-AGB mass-loss and superwind as assumed by the individual evolution models for (a) $M_i = 1.85\ M_\odot$ (top) and (b) $2.65\ M_\odot$ (bottom) – note the different mass-loss scales. The decreasing actual stellar mass is indicated. Large, brief changes of the mass-loss rate are a reaction to the thermal pulses by the dust-driven wind with its characteristic, strong temperature-sensitivity.

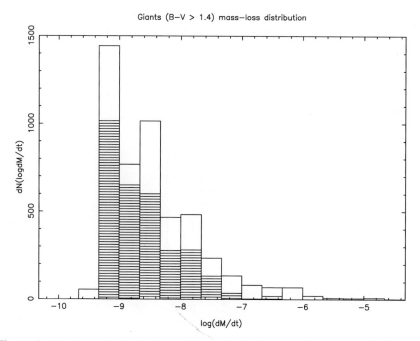

Figure 4: Mass-loss rate distribution of the synthetic giant sample – hatched: RGB giants; see text for more details.

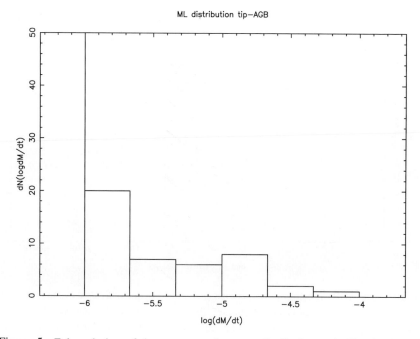

Figure 5: Enlarged view of the same mass-loss rate distribution as in Fig. 4: the dust-enshrouded tip-AGB stars with noticeable IR excess.

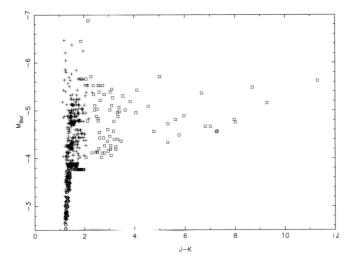

Figure 6: The resulting synthetic (M_{Bol}, J–K) diagram of the synthetic sample. Supposedly O-rich stars are marked by a cross, RGB stars by a circle, and C-rich stars by an square. See Table 3 for the properties of the dust-enshrouded, carbon-rich tip-AGB stars with J–K > 4.0.

Table 3: Properties of the 20 dust-enshrouded, carbon-rich tip-AGB objects in our synthetic sample with J–K > 4.0 seen in Figs. 5 and 6, organized by increasing J–K colour: J–K, mass-loss rate, initial mass, present mass, age, effective temperature and luminosity.

J–K	$\dot{M}[M_\odot/\text{yr}]$	$M_i/M_\odot$	$M_*/M_\odot$	Age/yrs	T_{eff}/K	$\log L_*/L_\odot$
4.07	$2.5 \cdot 10^{-6}$	1.83	1.40	$1.73 \cdot 10^9$	2563	3.870
4.11	$3.9 \cdot 10^{-6}$	2.47	2.00	$8.24 \cdot 10^8$	2636	4.058
4.56	$7.4 \cdot 10^{-6}$	1.95	1.51	$1.53 \cdot 10^9$	2544	3.923
4.78	$4.6 \cdot 10^{-6}$	1.54	0.65	$2.66 \cdot 10^9$	2928	3.713
5.01	$1.7 \cdot 10^{-5}$	2.76	2.17	$5.98 \cdot 10^8$	2565	4.172
5.32	$4.4 \cdot 10^{-6}$	1.21	0.84	$6.50 \cdot 10^9$	2552	3.621
5.33	$6.7 \cdot 10^{-6}$	1.56	1.14	$2.66 \cdot 10^9$	2525	3.777
5.63	$8.1 \cdot 10^{-6}$	1.74	0.66	$2.01 \cdot 10^9$	3022	3.812
5.77	$6.1 \cdot 10^{-6}$	1.31	0.92	$4.87 \cdot 10^9$	2524	3.683
5.97	$1.0 \cdot 10^{-5}$	1.67	1.21	$2.35 \cdot 10^9$	2476	3.843
6.82	$1.1 \cdot 10^{-5}$	1.67	0.73	$2.35 \cdot 10^9$	2731	3.754
6.67	$2.1 \cdot 10^{-5}$	2.55	1.64	$8.25 \cdot 10^8$	2497	4.031
7.02	$1.2 \cdot 10^{-5}$	1.41	0.94	$3.75 \cdot 10^9$	2438	3.754
7.26	$1.1 \cdot 10^{-5}$	1.48	0.76	$2.97 \cdot 10^9$	2542	3.709
7.29	$1.1 \cdot 10^{-5}$	1.48	0.75	$2.97 \cdot 10^9$	2542	3.715
7.97	$1.9 \cdot 10^{-5}$	1.48	0.92	$2.97 \cdot 10^9$	2403	3.810
8.01	$1.8 \cdot 10^{-5}$	1.68	0.81	$2.17 \cdot 10^9$	2483	3.791
8.70	$3.9 \cdot 10^{-5}$	2.46	1.38	$8.25 \cdot 10^8$	2377	4.080
9.30	$3.6 \cdot 10^{-5}$	2.16	1.11	$1.27 \cdot 10^9$	2377	3.949
11.31	$8.3 \cdot 10^{-5}$	2.66	1.31	$7.00 \cdot 10^8$	2333	4.137

4 Discussion

The idea of this work is to provide a consistent, theoretical approach to the galactic mass injection which describes the *collective* tip-AGB mass-loss of the moderate-mass stars in a stellar population, but which avoids the incompleteness of an object-by-object approach. We show that a synthetic tip-AGB stellar sample can be generated on a grid of evolution models with mass-loss, to serve as a direct and quantitative interpretation of a large (observed) IR data set. For this purpose, we derived relations for the mass-loss and IR-colour from the time-averaged output of hydrodynamic wind models. Using detailed computer simulations rather than observations, we are able to account for the most extreme cases of dust-enshrouded objects (J–K > 6), which contribute about half of all collective cool stellar wind mass-loss, but which are, at the same time, most likely to be missed out by an observed sample.

A very useful tool is the (J–K, M_{Bol}) diagram which allows us to compare observed and synthetic samples – especially with respect to the objetcs with a most extreme mass-loss. In our synthetic stellar sample of tip-AGB stars, which represents the solar neighbourhood stellar population, we find a small but crucial number of stars with extreme mass-loss which are likely to remain undetected in observed samples.

Unfortunately, a suitable observed sample of solar neighbourhood tip-AGB stars does not yet exist, mostly because the distances to appropriate objects are still just too badly known. The best complete tip-AGB star samples available to date are those of the Magellanic clouds (Loup et al. 1999). However, for a comparison in the (J–K, M_{Bol}) diagram between our synthetic sample and the observed LMC or SMC tip-AGB stars we must keep in mind that the metallicities do not match. Van Loon (2000) finds observational evidence for just a weak dependence of the mass-loss rate but a significant, approximately linear dependence on metallicity of the dust-to-gas ratio. Work on appropriate models for smaller metallicities is in progress but must remain the scope of a future publication.

In a qualitative comparison, the synthetic (J–K, M_{Bol}) diagram presented here resembles its observed LMC counterpart (Loup et al. 1999) quite well, with one exception: While the synthetic IR colours, in some extreme cases, reach out to J–K $\approx$ 12, only values up to about 5 have been observed in the Magellanic clouds. This should not come as a surprise, though, since (1) the lower metallicities in the LMC cause a lower dust-to-gas ratio with, accordingly, lower IR optical depth, and (2) the most dust-enshrouded objects (J–K > 6) are so much obscured by their circumstellar extinction that they would simply not be observable at the distance of the LMC, given the DENIS threshold (J < 16, K < 14). For example, the prototypical dust enshrouded carbon star IRC+10216 with a J–K of just 6 would, at the distance of the LMC, have J $\approx$ 20 and K $\approx$ 14.

A related question is the choice of reasonable IR opacities by the wind-model code (see Winters et al. 2000 for details). The good coincidence between IR colours and amplitudes of synthetic light curves of matching wind models on the one hand, and real objects on the other (Wachter et al. 2002), suggests to us that the adopted IR opacities (Preibisch et al. 1993) are appropriate.

As pointed out before, we have not been able to account for those tip-AGB objects with superwinds which have developed from massive stars, and which remained O-rich due to hot-bottom burning. The dust-enshrouded objects in our synthetic sample would compare best to a complete sample of C-rich tip-AGB stars in their superwind phase. Theoretical work on suitable wind models with an O-rich dust-formation chemistry is in progress (Jeong et al. 1999), but the number of the respective first generation wind models is still insufficient to derive representations of the respective mass-loss and IR properties.

Fortunately, the actual number of unaccounted stars may be very small: According to Blöcker (1999), hot-bottom burning is expected to occur only in stars with an initial mass of $M_i > 4$ to 5 $M_\odot$. Due to the steepness of the IMF, that should just be a small minority of objects ($< 15\%$). Also, the LMC tip-AGB sample of Loup et al. (1999) suggests a fraction just over 20% of O-rich objects among the dust-enshrouded tip-AGB stars (which is about 4 stars with respect to the 20 C-rich giants of group 4).

In our model grid, we adopted an upper mass-limit of $M_i \approx 3\ M_\odot$ for the stars which leave the tip-AGB with a C-rich superwind. Hence, there remains a mass range of $0.5 < \log M_i < 0.9$ for the unaccounted O-rich superwind objects (more massive stars lead to a SN). By comparison, the mass range of the 20 synthetic, supposedly C-rich stars of group (4) is $0.1 < \log M_i < 0.5$. With an IMF $\propto M_i^{-1.7}$, the number of O-rich objects in their superwind phase should then be 5 times smaller (even less, if the IMF is steeper in this mass range). If the O-rich dust-driven wind phases were indeed of similar duration as those of our 20 C-rich stars, this would correspond to just about 4 unaccounted tip-AGB, O-rich objects with $\dot{M} > 4 \cdot 10^{-6}\ M_\odot/\text{yr}$, consistent with Loup et al. (1999).

Certainly, a lot more work needs to be done to address the afore-mentioned unsolved problems. But in any case, the creation of synthetic stellar samples offers a direct way of comparing theoretical expectations with observational evidence. In combination with large, volume-limited samples of real stars with observed IR colour and C/O ratio, this approach will help to resolve open theoretical questions and, finally, will give a quantitative interpretation of IR observatinal data sets in terms of individual and collective stellar mass-loss.

References

Arndt, T.U., Fleischer, A.J., & Sedlmayr, E. 1997, A&A 327, 614

Bessell, M.S., Brett, J.M., Scholz, M., & Wood, P.R. 1989, A&AS 77, 1

Blöcker, T. 1999, in: Le Bertre, T., Lèbre, A., Waelkens, C. (eds.), Asymptotic Giant Branch Stars, Proc. IAU Symp. 191, ASP, 21

Cioni, M.-R.L., Marquette, J.-B., Loup, C., et al. 2001, A&A 377, 945

Fleischer, A.J., Gauger, A., & Sedlmayr, E. 1992, A&A 266, 321

Groenewegen, M.A.T., Whitelock, P.A. 1996, MNRAS 281, 1347

Habing, H.J. 1996, A&AR 7, 97

Jeong, K.S., Winters, J.M., Sedlmayr, E. 1999, in: Le Bertre, T., Lèbre, A., Waelkens, C. (eds.), Asymptotic Giant Branch Stars, Proc. IAU Symp. 191, ASP, 233

Le Bertre, T. 1997, A&A 324, 1059

Le Bertre, T. & Winters, J.M. 1998, A&A 334, 173

Le Bertre, T., Matsuura, M., Winters, J.M., et al. 2001, A&A 376, 997

Loup, C., Josselin, E., Cioni, M.-R., et al. 1999, in: Le Bertre, T., Lèbre, A., Waelkens, C. (eds.), Asymptotic Giant Branch Stars, Proc. IAU Symp. 191, ASP, 561

Omont, A., Ganesh, S., Alard, C., et al. 1999, A&A 348, 755

Parthasarathy, M. 1999, in: Le Bertre, T., Lèbre, A., Waelkens, C. (eds.), Asymptotic Giant Branch Stars, Proc. IAU Symp. 191, ASP, 475

Pols, O.R., Schröder, K.-P., Hurley, J.R., Tout, C.A., Eggleton, P.P. 1998, MNRAS 298, 525–537

Preibisch, Th., Ossenkopf, V., Yorke, H.W., Henning, Th. 1993, A&A 279, 577

Renzini, A. 1981, in: Iben Jr., I., Renzini, A. (eds.), Physical Processes in Red Giants, Reidel, Dordrecht, 431

Schröder, K.-P., Pols, O.R., Eggleton, P.P. 1997, MNRAS 285, 696

Schröder, K.-P., Winters, J.M., & Sedlmayr E. 1999, A&A 349, 898

Schröder, K.-P. & Sedlmayr, E. 2001, A&A 366, 913

Sedlmayr, E. 1994, in: Jørgensen, U.G. (ed.), Molecules in the Stellar Environment, Proc. IAU Coll. 146, Springer, Berlin, 163

Sedlmayr, E. & Winters, J.M. 1997, in: De Greve, J.P., Blomme, R., Hensberge, H. (eds.), Stellar Atmospheres: Theory and Observations, Proc. EADN Astrophys. School IX, Lec. Not. Phys. 497, 89

Sutton, E.C., Storey, J.W.V., Betz, A.L., Townes, C.H., Spears, D.L. 1977, ApJ 217, L97

van Loon, J.Th., Groenewegen, M.A.T., de Koter, A., et al. 1999, A&A 351, 559

van Loon, J.Th. 2000, A&A 354, 125

Wachter, A., Schröder, K.-P., Winters, J.M., et al. 2002, A&A 384, 452

Winters, J.M., Le Bertre, T., Jeong, K.S., Helling, Ch., Sedlmayr, E. 2000, A&A 361, 641

Zuckerman, B., Aller, L.H. 1986, ApJ 301, 772

The Dusty Sight of Galaxies:

ISOPHOT Surveys of
Normal Galaxies, ULIRGS, and Quasars

Ulrich Klaas

Max-Planck-Institut für Astronomie
Königstuhl 17, 69117 Heidelberg, Germany
klaas@mpia.de

Abstract

The FIR wavelength range is the realm of dust emission. The Infrared Space Observatory (ISO) offered a wide range of observing capabilities to characterize the properties of this solid state constituent of the ISM. The photometer ISOPHOT openend new wavelength windows and faint flux ranges for extragalactic FIR astronomy. Here we present highlights of studies of major galaxy classes, from nearby galaxies to $z \approx 1$ quasars. ISOPHOT provided FIR maps with unprecedented spatial resolution for nearby galaxies. With the ISOPHOT-S spectrometer PAH emission could be traced over large scales in galaxy disks. With ISOPHOT multi-filter photometry and complementary submm observations the most detailed 1 to 1000 µm spectral energy distribution (SED) templates of ultra-luminous infrared galaxies (ULIRGs) were established. The analysis of these SEDs provided the range of dust temperatures and emissivity indices as well as constraints for the size and opacity of the IR active regions. AGN and starburst heating can only be distinguished in the NIR to MIR part of the ULIRG SEDs. ISOPHOT multi-filter observations allowed for the first time the determination of the relative contributions by thermal and non-thermal radiation for a statistically significant sample of quasars. The new observations provided hints for evolution in PG quasars and clues for the unified scheme of radio galaxies and quasars. With the ISOPHOT 170 µm Serendipity Sky Survey the cold dust component in a large number of galaxies was quantified allowing a more accurate determination of the gas-to-dust mass ratio.

1 Introduction

When IRAS performed the first all sky survey in the infrared in 1983, one of the surprising results was the detection of more than 30 000 galaxies, some of which emit more than 90 % of their energy in the IR. It turned out that in most cases this IR emission is due to thermal radiation of dust.

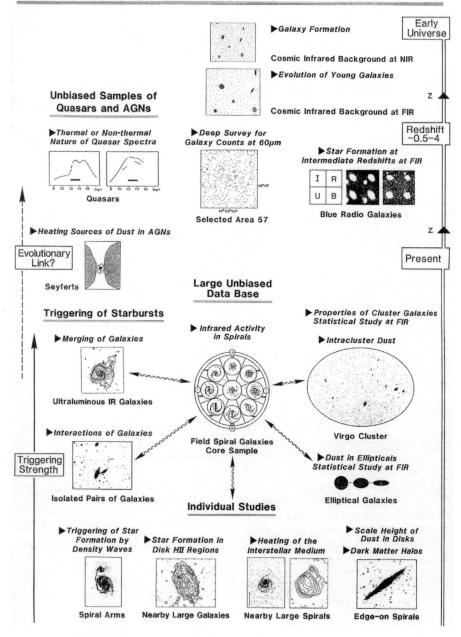

Figure 1: ISOPHOT studies of samples of galaxies as part of the ISO Central Programme defined several years before ISO's launch.

In 1995 ESA's Infrared Space Observatory ISO (Kessler et al. 1996) was launched giving the opportunity to observe many of these galaxies in much more detail. German institutes contributed to two of ISO's instruments, the Short Wavelength Spectrometer SWS and the photopolarimeter ISOPHOT. With ISOPHOT (Principal Investigator Dietrich Lemke, MPIA Heidelberg, Lemke et al. 1996) more than 4500 observations of extragalactic objects were performed during the 29 month mission.

When the ISOPHOT scientific consortium defined the guaranteed time observations as part of the ISO Central Programme, it was motivated by the findings of the IRAS mission to set up a comprehensive programme of extragalactic and cosmological studies. These ranged from investigations of nearby galaxies over large unbiased galaxy samples up to the earliest phases of galaxy formation as reflected in the diffuse cosmic infrared background light (Figure 1). All of these issues were addressed during the mission and here several highlights are presented.

2 ISOPHOT's Capabilities for Dust Observations

Although dust makes up only about 1 % of the mass of the interstellar medium, it can have a strong impact on observations: extinction affects mostly the UV and the optical wavelength regime, while in the IR and the submm the dust is practically transparent. In addition, the dust releases the absorbed energy in the IR by its self-emission. Figure 2 bottom shows a model of the dust composition and emission for the solar neighbourhood as proposed by Desert et al. (1990). Beside the big grains, which are in thermal equilibrium with temperatures around 20 K, a hotter dust component is made up by very small grains. A third component, a kind of macromolecule, the so-called Poly-Aromatic-Hydrocarbons (PAHs), shows typical emission features. All of these constituents show their emission in the IR and submm, while their heating sources emit in the UV and optical. Thus, dust causes an energy transfer between these two wavelength regimes.

From all this it is inferred that the best observing facilities for dust are in the IR, including the submm. ISOPHOT filters provided a continuous wavelength coverage between 3 and 200 μm, in particular covering the FIR peak of the big grain emission (Figure 2 top). With the wavelength extension beyond the IRAS 100 μm band, the very cold dust with temperatures around 15 K became visible for the first time, thus making large masses of dusty matter accessible to observations. The ISOPHOT-S dust spectrometer was designed to provide a sensitive instrument to measure the PAH features.

3 ISOPHOT Mapping of Nearby Galaxies

Figure 3 shows FIR maps of M31 (Haas et al. 1998) and the Small Magellanic Cloud (Wilke et al. 2003) with hitherto unprecedented spatial resolution. Hundreds of individual sources are detected in both galaxies which allows to study star formation in dependence of the position inside the galaxy, like distance from the centre, as well as the heating mechanisms in the arm and inter-arm regions of spiral galaxies.

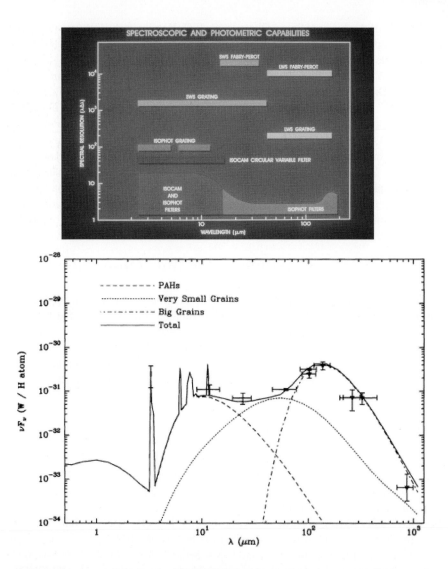

Figure 2: The top panel shows the wavelength coverage and spectral resolution of the ISO instruments. ISOPHOT had 25 filters covering the wavelength range 3.3 to 200 μm. In addition it contained a low-resolution grating spectrometer for the wavelength range 2.5 to 12 μm. The wavelength axis of this instrument capabilities panel is matched to the one of the bottom panel which displays the dust emission model of the solar neighbourhood as described by Desert et al. (1990).

The Dusty Sight of Galaxies

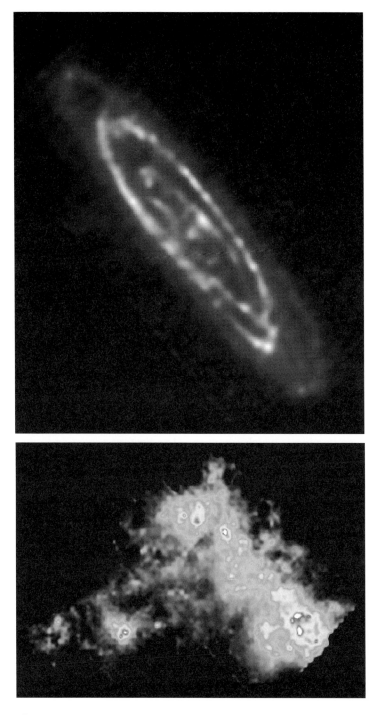

Figure 3: Top: ISOPHOT 170 μm map of M31 (Haas et al. 1998).
Bottom: ISOPHOT 170 μm map of the SMC (Wilke et al. 2003).

4 PAHs in the Disks of Galaxies

With the low-resolution spectrometer ISOPHOT-S it was possible to trace the existence of PAH emission over the whole disk of NGC 891 (Figure 4, Mattila et al. 1999). These spectra are identical with those from fields in the Milky Way containing diffuse emission. This allows to construct a template spectrum which is typical for the quiet areas of neutral molecular matter.

5 Power Sources of Ultra-luminous IR Galaxies

Ultra-luminous IR galaxies, ULIRGs, emit more than 10^{12} $L_\odot$ in the IR and are as luminous as quasars. For most of the objects the trigger of the high activity is tight interaction and merging of two gas-rich spirals. "Competing" heating sources are deeply embedded starbursts and active galactic nuclei (AGN). ISO has provided much better diagnostics for the issue "what powers ULIRGs?"

5.1 ULIRG SEDs

With ISOPHOT observations from 10 to 200 μm and follow-up observations in the submm we have obtained the most detailed 1 to 1000 μm SEDs for about 20 nearby ULIRGs. One can now look for statistically significant tracers of the central heating sources in the SEDs (Klaas et al. 2001).

Indeed, we find two major SED types which can be distinguished in the NIR-MIR (Figure 5):

1) One shows a power law ascent from 1 to 50 μm. This SED type is exclusively shown by objects which are classified as Seyferts according to their optical spectrum.

2) The second type shows a relatively flat SED from 1 out to 10 μm and then a relatively steep rise towards the FIR peak. This SED shape is exhibited by objects with all three narrow optical emission line excitation types, namely Seyfert 2, LINER and H II/starburst type spectra.

We concluded that the excess emission underneath the power-law SED (type 1) is due to a small amount of hot dust heated by the central AGN. The AGN is the main power source of the MIR luminosity.

In two-colour diagrams (Figure 6) the AGN contribution can be most clearly seen in the NIR colours, even better than in the previously used IRAS 25/60 colour. The diagrams also show that the majority of the nearby ULIRGs belongs to the "type 2" category.

However, the different optical spectral types are indistinguishable by their FIR colours. This suggests that powerful starbursts dominate the FIR luminosity in all nearby ULIRGs. Furthermore, in both diagrams H II/starburst and LINER-type galaxies are indistinguishable which argues for a starburst origin of the LINER emission.

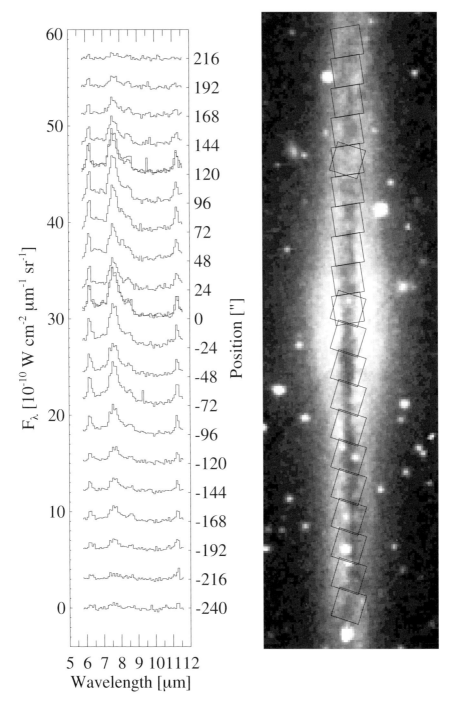

Figure 4: PAH spectra of the disk of NGC 891. 22 spectra were taken along the full extension of the disk and centered on the dust lane. The individual spectra are displayed on the left hand side (Mattila et al. 1999).

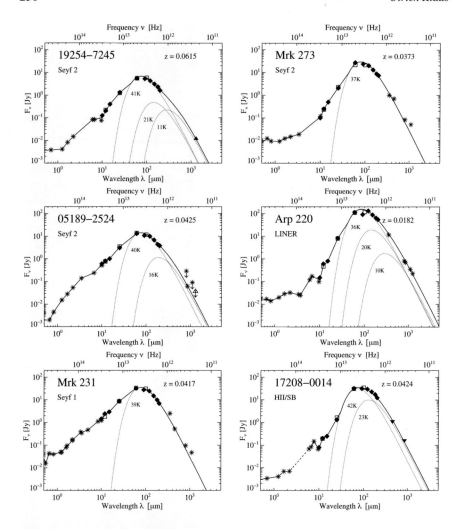

Figure 5: 1 to 1000 μm SEDs of ULIRGs (Klaas et al. 2001). Left column: Examples of SEDs with a NIR-MIR power-law shape. Right column: Examples of SEDs with a NIR-MIR flat shape. For each object the optical spectral type is indicated. The FIR part of the SED has been fitted by multiple modified blackbodies ($\varepsilon \propto \lambda^{-2}$) whose temperatures are indicated in the plot.

The relative ratio of MIR and FIR luminosities is less than 0.5 in all type 2 sources. For the few type 1 sources MIR and FIR luminosities have a similar amount. This implies that starbursts are the dominant power source in the type 2 ULIRGs and provide an equal amount of power as the AGN in type 1 ULIRGs.

Statistically there is an excellent correspondence of our SED shape classification with the classification of the ULIRG power source by PAH diagnostics, relating the PAH feature strength with the underlying continuum (Genzel et al. 1998, Lutz et al. 1998).

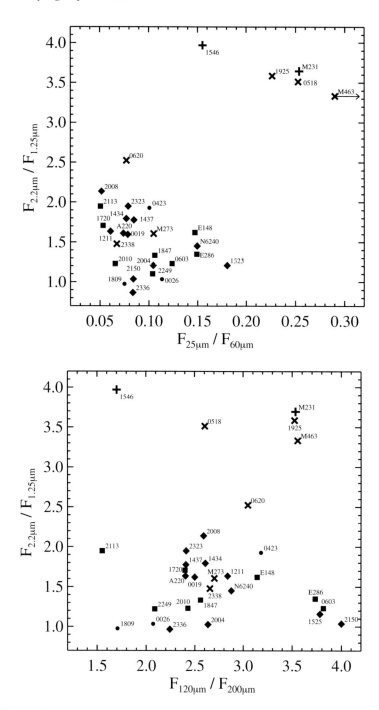

Figure 6: Two color diagrams of ULIRG SEDs (Klaas et al. 2001). The optical spectral type is indicated by the following symbols: plus: Seyfert 1 (2 objects), cross: Seyfert 2 (6), diamond: LINER (12), square: H II/starburst (8), dot: unclassified (3).

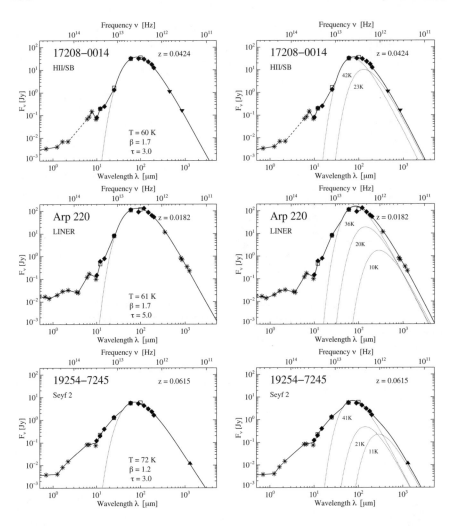

Figure 7: Temperature component fits to the FIR and submm SEDs of ULIRGs (Klaas et al. 2001). Left column: Single component fit with emissivity index β and optical depth $\tau_{100\,\mu m}$ as free parameters. Right column: Fit with multiple temperature components assuming $\tau_{100\,\mu m} \ll 1$ and $\beta = 2$ fixed.

5.2 Properties of the IR active regions

Another issue we can address with the SEDs is the temperature range of the dust and the transparency, size and mass of the dust clouds. The examples in the left column of Figure 7 demonstrate that it is possible to fit the FIR SEDs with a single source function $S_\lambda = B_\lambda(T)(1 - e^{-\tau_\lambda})$. In this case the fits usually give τ values indicating that the source is still optically thick at 100 μm. Alternatively, the FIR SEDs can be fitted with multiple modified blackbody spectra covering a temperature range between 10 and 40 K and with the emissivity index fixed to 2. These are, however, no unique solutions.

Which description is the more realistic one for the FIR emission regions? In the case of the single blackbody the emission is optically thick at 100 μm corresponding to an optical absorption A_V between 60 and 750 mag. The resulting brightness radii infer very compact sources between 100 and 400 pc. The dust masses are relatively small and range from a few 10^6 $M_\odot$ to a few 10^7 $M_\odot$. The dust temperatures are around 60 K (62 ± 7 K) and are significantly higher than for the warmest component in the multiple component case (38 ± 4 K). The resulting emissivity indices vary between 1.2 and 2 which would mean that the dust grain properties are different from object to object.

For the multiple blackbody case a size estimate can be made assuming that the optical depth of the FIR active component at 100 μm is around one. This yields radii which are 2 to 4 times larger than in the single blackbody case. Due to the existence of cold dust components the range of dust masses goes up to 10^9 $M_\odot$, more than ten times the highest dust mass for the single blackbody case. The mass of the FIR active dust is on average one quarter of the total dust mass, but with large variations $< M_{dust(FIRactive)} / M_{dust(tot)} >= 0.25 \pm 0.17$ (min/max: 0.02–0.50). Although the average dust temperature of around 40 K is considerably lower than for the single blackbody case, the strong dependence of the luminosity on the temperature with a power of 6 yields a consistent luminosity increase, compared to normal galaxies with 20 K dust temperature.

For those sources with published molecular hydrogen masses gas-to-dust mass ratios can be determined. For the single blackbody case anomalously high gas-to-dust mass ratios are obtained, while in the multiple blackbody case the ratios are much closer around the canonical value of the Milky Way. One weakness of the molecular hydrogen mass determination may however be that the standard $N(H_2)/I(CO)$ conversion factor is not applicable to the dense gas clouds in the central areas of ULIRGs.

Another check in the single versus multiple blackbody discussion is the comparison of the IR emission radii with sizes of molecular disks derived from CO interferometry (Table 1). For three out of four sources there is a better correspondence for the more transparent case, however, τ_{100} ranges very likely between 0.3 and 1 for the FIR active component.

Table 1: Sizes of molecular gas disks in ULIRGs derived from CO interferometry compared with the brightness radius for the optically thick single component FIR emission and the transparency radius for the multiple component FIR emission.

object	d_{COdisk} (″)	$2 \cdot r_b$ (″)	$2 \cdot r_\tau (\tau_{100}=1)$ (″)
Mrk231	0.9×0.8	0.9	1.0
Mrk273	$0.6 - 3.1$	0.56	1.6
17208-0014	1.8×1.6	0.64	2.2
23365+3604	1.0×0.9	0.24	1.0

Our conclusion is that the multiple component model is a better description for the ULIRGs.

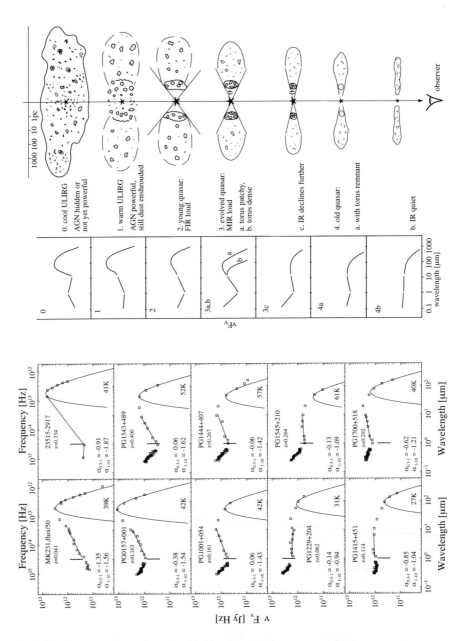

Figure 8: The left (bottom) panel shows examples of measured SEDs of PG quasars. The middle panel presents a sequence of SED models. The right (top) panel displays the derived evolutionary scenario. Mrk 231 and IRAS 23515–2917 are examples for SED shape 1 and PG 0157+001 and PG 1543+489 for SED shape 2. PG 1001+054 represents 3a, PG 1444+407 is 3a-b, PG 1229+204 and PG 1545+210 3c. PG 1700+518 is classified as SED type 4a and PG 1415+451 as type 4b (Haas et al. 2003).

The Dusty Sight of Galaxies

6 Evolution of Palomar-Green Quasars

Palomar-Green quasars are optically selected and are usually radio quiet. With ISOPHOT 64 out of 114 sources could be observed with a detection rate of higher than 80 %, three times higher than IRAS could achieve (Haas et al. 2003). Since these sources are optically blue, the dust must be arranged in a disk or torus seen face-on. Covering the wavelength range from the UV to the submm we find a variety of SED shapes (Figure 8) which cannot be explained due to aspect angle effects. We propose an evolutionary model starting from the ULIRG phase, when the source is still heavily enshrouded by dust. The central source breaks off the dust cocoon and the dust settles eventually into a torus/disk-like structure and finally the central source fades out. Individual SEDs can be ordered along this sequence (Figure 8). From phase 3 onwards the IR emission is mainly powered by the AGN. In some cases a patchy torus is needed to explain the relatively low dust temperatures.

7 Unified Scheme of AGN

The presence of dusty tori may cause different spectral appearance of the same type of source depending on the aspect angle. This is the issue of the unified scheme. For radio sources this is enforced by the direction of the relativistic jet (Figure 9). A face-on source will appear as a flat radio spectrum quasar whose SED is dominated by non-thermal synchrotron emission (e. g. FR 0234+28). The shallower the aspect angle becomes, the more the thermal bump of the torus will show up until at small angles we observe a radio galaxy with the torus seen edge-on (e. g. Cyg A).

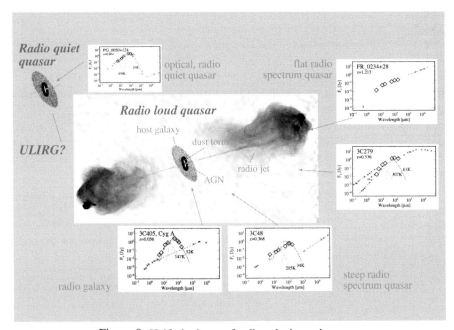

Figure 9: Unified scheme of radio galaxies and quasars.

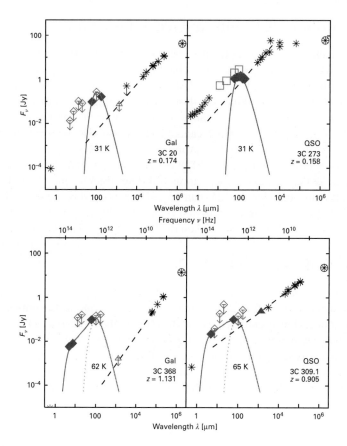

Figure 10: SEDs of matched pairs of 3C radio galaxies and quasars (Meisenheimer et al. 2001).

In order to provide further evidence for the unification we observed 10 pairs of 3C radio galaxies and quasars with ISOPHOT (Meisenheimer et al. 2001). The counterparts of each pair matched in luminosity at 178 MHz and redshift. Thermal dust emission is well detected above the synchrotron spectrum (Figure 10). For each pair the dust emission has similar properties. Of course, for the quasars the synchrotron emission is relatively stronger.

We found the following clues for a unified scheme: For the first time the sensitivity was sufficient to perform a pair-wise comparison of individual sources – beforehand only average SEDs of each type could be constructed. A balanced detection statistics was found. For 5 pairs both the quasar and the radio galaxy were detected, for 1 pair only the quasar and for 1 pair only the radio galaxy. The dust radiation is isotropic and balanced IR luminosities are found. Also the ratio of FIR power to core power, measured by the kinetic energy which is channeled via the jets into the radio lobes, is similar, at least for redshifts greater than 0.8. For smaller redshifts some of the radio galaxies show a low thermal power which could indicate exhaustion of the circumnuclear fuel.

8 ISOPHOT 170 μm Serendipitous Sky Survey

A serendipitous sky survey in a new wavelength window was performed during the slews of the satellite, making use of ISOPHOT's C200 camera (Bogun et al. 1996). 150 000° strip scans with a width of 3 arcmin resulted in a sky coverage of 15 % (Figure 11), thus increasing the efficiency of the mission.

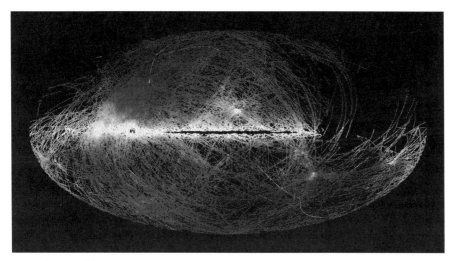

Figure 11: ISOPHOT 170 μm Serendipitous Sky Survey (Bogun et al. 1996).

The extraction of point sources detected with good signal-to-noise ratio by all four camera pixels yielded a first catalogue with 115 galaxies (Stickel et al. 2000). This represents already the largest homogeneous 170 μm galaxy sample. For this sample the average IR luminosity is a few 10^{10} $L_\odot$ and the average dust temperature 20 K, typical for normal spiral galaxies as our Milky Way (Figure 12). By inclusion of this cold dust component more realistic dust masses were derived, which are higher than previously derived from IRAS photometry up to 100 μm. With these new dust masses the average gas-to-dust mass ratio is close to the canonical value of the Milky Way.

9 Summary

Essential findings of the ISOPHOT extragalactic surveys presented here are:

- PAHs are ubiquitous and they are associated with colder and IR quiet regions. Spectra are similar in the local universe.

- Extending the wavelength regime out to 200 μm the existence of massive cold dust components was verified. Consequently lower gas-to-dust mass ratios which are similar to the one of the Milky Way are found.

- The predominant power sources of nearby ultra-luminous IR galaxies are starbursts. The sizes of the IR active regions are in the order of 1 to 2 kpc with optical depths of ≈ 1 at 100 μm. Beside the 40 K dust in the IR active regions there exist large amounts of colder dust with temperatures of ≤ 20 K.

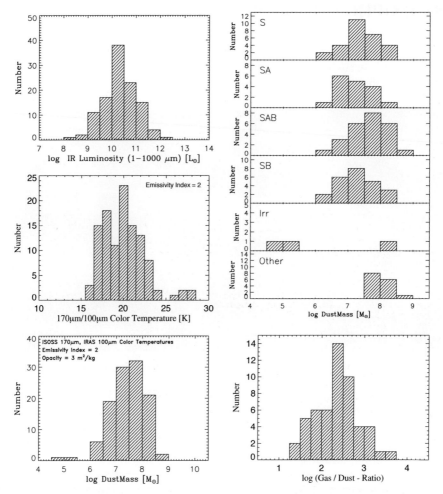

Figure 12: Statistical dust properties of normal galaxies observed by the ISOPHOT 170 μm Serendipitous Sky Survey (Stickel et al. 2000). Upper left: distribution of total IR luminosity (1–1000 μm); middle left: distribution of dust color temperatures derived from the 170 μm-to-100 μm flux ratio; lower left: distribution of dust masses derived from the IR luminosities and the color temperatures; upper right: distribution of dust masses depending on Hubble type; lower right: distribution of gas-to-dust mass ratios – the value for the Milky Way is around 170.

- More than 100 quasars were detected in more than one IR band. This allowed to trace the thermal component in their SEDs.
- The large data base of SEDs of radio quiet PG quasars suggests an evolutionary scheme of the dust distribution in these sources.
- The unified scheme could be verified for radio galaxies and radio quasars.
- The ISOPHOT Serendipitous Sky Survey delivers fluxes of galaxies in the new 170 μm band. A final catalogue with 2000 galaxies is under compilation.

10 Working with ISO Data

The ISO Legacy Archive (Kessler et al. 2000) was released in February 2002. An example of the Java based user interface (Arviset et al. 2000) can be seen in Figure 13. This archive user interface can be launched from the web site of the ISO Data Centre in Villafranca, http://www.iso.vilspa.esa.es, by selecting "General user access to ISO data". An overview of data products is also provided by the Simbad VizieR Service and by ADS. Ample documentation is available on the ISO Data Centre web site, too. Help in data reduction can be asked for by contacting the ISO Data Centre or one of the National Instrument Data Centres, e. g. the ISOPHOT Data Centre at MPIA.

Figure 13: Result of an ISO Data Archive query for observations of M51. Each observation is characterized by information on the observing mode, the position, the wavelength range, size of the field, and a graphical representation of the data product in form of an icon. Clicking on this icon a larger postcard representation of the data product will appear.

Acknowledgement

This highlight overview presents the work of many people working on ISOPHOT data. I am greatly indebted to the following colleagues:

- from Max-Planck-Institut für Astronomie in Heidelberg: Martin Haas, Hans Hippelein, Dietrich Lemke, Klaus Meisenheimer, Manfred Stickel, and Karsten Wilke
- from Astronomisches Institut der Universität in Bochum: Marcus Albrecht, Rolf Chini, and Sven A. H. Müller
- from ISO Data Centre in Villafranca, Spain: Bernhard Schulz
- from Joint Astronomy Centre in Hilo, Hawaii: Iain M. Coulson
- from University Observatory Helsinki: Kimo Lehtinen and Kalevi Mattila
- from Landessternwarte in Heidelberg: Max Camenzind
- from Max-Planck-Institut für Radioastronomie in Bonn: Frank Bertoldi
- from Harvard-Smithsonian Center for Astrophysics in Cambridge MA: Belinda Wilkes
- from Rutherford Appleton Laboratory in Chilton, Didcot: Phil Richards

The ISOPHOT Data Centre at MPIA is supported by Deutsches Zentrum für Luft- und Raumfahrt e. V. (DLR) with funds of Bundesministerium für Bildung und Forschung, grant no. 50 QI 0201. The results presented here are based on observations with the Infrared Space Observatory ISO, an ESA project with instruments funded by ESA Member States (especially the PI countries France, Germany, the Netherlands, and the United Kingdom) and with the participation of ISAS and NASA. Data reduction was performed with the ISOPHOT Interactive Analysis System (PIA) which is a joint development by the ESA Astrophysics Division and the ISOPHOT consortium led by MPIA.

References

Arviset, C., Dowson, J., Hernández, J., et al. 2000, ADASS IX, N. Manset, C. Veillet, and D. Crabtree (eds.), ASP Conf. Proc. 216, 191
Bogun, S., Lemke, D., Klaas, U., et al. 1996, A&A 315, L71
Desert, F. X., Boulanger, F., & Puget, J.L. 1990, A&A 237, 215
Genzel, R., Lutz, D., Sturm, E., et al. 1998, ApJ 498, 579
Haas, M., Lemke, D., Stickel, M., et al. 1998, A&A 338, L33
Haas, M., Klaas, U., Müller, S.A.H., et al. 2003, A&A 402, 87
Kessler, M.F., Steinz, J.A., Anderegg, M.E., et al. A&A 315, L27
Kessler, M.F., Müller, T.G., Arviset, C., García-Lario, P., Prusti, T. 2000, The ISO Handbook, Volume I: ISO-Mission Overview, Version 1.0, November 24, 2000, SAI/2000-035/dc, ESA publ., Villafranca
Klaas, U., Haas, M., Müller, S.A.H., et al. 2001, A&A 379, 823
Lemke, D., Klaas, U., Abolins, J., et al. 1996, A&A 315, L64
Lutz, D., Spoon, H.W.W., Rigopoulou, D., et al. 1998, ApJ 505, L103
Mattila, K., Lehtinen, K., Lemke, D. 1999, A&A 342, 643
Meisenheimer, K., Haas, M., Müller, S.A.H., et al. 2001, A&A 372, 719
Stickel, M., Lemke, D., Klaas, U., et al. 2000, A&A 359, 865
Wilke, K., Stickel, M., Haas, M., et al. 2003, A&A 401, 873

Abundance Evolution with Cosmic Time

James W. Truran

Department of Astronomy and Astrophysics and Enrico Fermi Institute
University of Chicago, 5640 S. Ellis Avenue, Chicago, IL 60637
truran@nova.uchicago.edu

Abstract

The primordial compositions of galaxies reflect the products of the cosmological Big Bang: hydrogen, deuterium, ^3He, ^4He, and ^7Li. Within galaxies, stars and supernovae play the dominant role both in synthesizing the elements from carbon through uranium and in returning heavy-element-enriched matter to the interstellar gas, from which new stars are formed. We review abundance patterns in field halo stars, globular clusters, dwarf spheroidal galaxies, and damped Lyman-α systems. We demonstrate that the often distinctive abundance patterns that characterize these low metallicity stellar populations are entirely consistent with expectations for nucleosynthesis in massive stars and associated Type II supernovae and identify abundance trends as a function of metallicity. We then discuss the constraints such abundance trends can impose on the star formation histories of galaxies and the Universe.

1 Introduction

Observations reveal that the mean level of heavy element abundances in Galactic matter increases monotonically with time. This is a forced consequence of the continuing cycle of star formation, stellar and supernova nucleosynthesis, and the return of heavy element enriched matter to the interstellar gas from which subsequent generations of stars are formed. This trend is displayed specifically in the "age-metallicity" relation for Galactic disk stars (Edvardsson et al. 1993; Feltzing, Holmberg, & Hurley 2001) in the form of an increasing Fe/H abundance ratio with time over the metallicity range $-1 \lesssim [Fe/H] \lesssim 0$, while Galactic halo stars are characterized by metallicities $-4 \lesssim [Fe/H] \lesssim -1$.

In contrast, the relative abundances of particular species can vary significantly. Such effects are most noticeable at early epochs ($\lesssim 10^8$–10^9 years), where contributions from massive stars of short lifetimes are dominant and incomplete mixing yields a significant degree of inhomogeneity. As we shall see, distinctive abundance patterns are characteristic of nucleosynthesis occurring in stars of different masses and corresponding lifetimes. For this reason, interesting constraints on the star formation histories of the halo, bulge, and disk components of our Galaxy may be contained in their abundance histories. Similarly, scrutiny of the absorption line systems associated with distant quasars and, most recently, gamma-ray bursts, allows us to probe and to constrain the star formation and abundance history of the high redshift Universe.

Significant progress in our understanding of the chemical evolution of stellar populations has occurred in recent years as a consequence of a wealth of new information on cosmic phenomena – spectroscopic and photometric properties of stars in our Galaxy and of gas clouds and galaxies at high redshifts – pouring in from new ground and space based observatories. A review of interesting composition trends derived from these studies is presented in section 3. The interpretation of this wealth of observational data is necessarily based upon our knowledge of stellar and supernova nucleosynthesis. In the next section, we identify critical nucleosynthesis products, their production sites, and the timescales on which they can be expected to enrich the interstellar media of galaxies. Discussion and conclusions are presented in section 5.

2 Nucleosynthesis and Production Timescales

We will confine our attention to three broad categories of stellar and supernova site with which specific nucleosynthesis products are understood to be associated: (1) the mass range $1 \lesssim M/M_\odot \lesssim 10$ of "intermediate" mass stars, for which substantial element production occurs during the asymptotic giant branch (AGB) phase of their evolution; (2) the mass range $M \gtrsim 10\ M_\odot$, corresponding to the massive star progenitors of Type II ('core collapse') supernovae; and (3) Type Ia supernovae, which are understood to arise as a consequence of the evolution of intermediate mass stars in close binary systems.

A knowledge of the production timescales – the effective timescales for the return of a star's nucleosynthesis yields to the interstellar gas – is also crucial to our use of abundance patterns to constrain star formation histories. The lifetime of a $10\ M_\odot$ star is approximately 5×10^7 years. We can thus expect all massive stars $M \gtrsim 10\ M_\odot$ to evolve on timescales $\tau_{\rm SNe\,II} < 10^8$ years, and to represent the first sources of heavy element enrichment of stellar populations. In contrast, intermediate mass stars (IMS) evolve on timescales $\tau_{\rm IMS} \gtrsim 10^8$–$10^9$ years. Finally, the timescale for SNe Ia product enrichment is a complicated function of the binary history of Type Ia progenitors (see, e.g., Livio 2000). Observations and theory suggest a timescale $\tau_{\rm SNeIa} \gtrsim 1.5$–$2 \times 10^9$ years.

With these timescales in mind, we will now review expectations for nucleosynthesis associated with each of these specific nucleosynthesis sites.

2.1 Intermediate Mass Stars

Intermediate mass stars ($1 \lesssim M/M_\odot \lesssim 10$) provide important contributions to galactic abundances primarily as a consequence of the occurrence of thermal pulses in their helium burning shells on the "asymptotic giant branch" (see, e. g. the review by Busso et al. 1999). The principle products are ^{12}C, ^{14}N, and the s-process neutron capture elements. Discussions of nucleosynthesis in low and intermediate mass stars have also been presented by Iben & Truran (1978) and Renzini & Voli (1981). Carbon production results from the incomplete burning of helium in the thermally pulsing shells of AGB stars, followed by the dredge-up of this shell matter into the envelope. "Carbon stars" are a product of such dredge-up. This environment also produces the bulk of the heavy s-process ("slow neutron capture") isotopes in the mass range A $\gtrsim$ 90, with the neutrons being provided by the $^{13}C(\alpha,n)^{16}O$ reaction. The main production of s-process isotopes occurs at the lower end of the mass range of intermediate mass stars (M $\approx$ 1–2 $M_\odot$), on a timescale of order $1-2 \times 10^9$ years.

2.2 Massive Stars and Type II Supernovae

Massive stars M $\gtrsim$ 10 $M_\odot$ represent the progenitors of Type II (or 'core collapse') supernovae. The combination of stellar burning stages and explosive nucleosynthesis in the supernova phase is understood to synthesize nuclei from oxygen to nickel and the r-process heavy neutron capture products.

Calculations of charged-particle nucleosynthesis in massive stars and Type II supernovae over the past decade (Woosley & Weaver 1995; Thielemann et al. 1996; Nomoto et al. 1997; Limongi et al. 2000) reveal one particularly significant distinguishing feature of the emerging abundance patterns: the elements from oxygen through calcium (and titanium) are overproduced relative to iron (peak nuclei) by a factor $\approx$ 2–3. This means that SNe II produce only approximately 1/3 to 1/2 of the iron in Galactic matter. (We note that, while all such models necessarily make use of an artificially induced shock wave via thermal energy deposition (Thielemann et al. 1996) or a piston (Woosley & Weaver 1995), the general trends obtained from such nucleosynthesis studies are expected to be valid.) As we shall see in the next section, these trends are reflected in the abundance patterns of metal deficient stars, for which typically [O-Ca/Fe] $\sim$ +0.3 to +0.5.

It now also seems most likely that the r-process synthesis of the heavy neutron capture elements in the mass region A $\gtrsim$ 130–140 occurs in an environment associated with massive stars. The association of the r-process site with massive star environments results from two factors: (1) the two most promising mechanisms for r-process synthesis – a neutrino heated "hot bubble" (Woosley et al. 1994; Takahashi et al. 1994) and neutron star mergers (Lattimer et al. 1977; Freiburghaus & Rosswog et al. 2001) – are both tied to environments associated with core collapse supernovae; and (2) observations of old stars (elaborated in the next section) confirm the early entry of r-process isotopes into Galactic matter.

2.3 Type Ia Supernovae

Type Ia supernovae are understood to represent the source of the 1/2 to 2/3rds of the iron-peak nuclei that are not produced in SNe II. The standard model for SNe Ia involves the growth of a white dwarf to the Chandrasekhar limiting mass in a close binary system (see the review of Type Ia progenitor models in Livio (2000)) and its subsequent incineration. Calculations of explosive nucleosynthesis associated with carbon deflagration models for Type Ia events (Thielemann et al. 1986; Iwamoto et al. 2000; Domínguez, Höflich, & Straniero 2001) predict that sufficient iron-peak nuclei ($\sim$ 0.6–0.8 $M_\odot$ of ^{56}Fe in the form of ^{56}Ni) are synthesized to explain both the powering of the light curves due to the decays of ^{56}Ni and ^{56}Co and the observed mass fraction of ^{56}Fe in Galactic matter. Estimates of the timescale for first entry of the ejecta of SNe Ia into the interstellar medium of our Galaxy yield $\approx 2 \times 10^9$ years, at a metallicity [Fe/H] ~ -1 (Kobayashi et al. 2000; Goswami & Prantzos 2000).

3 Composition Trends in Halo Stars

Spectroscopic abundance studies of the oldest stars in our Galaxy, down to metallicities [Fe/H] ≈ -4 to -3, have been most recently reviewed by Wheeler, Sneden, & Truran (1989) and McWilliam (1995). Such studies typically reveal abundance trends which can best be understood as reflecting the nucleosynthesis products of the massive stars and associated Type II supernovae that can be expected to evolve and to enrich the interstellar media of galaxies on rapid timescales ($\lesssim 10^8$ years). It is these trends in abundance patterns which we will review in this section.

3.1 The α-Elements

A strong and unambiguous signature of these massive stars is a ratio of the abundances of the so called α-elements (O, Ne, Mg, Si, S, Ar, and Ca) to that of iron-peak nuclei which is approximately 2–3 times the Solar System value (Wheeler et al. 1989; McWilliam 1995). The observed O/Fe and (Si-Ti)/Fe ratios fall gradually to the Solar value on longer timescales, as the iron-peak rich products of Type Ia supernovae are introduced at metallicities [Fe/H] ≈ -1. High (Si-Ca)/Fe ratios are also characteristic of globular cluster stars in our Galaxy (Brown, Burkert, & Truran 1991; Wheeler et al. 1989) and of nearby dwarf spheroidal galaxies (Shetrone et al. 2001).

For purposes of illustration, the trends in [O/Fe] and [Ca/Fe] as a function of [Fe/H] are shown in Figures 1 and 2, respectively. Following our discussion of the expected products of the evolution of massive stars and associated Type II supernovae in section 2, the observed trends with increasing metallicity can be understood in a straightforward manner. The ejecta of massive stars of short lifetimes dominate the nucleosynthesis contributions until such time (after approximately 1.5 to 2 billion years of Galactic evolution) as the iron-peak products of Type Ia are first introduced. This is seen to occur at a metallicity [Fe/H] ≈ -1, approximately the point at which the halo to disk 'transition' occurred. Subsequently, the [O/Fe] and [Ca/Fe] ratios fall monotonically to the Solar value. (Note here that we use [Ca/Fe] as representative of the intermediate mass alpha-particle elements – Ne, Mg, Si, S, Ar, Ca, and Ti).

In contrast to the α-chain elements, the even-Z elements in the iron peak region – specifically, Cr and Ni – are found to have been produced in approximately Solar proportions, relative to iron, at least down to metallicities [Fe/H] ≈ -3. This is consistent with the formation of these iron "equilibrium" peak nuclei in nuclear statistical equilibrium at high temperatures. The general consistency of the even-Z elemental abundance ratios with Solar system abundance ratios extends to zinc,

Abundance Evolution with Cosmic Time 265

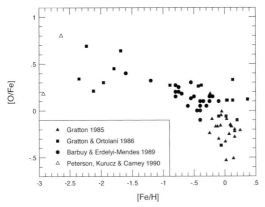

Figure 1: Observed evolution of the oxygen to iron abundance ratio with metallicity (Figure from Timmes 2003).

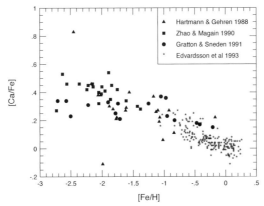

Figure 2: Observed evolution of the calcium to iron abundance ratio with metallicity (Figure from Timmes 2003).

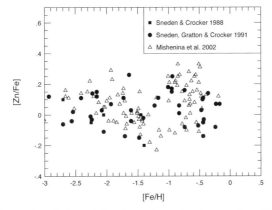

Figure 3: Observed evolution of the zinc to iron abundance ratio with metallicity (Figure from Timmes 2003).

as shown in Figure 3. The importance of zinc in this regard will become apparent when we discuss abundance data obtained from observations of Quasar and gamma-ray burst absorption line systems.

We should note here that the data regarding odd-Z nuclei, also reviewed by both Wheeler, Sneden, & Truran (1989) and McWilliam (1995), presents further challenges. The situation here is significantly more complicated. Early studies of explosive nucleosynthesis (Truran & Arnett 1971) first identified metallicity dependences of the relative abundances of odd-Z and even-Z nuclei: the ratios (odd-Z)/(even-Z) elements were predicted to decrease with decreasing metallicity. This is a straightforward consequence of the decreasing excess of neutrons relative to protons in the gas. In fact, these trends are indeed observed for odd-Z nuclei in the iron peak region: the ratios Mn/Fe, Co/Fe, and Cu/Fe are all found to decrease systematically in metal-poor halo stars back to [Fe/H] $\approx$ −3. The effects are less strong, however, for lighter odd-Z nuclei.

Observations of stars of the lowest metallicities (down to approximately [Fe/H] $\approx$ −4) reveal patterns which are not easily explained on the basis of current nucleosynthesis calculations (McWilliam et al. 1995a,b). Of particular interest, they find distinctive trends involving iron peak nuclei: the Cr/Fe and Mn/Fe ratios decrease with decreasing [Fe/H], while the Co/Fe ratio increases. These trends in [Cr/Fe], [Mn/Fe], and [Co/Fe] with [Fe/H] are not readily understandable in the context of current models for Type II supernova nucleosynthesis. Neither do they appear to be consistent with the abundance patterns characteristic of very massive stars (100–300 $M_\odot$) of low metallicity that can participate in the earliest stages of Galactic nucleosynthesis (Heger & Woosley 2002).

3.2 The Heavy Neutron-Capture Elements

The nuclei in the mass region extending from just beyond the iron peak through the actinides (60 $\lesssim$ A $\lesssim$ 238) are understood to have been formed by one of two processes of neutron capture: the s-process of (slow) neutron addition and the r-process of (rapid) neutron addition. As noted previously, intermediate mass stars are understood to be the site of formation of the heaviest s-process isotopes/elements while r-process synthesis is tied rather to an environment associated with massive stars, Type II supernovae, and their remnants.

In the heavy element region, both ground based and space based studies (see, e.g., the review by Truran et al. 2003) have confirmed that the abundances of the heavy neutron-capture products (A $\gtrsim$ 130–140) in the most extremely metal deficient halo fields stars ([Fe/H] $\lesssim$ −2.5) and globular cluster stars were formed in a robust r-process event (Truran 1981; Truran, Cowan, & Fields 2001). This is reflected, for example, in the abundance pattern for the metal poor but r-process enriched halo star BD+17°3248, shown in Figure 4. Note the striking agreement of the observed abundance pattern (solid circles and triangles) in the region from barium to lead with the Solar System r-process abundance pattern (solid line). These data provide strong confirmation of the association of the site of r-process nucleosynthesis with a massive star/Type II supernova environment, consistent with its introduction into the interstellar gas of our Galaxy at very early epochs.

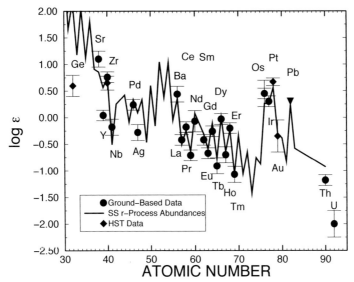

Figure 4: The neutron-capture elemental abundances for the halo star BD+17°3248 (Cowan et al. 2002), obtained from both ground-based and HST observations, are compared with the Solar System r-process abundance curve. The upper limit on the lead abundance is indicated by an inverted triangle (Figure from Truran et al. 2003).

Even more remarkable is the fact that this detailed agreement of the metal-poor star heavy-element abundance pattern with Solar System r-process abundances noted above is evident for all of the stars of [Fe/H] $\lesssim$ −2.5, for which data is available. This is evident in Figure 5, where the heavy element abundance patterns for three metal-poor but r-process-rich stars CS 22892–052 (Sneden et al. 2003), HD 115444 (Westin et al. 2000, and BD+17°3248 (Cowan et al. 2002) are compared with Solar System r-process abundances. The robustness of the r-process mechanism for the production of the heavy (A $\gtrsim$ 130–140) r-process elements is clearly reflected in the agreement of these patterns for low metallicity stars (which contain the nucleosynthesis products of at most a small number of progenitors) with the Solar pattern (which represents the integrated product of billions of years of Galactic chemical evolution).

3.3 Chemical Evolution Effects

We have seen that the most extremely metal deficient stars in our Galaxy have compositions that are generally consistent with predictions of the nucleosynthesis products of massive stars and Type II supernovae. The significant trends in Galactic chemical evolution with which we will be concerned in this paper are those involving the timing of first entry of the products of the other two broad classifications of nucleosynthesis contributors that we have identified: low mass stars (s-process nuclei) and Type Ia supernovae (iron peak nuclei).

The signatures of an increasing s-process contamination first appears at an [Fe/H] $\approx$ −2.5 to −2.0. The evidence for this is provided by an increase in the ratio of barium

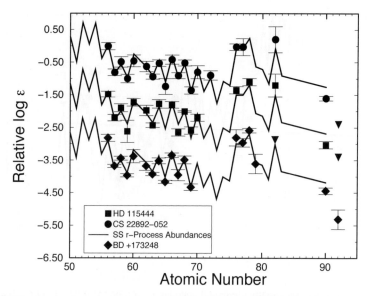

Figure 5: The heavy element neutron-capture elemental abundances for the three r-process rich halo stars CS 22892–052 ([Fe/H] = –3.1), HD 155444 ([Fe/H] = –3.0), and BD +17°3248 ([Fe/H] = –2.1) are compared with the Solar System r-process abundance curve (solid lines). The source data may be found in Sneden et al. (2003), Westin et al. (2000), and Cowan et al. (2002), respectively. Upper limits are indicated by inverted triangles (Figure from Truran et al. 2003).

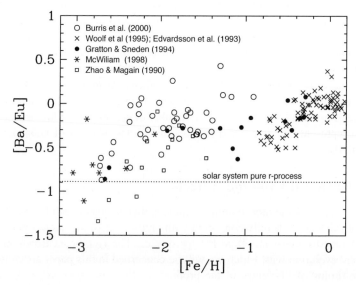

Figure 6: The history of the Ba/Eu ratio is shown as a function of metallicity [Fe/H]. This ratio reflects to a good approximation the ratio of s-process to r-process elemental abundances, and thus measures the histories of the contributions from these two nucleosynthesis processes to Galactic matter (Figure from Truran et al. 2003).

(the abundance of which in Galactic matter is dominated by s-process contributions) to europium (almost exclusively an r-process product). The ratio [Ba/Eu] is shown in Figure 6 as a function of [Fe/H] for a large sample of halo and disk stars. Note that at the lowest metallicities the [Ba/Eu] ratio clusters around the 'pure' r-process value ($[\text{Ba/Eu}]_{r-\text{process}} \approx -0.9$); at a metallicity [Fe/H] ≈ -2.5, the Ba/Eu ratio shows a gradual increase. In the context of our earlier review of nucleosynthesis sites, this provides observational evidence for the first input from AGB stars, on timescales perhaps approaching $\tau_{\text{IMS}} \gtrsim 10^9$ years.

In contrast, evidence for entry of the iron-rich ejecta of SNe Ia is seen first to appear at a metallicity [Fe/H] ≈ -1.5 to -1.0. This may be seen reflected in the abundance histories of [O/Fe] and [Ca/Fe] with [Fe/H], in Figures 1 and 2, respectively. This implies input from supernovae Ia on timescales $\tau_{\text{SNe Ia}} \gtrsim 1\text{–}2 \times 10^9$ years. It may be of interest that this seems to appear at approximately the transition from halo to (thick?) disk stars.

3.4 Inhomogeneity at Early Galactic Epochs

Significant star to star variations in abundance contributions from specific nucleosynthesis processes may be expected in the very early Galaxy, on timescales short with respect to mixing/homogenization timescales. Possible such effects have previously been discussed by e. g. Audouze and Silk (1994), with regard to the behaviors of stellar abundance patterns at the lowest metallicities discussed in the previous section. Perhaps the most dramatic evidence for such inhomogeneity, however, is provided by the increasing scatter identified in the ratio of r-process element abundances to iron, most pronounced at values below [Fe/H] ~ -2. The available data is displayed in Figure 7. Note that several of these stars are extremely r-process rich: CS 22892–052 (Sneden et al. 2003) and CS 31082–001 (Hill et al. 2002) both have [r-process/Fe] > 1.5. This makes apparent that not all early stars are sites for the formation of both r-process nuclei and iron. In the absence of other sources for either Fe or r-process nuclei, the logical conclusion to be drawn from the observed degree of scatter at the lowest metallicities is that only a small fraction of the massive star environments that produce iron also produce r-process elements. This issue is discussed further in e. g. the papers by Fields, Truran, & Cowan (2002) and Qian & Wasserburg (2001).

4 Abundance Probes of the High-z Universe

The abundance trends involving the α-elements are expected as well to be reflected in the composition of the high redshift Universe. By this we mean that, for systems at metallicities [Fe/H] $\lesssim -1$, the abundances in the oxygen-to calcium region should be elevated relative to iron by a factor $\approx 2\text{–}3$. Abundance determinations for Quasar absorption line systems (Lauroesch et al. 1996; Lu et al. 1996) reveal interesting trends in this regard. At redshifts out to at least $z \approx 3\text{–}3.5$, the intergalactic medium shows evidence for the presence of metals at levels $\sim 10^{-3}$ to 10^{-2} $Z_\odot$, indicating that massive star nucleosynthesis contributions have been realized. Damped

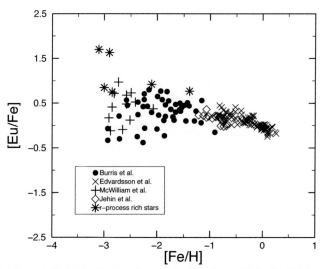

Figure 7: The ratio Eu/Fe is displayed as a function of metallicity [Fe/H] for a large sample of halo and disk stars. The increasing scatter apparent in [Eu/Fe] at lower metallicities reflects the increasing level of inhomogeneity of the halo gas at early Galactic epochs (Figure from Truran et al. 2003).

Lyman-α systems appear to be characterized by abundance patterns similar to those of Population II stars in our Galaxy.

A difficulty with reading this data arises from the fact that the elemental abundance determinations for the gas components of QSO absorbers (e. g. Si/Fe) may not reflect the corresponding total gas-plus-dust ratio (Kulkarni et al. 1997). This is why Zn (which is less prone to depletion) is often preferred over Fe as a tracer of abundance history (see, e. g., the nice review paper by Pettini (2002)). An alternative to Si/Fe as an indicator of such evolution is the S/Zn ratio. The increasingly available data (Figure 3) on metal deficient halo stars back to [Fe/H] $\approx$ –3 (Sneden, Gratton, & Crocker 1991; Mishenina et al. 2002; Primas 2002) strongly suggests that the abundance of zinc follows that of iron (an important clue to nucleosynthesis theorists). If indeed Zn traces Fe in total abundance, then the S/Zn ratio – involving relatively undepleted elements – can serve as a probe of explosive nucleosynthesis occurring in early stars and thus of the star formation histories of the observed damped Lyman-α systems.

Gamma-ray bursts now provide an alternative mode of study of the star formation and nucleosynthesis history of the high redshift Universe (Lamb and Reichart 2000). GRBs and their afterglows probe the compositions of the circumburst or interstellar media of their host galaxies. From a study of the optical transients associated with three gamma-ray bursts, Savaglio et al. (2003) find low [Fe/Zn], [Si/Zn], and [Cr/Zn] values, suggesting large dust depletions. This introduces uncertainty into the interpretation of the high values obtained for two of these systems ([Si/Fe] = +0.83 and +0.73) as signatures of massive star and Type II supernova nucleosynthesis. Future studies of such GRB afterburst environments can hopefully improve the situation, with the addition of data involving other interesting elements (e. g. sulfur).

5 Discussion

The Universe emerged from an exciting first three minutes with a composition consisting of the isotopes of hydrogen and helium and ^{7}Li. Further synthesis of heavy elements then awaited the formation and evolution of the first stars, some hundreds of millions of years later. The characteristics of the first stellar contributions to nucleosynthesis, whether associated with Population II or a Population III, reflect (as might be expected) the nucleosynthesis products of the evolution of massive stars (M $\gtrsim$ 10 M$_\odot$) of short lifetimes ($\tau \lesssim 10^8$ years).

Metal deficient stars ($-1.5 \lesssim$ [Fe/H] $\lesssim -3$) in our own Galaxy's halo show two significant variations with respect to Solar abundances: the elements in the mass range from oxygen to calcium are overabundant – relative to iron-peak nuclei – by a factor $\approx$ 2–3, and the abundance pattern in the heavy element region A $\gtrsim$ 60–70 closely reflects the r-process abundance distribution that is characteristic of Solar System matter, with no evidence for an s-process nucleosynthesis contribution. Both of these features are entirely consistent with nucleosynthesis expectations for massive stars and associated Type II supernovae.

The introduction of both the s-process products of the evolution of low mass stars (M $\approx$ 1–2 M$_\odot$) and of iron nuclei from Type Ia supernovae occurs on a significantly longer timescale. In our Galaxy, evidence for low mass star input is first seen at a metallicity $-2.5 \lesssim$ [Fe/H] $\lesssim -2$, while the products of Type Ia supernovae seem first to have appeared at $-1.5 \lesssim$ [Fe/H] $\lesssim -1$. Such trends are consistent as well with abundance determinations for both galactic globular clusters and nearby dwarf galaxies. They are also generally consistent with observations of the abundances and metallicity evolution (in the high redshift Universe) of damped Lyman-α systems.

The chemical evolution of Galactic matter is written in these abundance patterns. It follows that the abundance histories as a function of metallicity that we have reviewed in this paper should reflect the star formation histories of the various stellar populations of our Galaxy. We may also expect that the abundance history of the Universe itself, as revealed by abundances in other galaxies and in QSO and GRB absorption line systems, should provide comparable clues to the star formation histories of these populations as a function of time/redshift. How might we proceed to attempt to extract this information? Consider the following inferences from our discussions above for the case of our Galaxy:

- The abundances observed in halo field stars and globular cluster stars – representative of Population II stars in our Galaxy – generally exhibit Type II supernova signatures. This may be interpreted as suggesting either (i) a massive-star-dominated initial mass function or (ii) a short halo star formation and nucleosynthesis enrichment epoch.

- The duration of this phase of nucleosynthesis roughly follows from our knowledge of the lifetimes of the massive star progenitors of Type II supernovae: $\tau \lesssim \tau_{\text{SNe II}} < 10^8$ years.

- The approximate look-back time to this epoch can be determined by several methods. (1) The average age of the metal-poor globular clusters in our

Galaxy, from the main-sequence turnoff method is given by Chaboyer (2001) as 13.2 ± 1.5 Gyr. (2) A white dwarf luminosity function-based age for the globular cluster M4 is quoted by Hansen et al. (2002) as 12.7 ± 0.7 Gyr. (3) An exciting concomitant of the availability of accurate determinations of the abundances of r-process elements in the oldest halo stars is the ability to date individual stars. The long lived actinide nuclei ^{232}Th ($\tau_{1/2} = 1.4 \times 10^{10}$ years) and ^{238}U ($\tau_{1/2} = 4.5 \times 10^9$ years) are the nuclear chronometers of choice for this purpose. Thorium-uranium ages are now available for two ultra metal deficient stars: 14.0 ± 2.4 Gyr for CS 31082–001 (Cayrel et al. 2001; Hill et al. 2002) and 13.8 ± 4.0 Gyr for BD+17°3248 (Cowan et al. 2002). It seems clear from these consistent age estimates that the earliest star formation and nucleosynthesis events in our Galaxy occurred some 12–14 Gyr ago.

- The fact that the trends in metallicity in the halo stars persist through metallicities [Fe/H] ≈ -1 suggests that significant disk star formation activity may have been delayed $\gtrsim 1$–2 Gyr. This follows from the fact that both the introduction of s-process nuclei from AGB stars ($\tau_{\mathrm{IMS}} \gtrsim 10^8$–$10^9$ years) and of the iron-peak products of Type Ia supernovae ($\tau_{\mathrm{SNe\,Ia}} \gtrsim 1.5$–$2 \times 10^9$ years) occurred at metallicities below [Fe/H] ≈ -1.

- The age of the disk itself can be approximately determined by a study of the age-metallicity relation. Over the lifetime of the disk, the Fe/H ratio has increased by a factor ≈ 10, with significant scatter, while the ratios O/Fe and α-element/Fe have fallen from values 2–3 times Solar to their Solar values. Feltzing, Holmberg, & Hurley (2001) constructed an age-metallicity relation for 5828 dwarf and sub-dwarf stars from the Hipparcos Catalog, using evolutionary tracks to derive ages and Strömgren photometry to derive metallicities. Both the Edvardsson et al. (1993) and the Feltzing, Holmberg, & Hurley (2001) samples indicate ages back to approximately 12 billion years. Scrutiny of the Edvardsson et al. data reveals that there is increasing velocity dispersion with look-back time for the oldest 'disk' stars – perhaps suggesting an increasingly thick disk component (see, e. g. the discussion by Burkert, Truran, & Hensler 1992).

This simple summary of timescales and populations for our own Galaxy provides an illustration of the manner in which abundance studies can be utilized to trace the star formation histories of stellar populations. The increasingly available data on gas clouds and galaxies at high redshifts promises to allow us to trace in greater detail the chemical evolution of the Cosmos itself.

Acknowledgments

This research was supported in part by the DOE under contract DE-FG02-91ER40606 in Nuclear Physics and Astrophysics at the University of Chicago.

References

Audouze, J. & Silk, J. 1995, ApJL 451, 49

Brown, J.H., Burkert, A., & Truran, J.W. 1991, ApJ 376, 115

Burkert, A., Truran, J.W., & Hensler, G. 1992, ApJ 391, 651

Burris, D.L., Pilachowski, C.A., Armandroff, T., Sneden, C., Cowan, J.J., & Roe, H. 2000, ApJ 544, 302

Busso, M., Gallino, R., & Wasserburg, G.J. 1999, AAR&A 37, 239

Cayrel, R. et al. 2001, Nature 409, 691

Chaboyer, B.C. 2001, in "Astrophysical Ages and Time Scales", ed. T. von Hippel, C. Simpson, & N. Manset, ASP Conf. Ser. 245, 307

Cowan, J.J., & Roe, H. 2000, ApJ 544, 302

Cowan, J.J., Sneden, C., Burles, S., Ivans, I.I., Beers, T.C., Truran, J.W., Lawler, J.E., Primas, F., Fuller, G.M., Pfeiffer, B., & Kratz, K.-L. 2002, ApJ 572, 861

Domímquez, I., Höflich, P., & Straniero, O. 2001, ApJ 557, 279

Edvardsson, B., Andersen, J., Gustafsson, B., Lambert, D.L., Nissen, P.E., & Tomkin, J. 1993, A&A 275, 101

Feltzing, S., Holmberg, J., & Hurley, J.R. 2001, A&A 337, 911

Fields, B.D., Truran, J.W., & Cowan, J.J. 2002, ApJ 575, 845

Freiburghaus, C. & Rosswog, S. 1999, ApJ 525, 121

Goswami, A., & Prantzos, N. 2000, A&A 359, 191

Hansen, B.M.S. et al. 2002, ApJ 574, L155

Heger, A. & Woosley, S.E. 2002, ApJ 567, 532

Hill et al. 2002, A&A 387, 560

Iben, I. Jr., & Truran, J.W. 1978, ApJ 220, 980

Iwamoto, K., Brachwitz, F., Nomoto, K., Kishimoto, N., Hix, R., & Thielemann, F.-K. 1999, ApJS 125, 439

Kobayashi, C., Tsujimoto, T., & Nomoto, K. 2000, ApJ 539, 26

Kulkarni, V.P., Fall, S.M., & Truran, J.W. 1997, ApJ 484, L7

Lamb, D.Q. & Reichart, D.E. 2000, ApJ 536, 1

Lattimer, J.M., Mackie, F., Ravenhall, D.G., & Schramm, D.N. 1977, ApJ 213, 225

Lauroesch, J.T., Truran, J.W., Welty, D.E., & York, D.G. 1996, PASP 108, 641

Limongi, M., Straniero, O., & Chieffi, A. 2000, ApJS 129, 625

Livio, M. 2000, in "Type Ia Supernovae: Theory and Cosmology", ed. J.C. Niemeyer & J.W. Truran, Cambridge University Press, p. 33

Lu, L., Sargent, W.L.W., Barlow, T.A., Churchill, C.W., & Vogt, S.S. 1996, ApJS 107, 475

Madau, P., Della Valle, M., & Panagia, N. 1998, MNRAS 297, L17

McWilliam, A. 1995, AAR&A 35, 503

McWilliam, A., Preston, G.W., Sneden, C., & Searle, L. 1995, AJ 109, 2736

McWilliam, A., Preston, G.W., Sneden, C., & Searle, L. 1995, AJ 109, 2757

Mishenina, T.V., Kovtyukh, V.V., Soubiran, C., Travaglio, C., & Busso, M. 2002, A&A 396, 189

Nomoto, K., Hashimoto, M., & Tsujimoto, T. 1997, Nucl. Phys. A161, 79

Pettini, M. 2002, in "Cosmochemistry: The Melting Pot of Elements", Lectures given at the XIII Canary Islands Winter School of Astrophysics

Primas, F. 2002, private communication

Qian, Y.-Z. 2000, ApJL 534, 67

Qian, Y.-Z. & Wasserburg, G.J. 2001, ApJ 559, 925

Renzini, A., & Voli, M. 1981, A&A 94, 175

Savaglio, S., Fall, S.M., & Fiore, F. 2003, astro-ph/0203154

Shetrone, M.D., Côté, P., & Sargent, W.L.W. 2001, ApJ 548, 592

Sneden, C., Cowan, J.J., Burris, D.L., & Truran, J.W. 1998, ApJ 496, 235

Sneden, C., Cowan, J.J., Ivans, I.I., Fuller, G.M., Burles, S., Beers, T.C., & Lawler, J.E. 2000, ApJ 533, L139

Sneden, C., Cowan, J.J., Lawler, J.E., Ivans, I.I., Burles, S., Beers, T.C., Primas, F., Hill, V., Truran, J.W., Fuller, G.M., Pfeiffer, B., & Kratz, K.-L. 2003, ApJ, in press

Sneden, C. & Crocker, D.A. 1988, ApJ 335, 406

Sneden, C., Gratton, R.G., & Crocker, D.A. 1991, ApJ 246, 354

Takahashi, K., Witti, J., & Janka, H.-T. 1994, A&A 286, 857

Thielemann, F.-K., Nomoto, K., & Hashimoto, M. 1996, ApJ 460, 408

Thielemann, F.-K., Nomoto, K., & Yokoi, K. 1986, A&A 158, 17

Timmes, F.X. 2003, private communication

Truran, J.W. 1981, A&A 97, 391

Truran, J.W. & Arnett, W.A. 1971, Ap&SS 11, 430

Truran, J.W., & Cameron, A.G.W. 1971, Ap&SS 14, 179

Truran, J.W., Cowan, J.J., & Fields, B.D. 2001, Nucl. Phys. A688, 330

Truran, J.W., Cowan, J.J., Pilachowski, C.A., & Sneden, C. 2003, PASP 114, 1293

Westin, J., Sneden, C., Gustafsson, B., & Cowan, J.J. 2000, ApJ 530, 783

Wheeler, J.C., Sneden, C., & Truran, J.W. 1989, AAR&A 27, 279

Woosley, S.A., & Weaver, T.A. 1995, ApJS 101, 181

Woosley, S.A., Wilson, J.R., Mathews, G.J., Hoffman, R.D., & Meyer, B.S. 1994, ApJ 433, 229

Matter and Energy in Clusters of Galaxies as Probes for Galaxy and Large-Scale Structure Formation in the Universe

Hans Böhringer

Max-Planck-Institut für extraterrestrische Physik
D-85748 Garching, Germany
hxb@mpe.mpg.de

Abstract

Clusters of galaxies enclose a large enough volume to constitute ideal laboratories for a detailed study of a representative chunk of material of our Universe comprising a large galaxy population, dark matter, and intergalactic medium. X-ray observations allow a very precise investigation of the physical properties of the intracluster plasma which fills the entire cluster volume. With the knowledge of the intracluster medium properties we can probe the cluster structure, determine its total mass, and measure the baryon fraction in clusters and in the Universe as a whole. We can determine the abundance of heavy elements from O to Ni which originate from supernova explosions. From these metal abundances we can infer important information on the history of star formation in the cluster galaxy population and from the entropy structure of the intracluster medium we obtain constraints on the energy release during early star bursts. With the observational capabilities of the X-ray observatories XMM-Newton and Chandra this field of research is rapidly evolving. The present contribution gives an account of the current implications of the intracluster medium observations, but more importantly illustrates the prospects of this research for the coming years.

1 Introduction

Clusters of galaxies as the largest well defined and nearly virialized objects, are the largest well characterized astrophysical laboratories at our disposal for research, except for the Universe as a whole. The total masses of clusters range from less than 10^{14} to several 10^{15} $M_\odot$. (Below this mass scale groups of galaxies have quite similar downscaled properties as galaxy clusters and extend down to masses of several 10^{12} $M_\odot$. We will treat this size spectrum as continuous and discuss the differences of groups and clusters later in this paper). Hundreds to thousands of cluster galaxies are making up merely a few percent of this total mass while more mass is in the gaseous intracluster medium and the dominant fraction is made up by Dark Matter. Clusters are thus large mass associations allowing the study of a large representative sample of coeval galaxies and their large-scale environment.

Furthermore the formation of these large mass associations is a relatively well understood and well modeled process. They are formed through gravitational collapse of overdense regions in the primordial density fluctuation field and therefore their evolution and their properties are mostly determined by the action of gravitational forces. Since gravity acts on all forms of matter equally when they are assembled into the cluster laboratory, galaxy clusters contain to a good approximation a composition of matter well representative of the Universe as a whole (e.g. White et al. 1993). These properties of clusters make them ideal laboratories to study galaxy evolution in controlled large-scale environmental conditions.

In this contribution I will focus on the information we can gain on galaxy evolution from the study of the chemical composition and the thermodynamic structure of the intergalactic medium in groups and clusters of galaxies. Fig. 1 provides a sketch of the relevant scenario. From recent studies of the distant galaxy population we know that most activity in the galaxy formation process, measured by the rate of star formation, happened at redshifts between about $z = 1$ and 3 (e.g. Madau et al. 1998, Ferguson et al. 2000). In the denser environments of galaxy clusters these processes are expected to have occured even earlier at redshifts around $z = 3$ and before. The traces of what happened then can be found in the intracluster medium today: i) we observe that the intracluster medium is heavily polluted by metals (elements more massive than carbon) which must have been synthezised by supernovae in the cluster galaxies mostly in these early star formation epochs. We further observe in the entropy structure of groups and clusters of galaxies the effect of an early heating of the intergalactic medium which again is most likely the effect of the energy release in the star formation epochs.

For our understanding of cosmic galaxy formation, the study of the intracluster medium is therefore a very essential complement to the conventional study of the evolution of the stellar population in the galaxies done mostly by optical and infrared astronomy.

The intergalactic gas in clusters is a hot plasma with temperatures of several ten Million degrees which has its thermal radiation maximum in the soft X-ray regime (e.g. Sarazin 1986). It is a fortunate coincidence that it is the same wavelength region in which we can use the current technology of imaging X-ray telescopes and it is therefore through X-ray astronomy that we know most about this intracluster environment. The hot plasma is optically thin (except for the most exceptionally dense regions observed in certain emission lines) which makes the interpretation of the observations easy and the modeling is currently provided by several relatively well developed publicly available radiation codes (e.g. Mewe et al. 1985, 1995, Smith 2001). The observed spectra are mostly composed of bremsstrahlung and collisionally excited emission lines. The latter are relatively straight-forwardly used to measure the elemental composition of the intracluster plasma. Therefore this contribution is mostly based on information from X-ray observations. In this respect we are currently in the advantageous situation to have two new satellite X-ray observatories, ESA's XMM-Newton mission and NASA's Chandra mission, with greatly improved

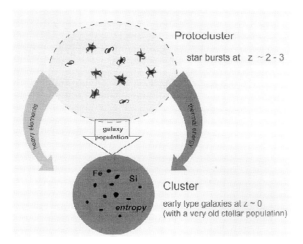

Figure 1: Schematic representation of a galaxy cluster used as a laboratory for the investigation of fossil records of processes that have occured during an early star burst phase of the cluster galaxy population.

spectral and imaging capabilities, which are providing a overwhelming new insight into the physics of the intracluster medium.

In this contribution I will use a scaling of physical parameters in different ways with the Hubble constant depending on the sources quoted. Therefore we will include a scaling parameter labeled with the Hubble constant used, such that e.g. $h_{70} = H_0/70 \,\mathrm{km\,s^{-1}\,Mpc^{-1}}$.

2 The mass fraction of the intergalactic gas

Before we start to look into more detail into the matter composition of the intracluster medium, let us investigate how much of the mass fraction of a cluster is in the observed intracluster gas – or in more general terms accounting for the intracluster medium together with the galaxies – what fraction of the mass is in baryonic form and how much is presumably non-baryonic Dark Matter.

The gas mass as well as the total mass can be determined from X-ray imaging and spectroscopy of the intracluster medium. With the usual assumption of spherical symmetry the X-ray surface brightness distribution which is proportional to the line of sight integral of the square of the gas density can be "deprojected" into the gas density distribution. Radial integration of the gas density profile yields the gas mass. The gas temperature is obtained from a spectral analysis of the X-ray emission usually conducted for concentric rings of the surface brightness distribution to obtain a radial gas temperature profile. From the known gas density and temperature distribution with the additional assumption of hydrostatic equilibrium the gravitational potential and the gravitational mass of the cluster can be determined. Fig. 2 shows the gas mass fractions determined for a sample 106 of the brightest galaxy clusters

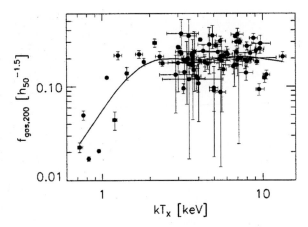

Figure 2: Gas mass fraction in units of $h_{50}^{-1.5}$ for 106 of the brightest clusters of galaxies detected by ROSAT in X-rays (from Reiprich 2001).

observed with ROSAT and ASCA (Reiprich 2001, Reiprich & Böhringer 2002). The data scatter around a value of 18–20 % (for $h = 0.5$) corresponding to 12 % (for $h = 0.7$). The error bars for these measurements originates mostly from the uncertain determination of the temperature and temperature profile based on ASCA observations.

Now the first detailed X-ray studies of galaxy clusters with XMM-Newton are becoming available providing much more detailed and accurate spatially resolved temperature information. One of the best recent examples of the determination of the gas mass fraction is shown in Fig. 3 for the galaxy cluster A1413 (Pratt & Arnaud 2002). Here the gas mass fraction has been determined out to a radius not far form the assumed virial radius of the cluster allowing to almost account for the mass distribution in the cluster as a whole. The resulting gas mass fraction is $0.2(\pm 0.02)h_{50}^{-1.5}$ consistent with the value given above. Similarly interesting results have been obtained with the Chandra observatory (e. g. Allen et al. 2001, Ettori et al. 2002) but with the lower sensitivity of Chandra these measurements do in general not extent to as large radii as for the A1413 result obtained with XMM-Newton.

With the reasonable assumption that most of the unseen matter is in non-baryonic form, probably in the form of yet unknow very weakly interacting elementary particles, we can determine the baryon fraction of the total cluster mass by summing the observed gas and galaxy mass. Taking the mass in the stellar population of the galaxies as estimated from the total light and a mass-to-light ratio for an old stellar population of about 5–8 in solar units to be about 2–3 % of the total mass we find a total baryon fraction of about 14–15 % $h_{70}^{-1.5}$.

If this is representative of the total matter composition of the Universe – as argued above (and shown in many simulations, e. g. White et al. 1993) – we can compare this number with the baryon fraction predicted from nucleosynthesis for the observed primordial deuterium to hydrogen ratio (e. g. O'Meara et al. 2001) together with the matter density of the Universe as provided by the cosmological "concordance" model. This model yields a baryon fraction of $\Omega_b h^2 = 0.020$ (e. g. Turner

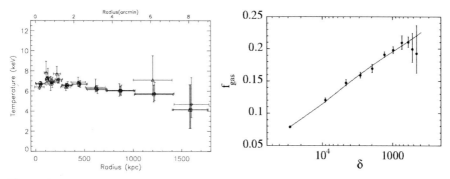

Figure 3: *Left panel:* Temperature profile of the cluster A1413 determined from XMM-Newton spectroscopic observations by Pratt & Arnaud (2002). This temperature profile stretches out to the largest radii in a cluster analysis reached so far, to 0.6 of the virial radius. *Right panel:* Gas mass fraction in A1314 from Pratt & Arnaud (2002).

2002). This is also consistent with the baryon fraction determined from recent analysis of microwave background observations which gives $\Omega_b h^2 = 0.019$ in good agreement with the above results (for $h = 0.7$ and $\Omega_m = 0.3$). This general consistency is a very nice confirmation of the results of the determination of the gas mass and gas mass fraction in galaxy clusters and we are therefore convinced that the gas mass fraction is a reasonably well determined parameter for galaxy clusters. This will also be important in the following when we determine the total masses of important elements in the intracluster medium from relative abundance measurements.

3 Observations of the metallicity of the intracluster medium before XMM-Newton

At typical cluster gas temperatures of several keV, the K-shell line of iron (Fe Lyman-α) is the most prominent line in the spectrum (this changes only for the plasma in colder groups of galaxies with temperatures below 2 keV). Therefore Fe was the first heavy element to be confirmed in X-ray spectra and for which most of the information was available before the era of XMM and Chandra. Before we come to the description of the more recent results I like to briefly review some of the information that was available before these two new instruments went into orbit.

Since 1993 the Japanese ASCA observatory provided good spectral capabilities for X-ray observations with spatial resolution. It allowed for the determination of element abundances mostly for the two elements iron and silicon. One of the interesting results was the finding of a central increase of the Fe abundance in clusters with so-called cooling flows (Matsumoto et al. 1996, Ezawa et al. 1997, Fukazawa et al. 1998, Finoguenov et al. 2000, 2001). More recent data obtained with the subsequently launched Italian X-ray satellite Beppo-SAX shown in Fig. 4 provide a survey of the relative abundance of iron in solar units as a function of cluster radius in a sample of galaxy clusters (De Grandi & Molendi 2001). The sample is split

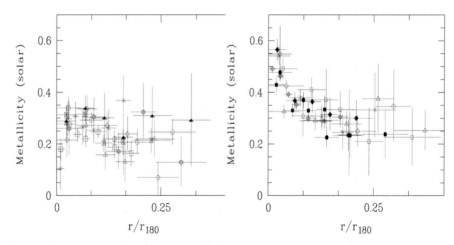

Figure 4: Iron abundance profiles of non-cooling flow clusters (*left*) and cooling flow clusters (*right*) as a function of the scaled radius expressed in units of the radius for a mean mass overdensity of 180 over the critical density of the Universe (De Grandi & Molendi 2001).

into two parts, one with galaxy clusters with pronounced centrally peaked X-ray surface brightness profiles with giant elliptical cD galaxies in the center, so-called cooling flow clusters, and clusters with shallower central surface brightness profiles. The latter group is also characterized by showing very frequently signs of ongoing cluster merging processes. One observes to distinct behaviors of the abundance profiles. While the "non-cooling flow" clusters show very flat abundance profiles with a typical abundance of about 0.2 solar on large-scales increasing up to a value of 0.3 solar in the center, the "cooling flow" clusters show a pronounced central abundance peak up to more than 0.5 solar (at the spatial resolution of the BeppoSAX X-ray telescope). De Grandi & Molendi further show that the abundance peak is roughly coincident with luminosity distribution of the stellar population in the cluster center dominated by the stars of the cD galaxy, as shown in Fig. 5.

These results lead to the possible implication that the central Fe abundance increase is connected to recent iron production in the central galaxies. Since the principle sources of the heavy elements are only the two types of supernovae, type II (core collapse supernovae) and type Ia (thermonuclear exploding white dwarfs) and only the supernovae of the tape Ia have been observed in the early type galaxies in the cluster centers, one is lead to the conclusion that the increased abundance of iron in the centers of these clusters comes from the enrichment by SN type Ia. This enhancement has to have occured since the time the structure of the cluster center with the cD in the middle has been established, that is since the time the cluster has settled after the last major cluster merger.

The hypothesis, that the abundance increase is due to more recent enrichment by SN type Ia can be tested by looking at the relative abundances of different elements produced in different ratios by the different types of supernovae. This has been done using ASCA results by Fukazawa et al. (1996, 1998) and Finoguenov et al. (2000, 2001). In Fig. 6 (Finoguenov et al. 2000) the ratio of the silicon abundance to the

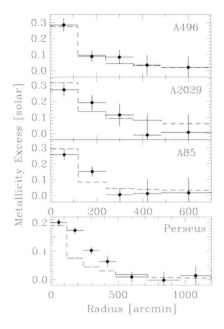

Figure 5: Comparison of the Fe abundance profile with the light profile of the cluster galaxies (De Grandi & Molendi 2001).

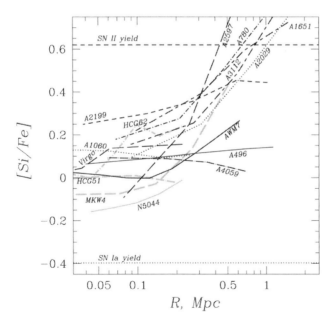

Figure 6: Radial profiles of the abundance ratio of Si to Fe in a sample of groups and clusters of galaxies (Finoguenov et al. 2000). Also shown is the expected abundance ratio from the yields of supernovae type Ia and type II.

iron abundance in solar units is plotted as a function of the cluster radius for a sample of galaxy systems ranging from rich groups to massive clusters. There is a clear trend for an increase in the Si to Fe ratio with cluster radius which is at least clearly observed for the more massive cluster systems (like A780, A1651, A2029, A2597, A3112). It is hardly seen in the less massive poor clusters and groups, however. Also shown in the figure are the classical abundance ratios which are believed to be produced by the two types of supernovae. The ratio of the Si to Fe yield for SN II is much higher than for SN Ia, and we observe that in the outskirts of the clusters the abundance pattern reflects quite well the yields of the core collapse supernovae. The observed abundance ratio moving towards the clusters center is than easily explained by an increased enrichment by SN Ia just as expected from the above speculations.

The importance of a late enrichment of cluster ellipticals was also pointed out in an ASCA study of the element abundances of Mg, Si, S, and Fe in six Virgo elliptical galaxies by Finoguenov and Jones (2000).

Models that tried to quantitatively explain the production of the observed iron abundances showed that simple calculations based on the chemical evolution of our own galaxy failed to reproduce the observed iron by a large factor. Two types of models were proposed to explain the observations: models with an increased production of Fe by type II supernovae by proposing an initial mass function (IMF) for the early star formation which leads to an enhanced occurance of these supernovae for a given stellar population (e. g. Arnaud et al. 1992) and models with an increased rate of type Ia supernovae in the past (e. g. Renzini et al. 1993).

A distinction between these two types of explanations will be possible if we have a detailed knowledge about the abundance distribution of different elements in a sample of clusters, which will be achieved with the observations with the X-ray observatories Chandra and XMM-Newton. Thus the observations previous to the launch of these two new observatories showed the great potential of heavy element abundance studies in the intracluster medium and highlighted the problem that it is not trivial to explain the large amount of metals found there. One can summarize this problem as follows. The above described observations indicated a universal iron abundance in the intracluster medium of about 0.3 of the solar value. We find at least in the rich clusters about five times more mass in the intracluster gas than in the galaxies. Thus there is as much iron in the gas as in a galaxy population with an iron abundance of two times the solar value – or put in other words, each galaxy has to have produce four times the solar abundance of iron. Therefore we know that two questions have to be inspected more closely with the new X-ray instruments: is the Fe abundance overestimated due to the neglection of gradients in the abundances (since the iron line is proportional to the density of iron and the electron density, the central regions are overrepresented in the integral spectrum – if the iron abundance is falling off with radius the iron abundance will thus be overestimated)? Can the abundance distribution of different heavy elements shed more light on the production process?

4 Metal abundance determination in the X-ray halo of M87 with XMM-Newton – a corner stone study

XMM-Newton with its high sensitivity, the capability to perform spatially resolved spectroscopy with an angular resolution of about 8 arcsec (e. g. 6.6 arcsec FWHM for the EPN detector; Jansen et al. 2001) and a spectral resolution of e. g. $\Delta E/E \sim 40$ around 6 keV for the EPN detector (Strüder et al. 2001) and slightly better resolution for the EMOS detectors (Turner et al. 2001) enables us to make a giant step forward in this analysis. At present the first detailed results are just becoming available indicating the potential of the instrument. The full impact will, however, only be seen, after a few years when larger cluster samples have systematically been studied as it has been done over years of research with the ROSAT, ASCA, and Beppo-SAX satellites with the results described above. Thus we will concentrate here on the results of a specific example which is also one of the best cases to be studied – the XMM-Newton observation of the X-ray halo of M87.

4.1 Abundance pattern in the halo of M87

The giant elliptical galaxy, M87, in the center of the more massive northern part of the Virgo cluster is the nearest X-ray luminous galaxy cluster center. It is the most luminous X-ray cluster source providing the highest signal-to-noise ratio for X-ray spectra and allows us to study the spatial structure of the cluster at the highest spatial resolution. In addition the temperature of the X-ray halo of M87 is relatively low with values in the range from 1 to 3 keV, a temperature range where the spectra are very rich in emission lines. For its importance M87 was made one of the verification targets of the XMM-Newton mission. Fig. 7 shows an X-ray image of the X-ray halo of M87 taken with the EPN detector of XMM. The physical size of the cluster region covered by the image extends to a radius of about 80 kpc (a distance of 17 Mpc for M87 is assumed here and throughout the paper). An inset shows the X-ray spectrum obtained from the central 2 arcmin region of the cluster. Besides the dominantly Bremsstrahlung continuum the emission lines of the nine astrophysically most important heavy elements are observed, providing the possibility to determine their abundances.

It is worth to recall that what we see in the X-ray spectrum as emission lines is the integral emission of the heavy elements in the entire volume of the intracluster medium. This is due to the fact that the plasma is optically thin (except for the central arcmin of the M87 observation in some emission lines where we observe resonant line scattering as discussed in Böhringer et al. (2001) and Matsushita et al. (2002a)). Thus we receive the emission equally from all the volume without having to resort to radiation transport. That the gas is so teneous further facilitates the spectral modeling, since we can safely assume that all excitations by electron collisions are deexcited by radiation and there is no density dependence of the spectral line intensities. Both effects make the modeling of the spectra very straightforward. The model depends only on atomic data and the temperature and emission measure distribution of the plasma. We further find very good evidence that thermal

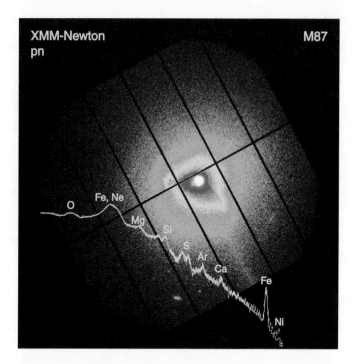

Figure 7: XMM-Newton EPN image of the X-ray halo of M87 with a spectrum of the central 2 arcmin region overlayed. The signature of the 9 most important heavy elements can be seen in the spectrum. The scale of the image is approximately 26 arcmin by 27 arcmin (Böhringer et al. 2001).

equilibrium prevails (e. g. between the electron temperature, the ion temperature, and the ionization structure). This is a better defined situation than in almost all other cases where abundances are determined from emission lines. In stars we see for example only the skin of the object and in many cases density corrections have to be applied and radiation transport calculations have to be performed. On the contrary, in the present case the iron line seen in Fig. 7 coming from the central 2 arcmin radius region in M87 is directly telling us about the presence of several Million solar masses of iron. Recalling these facts lets us better appreciate the reliability of the results on abundances observed in the intracluster and intragroup media.

Before one can resort to the determination of the element abundances the temperature structure of the gas in the M87 halo has to be known with high precision. Particularly in the relevant temperature range quoted above the Fe abundance is best determined from the L-shell lines, but the results are very sensitive to the assumed plasma temperature (see Fig. 8). Also for most of the other elements a temperature dependence has to be taken into account. The region observed with this central pointing on M87 is covering the region which was previously believed to harbour a classical cooling flow. Such cooling flows have been proposed to display a multi-temperature structure with a wide range of temperatures (e. g. Nulsen 1986, Thomas et al. 1987, Fabian 1994). An analysis of the observed spectra based on a multi-

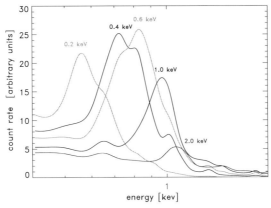

Figure 8: Simulations of Fe L line feature observations with the XMM EPN detector as a function of temperature (Böhringer et al. 2002).

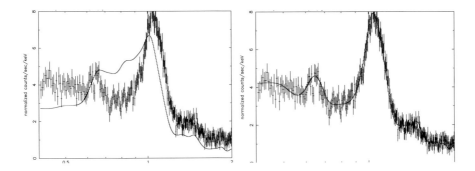

Figure 9: The observed Fe L line feature in the X-ray halo of M87 (around a radius of 2 arcmin) compared to a single temperature (*right*) and a classical cooling flow model (*left*; Böhringer et al. 2002).

temperature model may provide very different results from a simple analysis assuming a single temperature for the plasma (e. g. Buote 1999). Therefore the first step of our analysis of the data was a careful determination of the temperature structure of the plasma in the M87 halo (Matsushita et al. 2002a). It turns out that the classical cooling flow picture does not apply to the halo of M87 (Böhringer et al. 2001, 2002, Moldendi et al. 2001a,b, 2002) and that the gas is very close to be composed of a single temperature phase locally, but shows a monotonic decrease in temperature towards the center (Matsushita et al. 2002a).

This is illustrated in Figs. 8 and 9. Fig. 8 shows, that while the feature of the Fe L-shell line blend is very variable with temperature and yields abundances dependent on the assumed temperature as explained above, it is also a very good thermometer, since the peak of the line blend shifts significantly with temperature due to the changes of the mean ionisation level. If the gas is locally characterized by a narrow temperature interval we consequently would expect a relatively narrow line blend

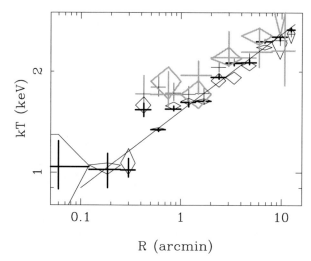

Figure 10: Temperature profile of the intracluster in M87 obtained from the analysis of deprojected spectra in concentric rings around the center of M87 (Matsushita et al. 2002a). The temperature has been determined from the complete spectrum (thick crosses), the Fe-L line region (black diamonds), the spectrum above 1.6 keV (grey crosses), and in the 2.3–2.7 keV region comprising lines of Si and S (grey diamonds).

feature, while for a broad temperature structure as predicted for a classical cooling flow, a broad feature should appear.

This distinction is made in Fig. 9 where we compare the locally single temperature case in the right panel with the classical cooling flow case in the left panel. The observations clearly favour the locally isothermal plasma (for details see Böhringer et al. 2002). This is very good news for our application. Since the spectra obtained from concentric rings turn out to be simply modeled by an ionization equilibrium isothermal plasma their interpretation can be achieved in the most straightforward way. This scenario has been checked in very much detail by making sure that spectral temperatures measured in at least four different ways, by line ratios as well as continuum shape, provide consistent results by Matsushita et al. (2002a) and we therefore believe that the resultant analysis of the spectra to obtain elemental abundances is correct. The resulting temperature profile is shown in Fig. 10. The results are also quite nicely consistent with the temperature profile anticipated by Matsumoto et al. (1996) from the less precise ASCA data.

In Fig. 7 we note that the surface brightness distribution is not perfectly symmetric. The are two "arms" of enhanced surface brightness stretching out from the center to radii of about 4–5 arcmin. These enhancements are spatially coincident with the radio lobes of M87 and have been interpreted as the interaction effects of the radio lobes with the intracluster gas and the gas temperature has been found to be lower and the gas density higher in these interaction regions (Böhringer et al. 1995, Churazov et al. 2001). This explains the higher surface brightness. Belsole et al. (2001) have actually used the Fe L-line thermometer to confirm the low temperature of the lobe gas with the XMM data. In the following analysis we are therefore carefully

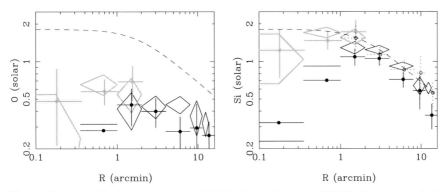

Figure 11: O and Si abundance profiles in M87 (Matsushita et al. 2002b). The grey symbols came from a two-temperature model (grey symbols) for the central region where the disturbance of the central AGN may not be perfectly subtracted. These results are more reliable than the simple one-temperature treatment (black symbols). The data are obtained from the two XMM detectors EPN (crosses) and EMOS (diamonds).

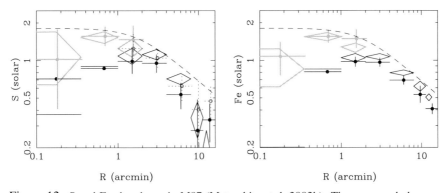

Figure 12: S and Fe abundance in M87 (Matsushita et al. 2002b). The grey symbols came from a two-temperature model (grey symbols) for the central region where the disturbance of the central AGN may not be perfectly subtracted. These results are more reliable than the simple one-temperature treatment (black symbols). The data are obtained from the two XMM detectors EPN (crosses) and EMOS (diamonds).

excising these regions and report only on the properties of the undisturbed intracluster medium.

Abundance profiles determined for the elements O, Si, S, and Fe are shown in Figs. 11 and 12 (Matsushita et al. 2002b). Note that for the derivation of these results the observed spectra have been deprojected, that is for the spectra observed in each ring the contribution of emission from the outer regions have been subtracted such that the analysed spectrum reflects purely the emission of the innermost projected three-dimensional shell. We clearly see a different behavior of the different elements. While O shows a relatively flat profile, the profiles of Si and Fe decrease significantly with radius (in approximately the same way) and S seems to fall off

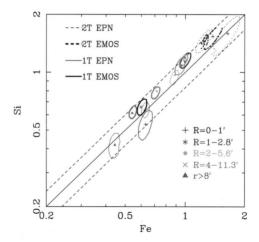

Figure 13: Si to Fe abundance ratio in M87 for different radial regions (different shading) and different ways of data analysis (different line styles; Matsushita 2002b).

even more steeply. This is partly what we could have expected: oxygen which is produced essentially only by SN type II at early epochs and which should be well mixed shows the more even distribution. The strong increase in Fe is confirming theearlier results (e. g. Matsumoto et al. 1996) but we now can see the profile at higher spatial resolution and note that it is even more peaked than inferred so far from the lower resolution ASCA data.

It comes somewhat as a surprise, however, that the profiles of Fe and Si are practically running in parallel. This is further illustrated in Fig. 13 where the Si to Fe ratio is shown for different concentric shells and different ways of data interpretation (Matsushita et al. 2002b). As a continuation of the results shown in Fig. 6 and the accompanying reasoning we should have expected a stronger increase of Fe compared to Si, because in the classical models the SN type Ia which are responsible for the central increase of the abundance profiles, should produce predominantly iron and less Si as e. g. assumed in the model constraints shown in Fig. 6 as horizontal lines.

4.2 Contribution of the different types of supernovae to the abundance pattern

To investigate this further the behavior of all the elements was studied by Finoguenov et al. (2001). In this study the data where grouped into two radial bins (where the temperature variation is affecting the spectra not so strongly) from 1 to 3 arcmin (excluding well the contribution of the central AGN and the radio lobe regions to the X-ray emission) and from 8 to 16 arcmin (the outermost region in this observation). The abundance patterns in solar units are shown in Fig. 14. These data reflect what is seen in the profiles shown in Figs. 11 and 12: the decrease in oxygen is considerably less than that of iron.

To deconvolve the contributions of the two different types of supernovae the following exercise is made. Under the assumption that the central abundance increase

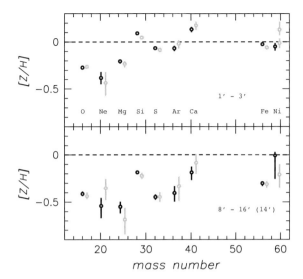

Figure 14: Abundance pattern in M87 for two different radial regions of 1–3 arcmin and 8–16 arcmin (Finoguenov et al. 2002a). The pairs of symbols show the separate results for the two XMM-Newton detectors EPN and EMOS.

is entirely due to SN type Ia one can subtract the outer abundance pattern from the inner one to obtain essentially the relative yields of the responsible SN type Ia. Alternatively on can use oxygen as a tracer of the contribution of SN type II and subtract from the inner abundances a weighted outer abundance pattern such that the O abundance is zero. (see Finoguenov et al. 2002a for details). The results of the latter procedure is shown in Fig. 13 where we also show the residual abundance pattern from SN type II. Note that these abundances have been derived with an earlier version of the M87 data reduction involving less perfect versions of the radiation codes and the XMM-Newton detector response matrices compared to the superceeding results published in Matsushita et al. (2002b); but even though the numerical values of the results have slightly changed, the changes are small and do not alter any of the conclusions.

These deconvolved abundance patterns are compared to theoretical models in the Figure. For the SN type Ia yields models by Nomoto et al. (1997a, see also Iwamoto et al. 1999) are used for the comparison. The model with the lowest yields for the α-elements compared to the Fe group elements is the classical Nomoto et al. (1984) W7 model of a fast SN Ia explosion which has been widely used for chemical evolution models. The relative yields inferred from the present observations show much larger α-element yields more in line with the slow deflagration-detonation models featuring a less complete burning of the α-elements. One also notes a second order result that the two patterns of SN type Ia for the inner and outer region do not perfectly agree, but the outer region seems to feature relatively lower yields for the α-elements. These may be an artefact of the imperfect subtraction procedure used in our analysis and one should treat this conclusion with care. However, this result may well be real since it is also a consequence of the finding shown in Fig. 13 of a

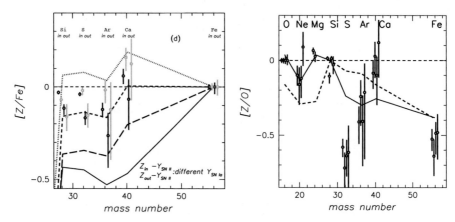

Figure 15: The contribution of SN type Ia (*left*) and SN II (*right*) to the abundance pattern in M87 (Finoguenov et al. 2002a). Also shown is a comparison with nuclear synthesis model calculations. *Left panel:* the lowest model curve is the classical SN Ia model of a fast explosion by Nomoto et al. 1984, the other three curves are slow deflagration-detonation models with the higher curves having decreasing deflagration speeds and less complete burning (Nomoto et al. 1997). *Right panel:* Shows a comparison with SN II models by Nomoto et al. 1997 (upper curve) and Woosley and Weaver (lower curve).

constant Fe to Si ratio. Thus, alternatively the result may fit into the new, tentative overall interpretation of the observations of a secular variation of the nature of type Ia SN explosions as outlined below.

A look at the yield pattern for SN type II shows that it agrees reasonably well with the prediction given e. g. in the review by Nomoto et al. (1997b) and a little less with with earlier results by Woosley and Weaver (1995). An exception is the bad match of the sulfur abundance for which we have no explanation at this moment.

4.3 Implications from the XMM-observations of element abundances

How do these results fit into the overall scheme of the chemical evolution of galaxies? In Fig. 14 a comparison of the element abundance distribution of stars in our Milky Way is compared to the M87 results (Matsushita et al. 2002b). The ratio of iron to silicon is plotted over the ratio of iron to oxygen which may be taken as a population age indicator. Two model lines for the evolution of the silicon to iron ratio are overplotted in this graph one for an increase of the abundances with a Si to Fe ratio of ~ 4 and with a ratio of 1.25. First of all we note that the M87 results are just a continuation of the Milky Way data in the sense that they reflect a more aged stellar population, which is to be expected for the old stellar population of the giant elliptical M87. It further appears that the data seem to follow at early evolution the upper theoretical line while at later evolutionary stages the data points approach the lower line indicating a change in the abundance pattern of the later enrichment (Matsushita 2002b).

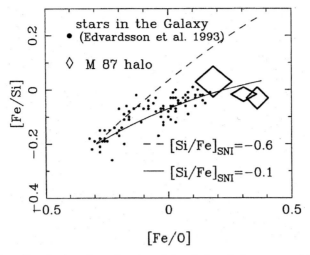

Figure 16: Fe to Si ratio plotted as a function of chemical evolution measured by the ratio of Fe to O from Matsushita et al. (2002b). The data points marking the abundances of galactic stars (from Edvardsson et al. (1993) and the results for M87 (diamonds) are compared. The M87 data (coming from a very old stellar population) appear as a continuation of the evolution traced by the galactic stars. The two SN type Ia enrichment model curves are further explained in the text.

One possible overall interpretation of these results is that there is a change in the nature and the yields of the supernovae contributing to the late enrichment of the interstellar or intergalactic medium. This would be a very dramatic result, because it would indicate a change in the physics of SN Ia explosions with stellar population, a change of a process which is used as a standard candle for cosmological tests. These variations in peak luminosity are presumably corrected by means of the empirical peak luminosity-light curve decay time correlation (see Hamuy et al. 1996). The yield pattern of the SN Ia supernovae is actually directly related to their total luminosity, since the luminosity is practically proportional to the amount of radioactive Ni synthesized in the explosion. Thus the SN with a less complete burning of α-elements will result in light curves with lower maxima.

Preliminary results from other elliptical galaxies and galaxy groups seem to confirm the results obtained from M87. A further test of this picture is the study of the Ni abundances, for which we have only uncertain data from the M87 observations but which can be studied much better in slightly hotter clusters.

In Fig. 17 we show the comparison of the Mg to O ratio for the intracluster medium of M87 and galactic stars (Matsushita et al. 2002b). The ratio observed in M87 fits well to our understanding of chemical evolution in our galaxy. Both elements are essentially only produced by type II SN and thus the ratio reflect the yields of this type of supernovae. In Fig. 18 (Matsushita et al. 2002b) we compare the Mg abundance of the intracluster medium with that of the stellar population as determined from optical spectra. In the past such a comparison was believed to produce contradicting results since the abundance of Fe in the intracluster medium was found

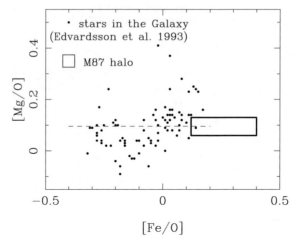

Figure 17: The oxygen to magnesium abundance ratio in M87 compared to that of galactic stars (Matsushita et al. 2002b).

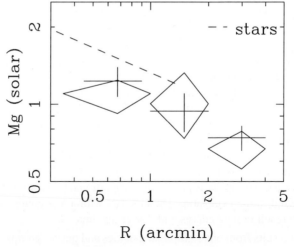

Figure 18: The magnesium abundance profile in the intracluster medium compared to the of the stellar population of M87 (Matsushita et al. 2002b).

to be lower than the Mg abundance of the central galaxy, if both abundances are expressed in solar units. Here we have two advantages: the abundances can be compared for the first time for the same element and with overlapping spatial resolution (see also discussion in Finoguenov and Jones 2000). We find that the discrepancy has almost vanished and the small differences can easily be due to some redistribution of the intracluster medium during the enrichment.

We can also now turn back to the problem of explaining the total iron abundance in the intracluster medium. The M87 observation implies that the Fe abundance that comes from SN II is about 0.15 solar, the ASCA results shown above and some recent XMM observations indicate that the iron abundance decreases with radius to values as low as 0.15 to 0.2 of the solar value. These values obtained at the outer radii accounts actually for most of the intracluster medium matter. Taking this value of the iron abundance that comes from type II SN, using a gas mass to light ratio of about 34 (see e.g. Arnaud et al. 1992), an overall mass to light ration for clusters of $\sim 200\ h_{70}$, and taking the stellar contribution to the iron abundance into account we arrive at a value of about 0.015–0.02 for the iron mass (from SN II) to light ration. A theoretical calculation using a Salpeter IMF gives a value around 0.009 for the iron mass to light ratio assuming an iron yield of 0.07 solar masses per supernova. Thus the new results show a decreased discrepancy for the explanation of the Fe production by SN II compared to the earlier studies, but note that we have already subtracted the centrally concentrated contribution by SN Ia (which is the smaller contribution). The gap is only up to a factor of two. More detailed observations are needed to confirm this result and to determine whether a different IMF for the early star burst phase is still necessary to explain the high metal abundance in the intracluster medium.

5 The entropy structure of groups and clusters of galaxies

The formation of dark matter halos from primordial overdensities is a self similar process. The structure of the groups and clusters of galaxies which are embedded in these dark matter halos is essentially determined by the self-similar nature of the dark matter halo structure and many of the properties of clusters are well characterized by self-similar scaling relations. E.g. the self-similar scenario predicts that the intracluster plasma temperature is proportional to the cluster mass with a power of 2/3 which is approximately observed (e.g. Finoguenov et al. 2001b). A closer look at the X-ray appearance of groups and clusters shows, however, that the predicted scaling relations are not always very well observed – the temperature – X-ray luminosity (bolometric luminosity) relation is a good example for this. The observed relation with a luminosity roughly proportional to the third power of the temperature is steeper than the self-similar prediction with a power of two (e.g. Navarro et al. 1996).

The possible reason for this is revealed in the diagnostics pioneered by Ponman et al. (1999). As shown in Fig. 19 the scaled gas density in groups (the X-ray temperature of the system is here used as a measure of the system mass) is significantly lower than those of more massive clusters. This implies a higher entropy of the group's intracluster medium as shown in Fig. 20. There the entropy of the intracluster medium (in the central region outside the so-called cooling flow zone) is plotted as a function of intracluster medium temperature shown in Fig. 20. The hot gas temperature and the entropy is in general the result of shock heating of the gas when the cluster is subsequently assembled by a series of substructure mergers. The shock strength should thereby be proportional to the depth of the gravitational potential in which the accreted matter settles which is itself proportional to the temperature.

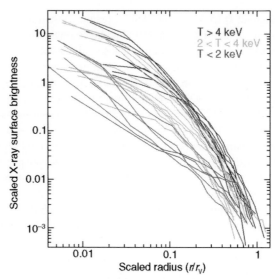

Figure 19: Scaled surface brightness profiles of groups and clusters of galaxies (Ponman et al. 1999). The surface brightness profiles are scaled according to the self-similar model such that they should coincide. The lower surface brightness of the low temperature systems (groups and poor clusters) constitutes a breaking of the self similar relation indicating that the central gas density is lower in the low mass systems than expected from self-similarity.

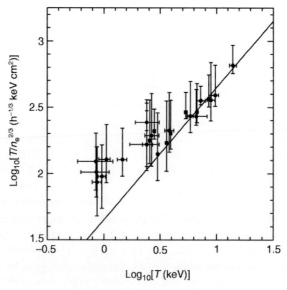

Figure 20: Central entropy (defined as $S = T/n_e^{2/3}$) of the intracluster medium in groups and clusters of galaxies (Ponman et al. 1999). In the self-similar model and with the assumption that the entropy is produced in the cluster accretion and merger shocks the entropy should increase with the depth of the potential well and the temperature of the system as indicated by the solid line. We clearly note an entropy excess for the low mass systems.

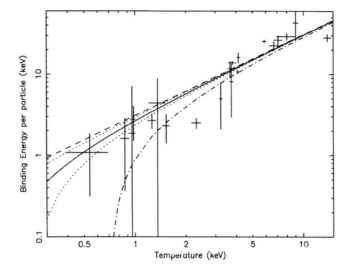

Figure 21: Gravitational binding energy of the intracluster medium protons in groups and cluster of galaxies (Lloyd-Davies et al. 2000). For the low mass systems where the effect of excess energy is strongest we note a decrease binding energy which corresponds to a specific energy input of about 0.5 keV per proton as indicated by the solid line.

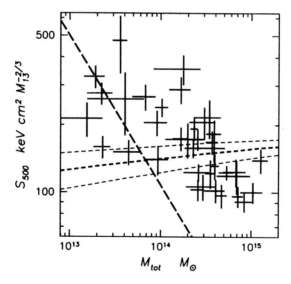

Figure 22: Entropy in clusters and groups measured at an outer radius, r_{500}, at a mean matter overdensity of the cluster of 500 over the critical density of the Universe scaled by the mass of the system with $M^{-2/3}$ in units of 10^{13} $M_\odot$. Self-similar shock heating is indicated by the dashed lines. A clear excess is again observed for the low temperature systems. The entropy values in unscaled units are higher than for the central regions with values as high as 300–500 keV cm^{-2}.

Thus we expect that the gas entropy (expressed here with the specific definition of $S = T/n^{2/3}$) is proportional to the shock strength and thus to the temperature. This is observed for the hotter, more massive clusters in Fig. 15 where the shock heating dominates, but for the lower temperature systems an "entropy floor" is observed (Ponman et al. 1999). This entropy floor is indicative of an additional heating mechanism which affects specifically the low temperature systems. Such an effect is easily explained if the heat input is proportional to the gas mass, that is a mechanism that provides a source of fixed specific thermal energy. While the energy input by shock heating increases with the depth of the potential this fixed amount is less noticible in the more massive clusters which are shock heated to high temperatures. But it contributes significantly or even dominantly in the low temperature systems (see Fig. 22). The entropy in the central regions amounts to typical values of 100 to 200 keV cm^2. It has recently been shown that the entropy excess can also be followed into the outer regions (Finoguenov et al. 2002b) where even higher values are noted.

One obvious source of a fixed amount of specific energy is the energy released by those supernovae which also contribute to the enrichment of the intracluster medium. From the above considerations on the metal enrichment of the intracluster medium we can also calculate the energy that is released by the same supernovae. Assuming that not much energy is dissipated locally in the galaxies where the supernovae explode but that most of it is transported out by a galactic wind, one finds that about 0.5 to 0.75 keV per proton is available (see e. g. also Finoguenov et al. 2001a) for an early heating of the intergalactic medium which could result in the observed entropy excess. Lloyd-Davies et al. (2000) have attempted to determine this extra energy by measuring the gravitational unbinding of the gas in groups under the assumption that the preheating energy is used to increase the pressure and thus levitate some of the intracluster gas. The results shown in Fig. 21 indicate a preheating energy of about 0.5 keV per proton.

Knowing the energy and entropy excess we can actually calculate at which environmental density this entropy excess was created (under the assumption that no energy is meanwhile dissipated). One finds that this environmental density for $\Delta S \sim$ 100–200 keV cm^{-2} is about $1-2 \cdot 10^{-4}$ cm^{-3}. This gas density is reached in the cosmic mean gas density at redshifts of about 6 to 7 and for an object that is at turn-around and starting to collapse at a redshift of 3 to 4. Since we expect that most of the SN II enrichment has happened before the cluster formed, we have to conclude that most of the enrichment happened before reshift of 3, but after redshift of 6. This is earlier than most of the star formation in the field, but this is to be expected for the cluster environment where due to the higher matter density galaxy formation should proceed earlier.

This gives only a very rough picture of what might have happened. So far we have excluded any dissipational process, that is in particular radiative cooling which is bound to play some role. Therefore the modeling of the observed entropy excess is still a controversial issue in the literature. Some authors have shown that the amount of energy needed at the typical time of most of the star formation to produce the entropy excess is much higher than what can be expected from the energy release of supernovae, even if all this energy is dumped into the intracluster medium with

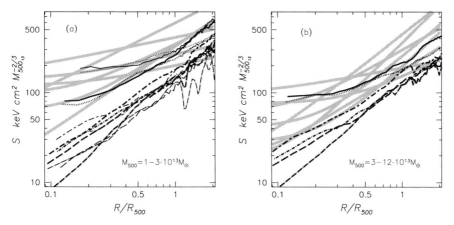

Figure 23: Observed and simulated entropy profiles of groups (left) and poor clusters (right) of galaxies where the system mass is indicated in the panels (Finoguenov et al. 2002d). The observations are denoted by the grey lines. Different model simulations are shown. The models shown by solid and dotted lines involve an energy input of about 0.75 keV per particle at a redshift of about 3 which reproduces the observed profiles best.

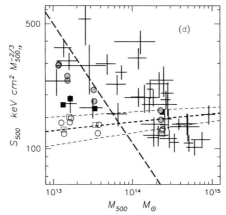

Figure 24: Same data as Fig. 22 but with results from simulations added (Finoguenov et al. 2003). The grey data points show the results from simulations with energy input at redshift around 3 while the open points show the results for energy input at $z \sim 9$. The black squares show results where only cooling is included.

negligible spontaneous energy dissipation (e. g. Wu et al. 2000, Borgani et al. 2002). Voit and Brian (2000) as well as others (Bryan 2000, Pearce et al. 2000, Muanwong et al. 2001, 2002, Wu and Xue 2002, Voit et al. 2002) have shown that the cooling and condensation of the coldest gas in groups and clusters can produce the observational effect. But it would imply a much to high degree of conversion of gas into stars (e. g. Balogh et al. 2001). Thus recent models have been more successful by combing the effect of cooling and feedback heating. Fig. 23 shows for example a successful modeling by Finoguenov et al. (2003) where the observed entropy profiles of groups, poor clusters, and rich clusters are compared to the simulations. It is clear that the critical comparison with observations is that for the groups where the effect is strongest. It is shown that models with a homogeneous specific energy input of 0.75 keV per particle at a redshift of $z \sim 3$ with additional cooling included in the simulations provides a good representation of the observations. Fig. 24 also shows that the high excess entropy observed in the outskirts of groups can be reproduced by this model, while a much earlier energy release at a redshift if 9 (which also inhibits star formation very effectively) fails to reproduce the high observed levels.

Alternatively models of heat input by AGN are discussed, but here the prescription are less constraint and the models have not been so successful yet. Again in this field of research XMM-Newton and Chandra are providing a new degree of precision. To have better model constraints we need more accurate entropy profiles. The present data mostly obtained with the poor spatial resolution of ASCA are relatively inaccurate. The data base for these studies from the XMM and Chandra observations is just becoming available. The still more speculative results presented here can than be pinned down with sufficient accuracy to allow a detailed comparison of models and observations. It is fortunate and stimulating that the precision and predictive capabilities of N-body/hydrodynamic modeling is improving right now to a state where results for the thermal structure of the intracluster medium can be reliably predicted.

6 Summary and conclusions

With X-ray observations we are now starting to gain precise knowledge about the structure of galaxy clusters and the physical state of the intracluster medium. We find that the intracluster medium comprises about 12 % of the total cluster mass and we find that it is surprisingly strongly enriched with metals. Nevertheless we are approaching an explanation for the high metal abundance to within a factor of two. Further more detailed studies have to show if it is necessary to resort to models with an unusual initial mass function for the star formation in the cluster ellipticals or not. The abundance pattern of the now 9 observed heavy elements allows to distinguish the contribution by type Ia and type II supernovae and to simultaneously test the nuclear synthesis model yields of these supernovae. We can clearly see the late enrichment by SN Ia in the centers of clusters. In the case of e. g. M87 the Fe enrichment in the center can easily be explained by the observed SN Ia rate in ellipticals and an enrichment time of about 2 Gyr. The abundance pattern also seems to favour models of slow supernova type Ia explosions (delayed-deflagration models) with incomplete burning of the α-elements in this aged stellar population environment. This should have far reaching consequences for chemical evolution modeling.

The entropy structure in groups and clusters of galaxies provides important information about the energy injection during the early star burst phase (where also most of the metals seen in the intracluster medium have been produced). These observations have triggered a large effort of hydrodynamical modeling of the star formation effect on the evolution of the intracluster medium. The first implications seem to be that most of the star formation in clusters happened a little earlier than redshift 3. In general these efforts will also greatly help to understand the very important feedback mechanisms controlling the galaxy formation and evolution in general.

The detailed solution of most of the questions rised in this contribution are still to come. But with the capabilities of the new X-ray observatories XMM-Newton and Chandra and a few more years of observation time we have all the means to solve most of these problems. Thus we are witnessing a great time for X-ray astronomy, where this branch of observational astronomy is contributing very substantially to the major questions of astrophysics and cosmology.

Acknowledgements

I like to thank my collaborators Kyoko Matsushita, Alexis Finoguenov, Ulrich Briel, Stefano Borgani for the fruitful joint work and the many discussions. I also like to thank the XMM-Newton team at MPE and outside for making these beautiful results possible and for supporting the data reduction of the XMM-Newton observations This paper is based to a large part on results obtained with XMM-Newton, an ESA science mission with instruments and contributions funded by ESA Member States and the USA (NASA). The XMM-Newton project is supported by the Bundesministerium für Bildung und Forschung, Deutsches Zentrum für Luft und Raumfahrt, The Max-Planck Society and the Haidenhain-Stiftung.

References

Allen, S.W., Schmidt, R.W., Fabian, A.C. 2002, MNRAS 334, L11

Arnaud, M., Rothenflug, R., Boulade, O., Vigroux, L., Vangioni-Flam, E. 1992, A&A 254, 49

Balogh, M.L., Pearce, F.R., Bower, R.G. & Kay, S.T. 2001, MNRAS 326, 1228

Belsole, E., Sauvageot, J.L., Böhringer, H., et al. 2000, A&A 365, L188

Böhringer, H., Nulsen, P.E.J., Braun, R., Fabian, A.C. 1995, MNRAS 274, L67

Böhringer, H., Belsole, E., Kennea, J., et al. 2001, A&A 365, L18

Böhringer, H., Matsushita, K., Ikebe, Y., et al. 2002, A&A 382, 804

Borgani, S., Governato, F., Wadsley, J., Manci, N., Tozzi, P., Lake, G., Quinn, P., Stadel, J. 2001, ApJ 559, L71

Bryan, G.L. 2000, ApJ 544, L1

Buote, D.A. 1999, MNRAS 309, 685

Churazov, E., Brüggen, M., Kaiser, C.R., Böhringer, H., Forman, W. 2001, ApJ 554, 261

De Grandi, S., Molendi, S. 2001,

Edvardsson, E., Andersen, J. Gustafsson, B., et al. 1993, A&A 275, 101

Ettori, S., Tozzi, P., Rosati, P. 2002, A&A, in press

Ezawa, H., Fukazawa, Y., Makishima, K., Ohashi, T., Takahara, F., Xu, H., Yamasaki, Y. 1997, ApJ 490, L33

Fabian, A.C. 1994, ARA&A 32, 277

Ferguson, H.C., Dickinson, M., Williams, R. 2000, ARAA 38, 667

Finoguenov, A., Arnaud, M., David, L.P. 2001a, ApJ 555, 191

Finoguenov, A., Borgani, S., Tornatore, L., Böhringer, H. 2003, A&A, in press, astro-ph/0212450

Finoguenov, A., David, L.P., Ponman, T.J. 2000, ApJ 514, 188

Finoguenov, A., Jones, C. 2000, ApJ 539, 603

Finoguenov, A., Jones, C., Böhringer, H., Ponmann, T.J. 2002b, ApJ 578, 74

Finoguenov, A., Matsushita, K., Böhringer, H., Ikebe, Y., Arnaud, M. 2002a, A&A 381, 21

Finoguenov, A., Reiprich, T.H., Böhringer, H. 2001b, A&A 368, 749

Fukazawa, Y., Makishima, K., Matsushita, K., Yamasaki, N., Ohashi, T., Mushotzky, R.F., Sakima, Y., Tsusaka, Y., Yamashita, K. 1996, PASJ 48, 395

Fukazawa, Y., Makishima, K., Tamura, T., Ezawa, H., Xu, H., Ikebe, Y., Kikuchi, K., Ohashi, T. 1998, PASP 50, 187

Hamuy, M., Phillips, M.M., Schommer, R.A., Suntzeff, N.B. 1996, AJ 112, 2391

Iwamoto, K., Brachwitz, F., Nomoto, K., Kishimoto, N., Umeda, H., Hix, W.R., Thielemann, F.K. 1999, ApJ SS 125, 439

Jansen, F., Lumb, D., Altieri, B., et al. 2001, A&A 365, L1

Lloyd-Davies, E.J., Ponman, T.J., Cannon, D.B. 2000, MNRAS 315, 689

Madau, P., Ferguson, H.C., Dickinson, M., Giavalisco, M., Steidel, C.C., Fruchter, H.S. 1996, MNRAS 283, 1388

Madau, P., Pozetti, L., Dickinson, M. 1998, ApJ 498, 106

Matsumoto, H., Koyama, K., Awaki, H., Tomida, H., Tsuru, T. 1996, PASP 48, 201

Matsushita, K., Belsole, E., Finoguenov, A., Böhringer, H. 2002a, A&A 386, 77

Matsushita, K., Finoguenov, A., Böhringer, H. 2002b, A&A, in press, astro-ph/0212069

Mewe, R., Gronenschild, E.H.B.M., Oord, G.H.J. 1985, A&SA Suppl. 62, 197

Mewe, R., Kaastra, J.S., Liedahl, D.A. 1995, LEGACY, J. HEASARC at NASA GSCF 6, 16

Molendi, S., Pizzolato, F. 2001, ApJ 560, 194

Molendi, S., Gastaldello, F. 2001, A&A 375, L14

Molendi, S., 2002, ApJ, in press, astro-ph/0207545

Muanwong, O., Thomas, P.A., Kay, S.T., Pearce, F.R., Couchman, H.M.P. 2001, ApJ 552, L27

Muanwong, O., Thomas, P.A., Kay, S.T., Pearce, F.R. 2002, MNRAS 336, 527

Navarro, J.F., Frenk, C.S., White, S.D.M. 1995, MNRAS 275, 720

Navarro, J.F., Frenk, C.S., White, S.D.M. 1997, ApJ 490, 493

Nomoto, K., Thielemann, F.-K., Yokoi, K. 1984, ApJ 286, 644

Nomoto, K., Iwamoto, K., Nakasato, N., Thielemann, F.-K., Brachwitz, F., Tsyjimoto, T., Kubo, Y., Kishimoto, N. 1997, Nucl. Phys. A621, 467

Nulsen, P.E.J. 1986, MNRAS 221, 377

O'Meara, J.M., Tytler, D., Kirkman, D., et al. 2001, ApJ 552, 718

Pearce, F.R., Thomas, P.A., Couchman, H.M.P., Edge, A.C. 2000, MNRAS 317, 1029

Ponman, T.J., Cannon, D.B., Navarro, J.F. 1999, Nature 397, 135
Pratt, G.W., Arnaud, M. 2002, A&A 394, 375
Reiprich, T.H. 2001, Ph.D. Thesis, Ludwig-Maximilians-Universität München
Reiprich, T.H., Böhringer, H. 2002, ApJ 567, 716
Renzini, A., Ciotti, L., D'Ercole, A., Pellegrini, S. 1993, ApJ 419, 52
Renzini, A. 1997, ApJ 488, 35
Sarazin, C.L. 1986, Rev. Mod. Phys. 58, 1
Smith, R.K., Brickhouse, N.S., Liedahl, D.A., Raymond, J.C. 2001, ApJ 556, L91
Strüder, L., Briel, U., Dennerl, K., et al. 2001, A&A 365, L18
Thomas, P.A., Fabian, A.C., Nulsen, P.E.J. 1987, MNRAS 228, 973
Turner, M.J.L., Abbey, A., Arnaud, M., et al. 2001, A&A 365, L27
Turner, M.S. 2002, astro-ph/0202007
Viot, M., Bryan, G.L. 2001, Nature 414, 425
Voit, M., Bryan, G.L., Balogh, M.L., Bower, R.G. 2002, ApJ 576, 601
White, S.D.M., Navarro, J.F., Evrard, A.E., Frenk, C.S. 1993, Nature 366, 429
Woosley, S.E., Weaver, T.A. 1995, ApJ 466, 114
Wu, X.P., Fabian, A.C., Nulsen, P.E.J. 2000, MNRAS 318, 889
Wu, X.P., Xue, Y.-J. 2002, ApJ 569, 112; ApJ 572, L19

Index of Contributors

Bergman, Per	171	Kerschbaum, Franz	171
Böhringer, Hans	275	Klaas, Ulrich	243
		Klessen, Ralf S.	23
Christlieb, Norbert	171	Kurtz, Stan	85
Dorschner, Johann	171	Mutschke, Harald	171
Franco, José	85	Olofsson, Hans	171
García-Segura, Guillermo	85	Pauldrach, A. W. A.	133
González Delgado, David	171	Posch, Thomas	171
Hanslmeier, Arnold	55	Schöier, Fredrik	171
Helling, Christiane	115	Schröder, Klaus-Peter	227
Hüttemeister, Susanne	207		
		Townes, Charles H.	1
Jäger, Cornelia	171	Truran, James W.	261

General Table of Contents

Volume 1 (1988): Cosmic Chemistry

Geiss, J.: Composition in Halley's Comet:
Clues to Origin and History of Cometary Matter 1/1

Palme, H.: Chemical Abundances in Meteorites 1/28

Gehren, T.: Chemical Abundances in Stars 1/52

Omont, A.: Chemistry of Circumstellar Shells 1/102

Herbst, E.: Interstellar Molecular Formation Processes 1/114

Edmunds, M.G.: Chemical Abundances in Galaxies 1/139

Arnould, M.: An Overview of the Theory of Nucleosynthesis 1/155

Schwenn, R.: Chemical Composition and Ionisation States of the
Solar Wind – Plasma as Characteristics of Solar Phenomena 1/179

Kratz, K.-L.: Nucear Physics Constraints to Bring the Astrophysical
R-Process to the "Waiting Point" 1/184

Henkel, R., Sedlmayr, E., Gail, H.-P.: Nonequilibrium Chemistry
in Circumstellar Shells 1/231

Ungerechts, H.: Molecular Clouds in the Milky Way: the Columbia-Chile
CO Survey and Detailed Studies with the KOSMA 3 m Telescope 1/210

Stutzki, J.: Molecular Millimeter and Submillimeter Observations 1/221

Volume 2 (1989)

Rees, M.J.: Is There a Massive Black Hole in Every Galaxy?
(19th Karl Schwarzschild Lecture 1989) 2/1

Patermann, C.: European and Other International Cooperation
in Large-Scale Astronomical Projects 2/13

Lamers, H.J.G.L.M.: A Decade of Stellar Research with IUE 2/24

Schoenfelder, V.: Astrophysics with GRO 2/47

Lemke, D., Kessler, M.: The Infrared Space Observatory ISO 2/53

Jahreiß, H.: HIPPARCOS after Launch!?
The Preparation of the Input Catalogue 2/72

Ip, W.H.: The Cassini/Huygens Mission 2/86

Beckers, J.M.: Plan for High Resolution Imaging with the VLT 2/90

Rimmele, Th., von der Luehe, O.: A Correlation Tracker
for Solar Fine Scale Studies 2/105

Schuecker, P., Horstmann, H., Seitter, W.C., Ott, H.-A., Duemmler, R.,
Tucholke, H.-J., Teuber, D., Meijer, J., Cunow, B.:
The Muenster Redshift Project (MRSP) 2/109

Kraan-Korteweg, R.C.: Galaxies in the Galactic Plane 2/119

Meisenheimer, K.: Synchrotron Light from Extragalactic Radio Jets
and Hot Spots 2/129

Staubert, R.: Very High Energy X-Rays from Supernova 1987A 2/141

Hanuschik, R.W.: Optical Spectrophotometry
of the Supernova 1987A in the LMC 2/148

Weinberger, R.: Planetary Nebulae in Late Evolutionary Stages 2/167

Pauliny-Toth, I.I.K., Alberdi, A., Zensus, J A., Cohen, M.H.:
Structural Variations in the Quasar 2134+004 .... 2/177

Chini, R.: Submillimeter Observations
of Galactic and Extragalactic Objects 2/180

Kroll, R.: Atmospheric Variations in Chemically Peculiar Stars 2/194

Maitzen, H.M.: Chemically Peculiar Stars of the Upper Main Sequence 2/205

Beisser, K.: Dynamics and Structures of Cometary Dust Tails 2/221

Teuber, D.: Automated Data Analysis 2/229

Grosbol, P.: MIDAS 2/242

Stix, M.: The Sun's Differential Rotation 2/248

Buchert, T.: Lighting up Pancakes –
Towards a Theory of Galaxy-formation 2/267

Yorke, H.W.: The Simulation of Hydrodynamic Processes
with Large Computers 2/283

Langer, N.: Evolution of Massive Stars
(First Ludwig Biermann Award Lecture 1989) 2/306

Baade, R.: Multi-dimensional Radiation Transfer
in the Expanding Envelopes of Binary Systems 2/324

Duschl, W.J.: Accretion Disks in Close Binarys 2/333

Volume 3 (1990): Accretion and Winds

Meyer, F.: Some New Elements in Accretion Disk Theory 3/1

King, A.R.: Mass Transfer and Evolution in Close Binaries 3/14

Kley, W.: Radiation Hydrodynamics of the Boundary Layer
of Accretion Disks in Cataclysmic Variables 3/21

Hessman, F.V.: Curious Observations of Cataclysmic Variables 3/32

Schwope, A.D.: Accretion in AM Herculis Stars 3/44

Hasinger, G.: X-ray Diagnostics of Accretion Disks 3/60

Rebetzky, A., Herold, H., Kraus, U., Nollert, H.-P., Ruder, H.:
Accretion Phenomena at Neutron Stars 3/74

Schmitt, D.: A Torus-Dynamo for Magnetic Fields
in Galaxies and Accretion Disks 3/86

Owocki, S.P.: Winds from Hot Stars 3/98

Pauldrach, A.W.A., Puls, J.: Radiation Driven Winds
of Hot Luminous Stars. Applications of Stationary Wind Models 3/124

Puls, J., Pauldrach, A.W.A.: Theory of Radiatively Driven Winds
of Hot Stars: II. Some Aspects of Radiative Transfer 3/140

Gail, H.-P.: Winds of Late Type Stars 3/156

Hamann, W.-R., Wessolowski, U., Schmutz, W., Schwarz, E., Duennebeil, G., Koesterke, L., Baum, E., Leuenhagen, U.: Analyses of Wolf-Rayet Stars 3/174

Schroeder, K.-P.: The Transition of Supergiant CS Matter from Cool Winds to Coronae – New Insights with X AUR Binary Systems 3/187

Dominik, C.: Dust Driven Mass Lost in the HRD 3/199

Montmerle, T.: The Close Circumstellar Environment of Young Stellar Objects 3/209

Camenzind, M.: Magnetized Disk-Winds and the Origin of Bipolar Outflows 3/234

Staude, H.J., Neckel, Th.: Bipolar Nebulae Driven by the Winds of Young Stars 3/266

Stahl, O.: Winds of Luminous Blue Variables 3/286

Jenkner, H.: The Hubble Space Telescope Before Launch: A Personal Perspective 3/297

Christensen-Dalsgaard, J.: Helioseismic Measurements of the Solar Internal Rotation 3/313

Deiss, B.M.: Fluctuations of the Interstellar Medium 3/350

Dorfi, E.A.: Acceleration of Cosmic Rays in Supernova Remnants 3/361

Volume 4 (1991)

Parker, E.N.: Convection, Spontaneous Discontinuities, and Stellar Winds and X-Ray Emission (20th Karl Schwarzschild Lecture 1990) 4/1

Schrijver, C.J.: The Sun as a Prototype in the Study of Stellar Magnetic Activity 4/18

Steffen, M., Freytag, B.: Hydrodynamics of the Solar Photosphere: Model Calculations and Spectroscopic Observations 4/43

Wittmann, A.D.: Solar Spectroscopy with a 100×100 Diode Array 4/61

Staude, J.: Solar Research at Potsdam: Papers on the Structure and Dynamics of Sunspots 4/69

Fleck, B.: Time-Resolved Stokes V Polarimetry of Small Scale Magnetic Structures on the Sun 4/90

Glatzel, W.: Instabilities in Astrophysical Shear Flows 4/104

Schmidt, W.: Simultaneous Observations with a Tunable Filter and the Echelle Spectrograph of the Vacuum Tower Telescope at Teneriffe 4/117

Fahr, H.J.: Aspects of the Present Heliospheric Research 4/126

Marsch, E.: Turbulence in the Solar Wind 4/145

Gruen, E.: Dust Rings Around Planets 4/157

Hoffmann, M.: Asteroid-Asteroid Interactions – Dynamically Irrelevant? 4/165

Aschenbach, B.: First Results from the X-Ray Astronomy Mission ROSAT 4/173

Wicenec, A.: TYCHO/HIPPARCOS A Successful Mission! 4/188

Spruit, H.C.: Shock Waves in Accretion Disks 4/197
Solanki, S.K.: Magnetic Field Measurements on Cool Stars 4/208
Hanuschik, R.W.: The Expanding Envelope of Supernova 1987A
    in the Large Magellanic Cloud
    (2nd Ludwig Biermann Award Lecture 1990) 4/233
Krause, F., Wielebinski, R.: Dynamos in Galaxies 4/260

Volume 5 (1992): Variabilities in Stars and Galaxies

Wolf, B.: Luminous Blue Variables; Quiescent and Eruptive States 5/1
Gautschy, A.: On Pulsations of Luminous Stars 5/16
Richter, G.A.: Cataclysmic Variables – Selected Problems 5/26
Luthardt, R.: Symbiotic Stars 5/38
Andreae, J.: Abundances of Classical Novae 5/58
Starrfield, S.: Recent Advances in Studies of the Nova Outburst 5/73
Pringle, J.E.: Accretion Disc Phenomena 5/97
Landstreet, J.D.: The Variability of Magnetic Stars 5/105
Baade, D.: Observational Aspects of Stellar Seismology 5/125
Dziembowski, W.: Testing Stellar Evolution Theory
    with Oscillation Frequency Data 5/143
Spurzem, R.: Evolution of Stars and Gas in Galactic Nuclei 5/161
Gerhard, O.E.: Gas Motions in the Inner Galaxy
    and the Dynamics of the Galactic Bulge Region 5/174
Schmitt, J.H.M.M.: Stellar X-Ray Variability
    as Observed with the ROSAT XRT 5/188
Notni, P.: M82 – The Bipolar Galaxy 5/200
Quirrenbach, A.: Variability and VLBI Observations
    of Extragalactic Radio Surces 5/214
Kollatschny, W.: Emission Line Variability in AGN's 5/229
Ulrich, M.-H.: The Continuum of Quasars and Active Galactic Nuclei,
    and Its Time Variability 5/247
Bartelmann, M.: Gravitational Lensing by Large-Scale Structures 5/259

Volume 6 (1993): Stellar Evolution and Interstellar Matter

Hoyle, F.: The Synthesis of the Light Elements
    (21st Karl Schwarzschild Lecture 1992) 6/1
Heiles, C.: A Personal Perspective of the Diffuse Interstellar Gas
    and Particularly the Wim 6/19
Dettmar, R.-J.: Diffuse Ionized Gas and the Disk-Halo Connection
    in Spiral Galaxies 6/33
Williams, D.A.: The Chemical Composition of the Interstellar Gas 6/49
Mauersberger, R., Henkel, C.: Dense Gas in Galactic Nuclei 6/69
Krabbe, A.: Near Infrared Imaging Spectroscopy of Galactic Nuclei 6/103

Dorschner, J.: Subject and Agent of Galactic Evolution 6/117

Markiewicz, W.J.: Coagulation of Interstellar Grains in a
    Turbulent Pre-Solar Nebula: Models and Laboratory Experiments 6/149

Goeres, A.: The Formation of PAHs in C-Type Star Environments 6/165

Koeppen, J.: The Chemical History of the Interstellar Medium 6/179

Zinnecker, H., McCaughrean, M.J., Rayner, J.T., Wilking, B.A.,
    Moneti, A.: Near Infrared Images of Star-Forming Regions 6/191

Stutzki, R.: The Small Scale Structure of Molecular Clouds 6/209

Bodenheimer, P.: Theory of Protostars 6/233

Kunze, R.: On the Impact of Massive Stars on their Environment –
    the Photoevaporation by H II Regions 6/257

Puls, J., Pauldrach, A.W.A., Kudritzki, R.-P., Owocki, S.P., Najarro, F.:
    Radiation Driven Winds of Hot Stars – some Remarks on Stationary
    Models and Spectrum Synthesis in Time-Dependent Simulations
    (3rd Ludwig Biermann Award Lecture 1992) 6/271

Volume 7 (1994)

Wilson, R.N.: Karl Schwarzschild and Telscope Optics
    (22nd Karl Schwarzschild Lecture 1993) 7/1

Lucy, L.B.: Astronomical Inverse Problems 7/31

Moffat, A.F.J.: Turbulence in Outflows from Hot Stars 7/51

Leitherer, C.: Massive Stars in Starburst Galaxies
    and the Origin of Galactic Superwinds 7/73

Mueller, E., Janka, H.-T.:
    Multi-Dimensional Simulations of Neutrino-Driven Supernovae 7/103

Hasinger, G.: Supersoft X-Ray Sources 7/129

Herbstmeier, U., Kerp, J., Moritz, P.:
    X-Ray Diagnostics of Interstellar Clouds 7/151

Luks, T.: Structure and Kinematics of the Magellanic Clouds 7/171

Burkert, A.: On the Formation of Elliptical Galaxies
    (4th Ludwig Biermann Award Lecture 1993) 7/191

Spiekermann, G., Seitter, W.C., Boschan, P., Cunow, B., Duemmler, R.,
    Naumann, M., Ott, H.-A., Schuecker, P., Ungruhe, R.:
    Cosmology with a Million Low Resolution Redshifts:
    The Muenster Redshift Project MRSP 7/207

Wegner, G.: Motions and Spatial Distributions of Galaxies 7/235

White, S.D.M.: Large-Scale Structure 7/255

Volume 8 (1995): Cosmic Magnetic Fields

Trümper, J.E.: X-Rays from Neutron Stars
    (23rd Karl Schwarzschild Lecture 1994) 8/1

Schuessler, M.: Solar Magnetic Fields 8/11

Keller, Ch.U.: Properties of Solar Magnetic Fields from Speckle Polarimetry
(5th Ludwig Biermann Award Lecture 1994) 8/27

Schmitt, D., Degenhardt, U.:
Equilibrium and Stability of Quiescent Prominences 8/61

Steiner, O., Grossmann-Doerth, U., Knoelker, M., Schuessler, M.:
Simulation oif the Interaction of Convective Flow
with Magnetic Elements in the Solar Atmosphere 8/81

Fischer, O.: Polarization by Interstellar Dust –
Modelling and Interpretation of Polarization Maps 8/103

Schwope, A.D.: Accretion and Magnetism – AM Herculis Stars 8/125

Schmidt, G.D.: White Dwarfs as Magnetic Stars 8/147

Richtler, T.: Globular Cluster Systems of Elliptical Galaxies 8/163

Wielebinski, R.: Galactic and Extragalactic Magnetic Fields 8/185

Camenzind, M.: Magnetic Fields and the Physics of Active Galactic Nuclei 8/201

Dietrich, M.:
Broad Emission-Line Variability Studies of Active Galactic Nuclei 8/235

Böhringer, H.: Hot, X-Ray Emitting Plasma, Radio Halos,
and Magnetic Fields in Clusters of Galaxies 8/259

Hopp, U., Kuhn, B.:
How Empty are the Voids? Results of an Optical Survey 8/277

Raedler, K.-H.: Cosmic Dynamos 8/295

Hesse, M.: Three-Dimensional Magnetic Reconnection
in Space- and Astrophysical Plasmas and its Consequences
for Particle Acceleration 8/323

Kiessling, M.K.-H.: Condensation in Gravitating Systems as Pase Transition 8/349

Volume 9 (1996): Positions, Motions, and Cosmic Evolution

van de Hulst, H.:
Scaling Laws in Multiple Light Scattering under very Small Angles
(24th Karl Schwarzschild Lecture 1995) 9/1

Mannheim, K.: Gamma Rays from Compact Objects
(6th Ludwig Biermann Award Lecture 1995) 9/17

Schoenfelder, V.:
Highlight Results from the Compton Gamma-Ray Observatory 9/49

Turon, C.: HIPPARCOS, a new Start
for many Astronomical and Astrophysical Topics 9/69

Bastian, U., Schilbach, E.:
GAIA, the successor of HIPPARCOS in the 21st century 9/87

Baade, D.: The Operations Model for the Very Large Telescope 9/95

Baars, J.W.M., Martin, R.N.: The Heinrich Hertz Telescope –
A New Instrument for Submillimeter-wavelength Astronomy 9/111

Gouguenheim, L., Bottinelli, L., Theureau, G., Paturel, G.,
Teerikorpi, P.: The Extragalactive Distance Scale
and the Hubble Constant: Controversies and Misconceptions 9/127

Tammann, G.A.: Why is there still Controversy on the Hubble Constant? 9/139

Mann, I.: Dust in Interplanetary Space:
 a Component of Small Bodies in the Solar System 9/173

Fichtner, H.: Production of Energetic Particles at the Heliospheric Shock –
 Implications for the Global Structure of the Heliosphere 9/191

Schroeder, K.-P., Eggleton, P.P.: Calibrating Late Stellar Evolution
 by means of zeta AUR Systems – Blue Loop Luminosity
 as a Critical Test for Core-Overshooting 9/221

Zensus, J.A., Krichbaum, T.P., Lobanov, P.A.:
 Jets in High-Luminosity Compact Radio Sources 9/221

Gilmore, G.: Positions, Motions, and Evolution
 of the Oldest Stellar Populations 9/263

Samland, M., Hensler, G.: Modelling the Evolution of Galaxies 9/277

Kallrath, J.: Fields of Activity for Astronomers and Astrophysicists
 in Industry – Survey and Experience in Chemical Industry – 9/307

Volume 10 (1997): Gravitation

Thorne, K.S.: Gravitational Radiation – a New Window Onto the Universe
 (25th Karl Schwarzschild Lecture 1996) 10/1

Grebel, E.K.: Star Formation Histories of Local Group Dwarf Galaxies
 (7th Ludwig Biermann Award Lecture 1996 (i)) 10/29

Bartelmann, M.L.: On Arcs in X-Ray Clusters
 (7th Ludwig Biermann Award Lecture 1996 (ii)) 10/61

Ehlers, J.: 80 Years of General Relativity 10/91

Lamb, D.Q.: The Distance Scale To Gamma-Ray Bursts 10/101

Meszaros, P.: Gamma-Ray Burst Models 10/127

Schulte-Ladbeck, R.: Massive Stars – Near and Far 10/135

Geller, M.J.: The Great Wall and Beyond –
 Surveys of the Universe to $z < 0.1$ 10/159

Rees, M.J.: Black Holes in Galactic Nuclei 10/179

Mueller, J., Soffel, M.: Experimental Gravity and Lunar Laser Ranging 10/191

Ruffert, M., Janka, H.-Th.: Merging Neutron Stars 10/201

Werner, K., Dreizler, S., Heber, U., Kappelmann, N., Kruk, J., Rauch, T.,
 Wolff, B.: Ultraviolet Spectroscopy of Hot Compact Stars 10/219

Roeser, H.-J., Meisenheimer, K., Neumann, M., Conway, R.G., Davis, R.J.,
 Perley, R.A.: The Jet of the Quasar 3C 273/ at High Resolution 10/253

Lemke, D.: ISO: The First 10 Months of the Mission 10/263

Fleck, B.: First Results from SOHO 10/273

Thommes, E., Meisenheimer, K., Fockenbrock, R., Hippelein, H.,
 Roeser, H.-J.: Search for Primeval Galaxies
 with the Calar Alto Deep Imaging Survey (CADIS) 10/297

Neuhaeuser, R.: The New Pre-main Sequence Population
 South of the Taurus Molecular Clouds 10/323

Volume 11 (1998): Stars and Galaxies

Taylor, J.H. jr.: Binary Pulsars and General Relativity
(26th Karl Schwarzschild Lecture 1997 – *not published*) 11/1

Napiwotzki, R.: From Central Stars of Planetary Nebulae to White Dwarfs
(8th Ludwig Biermann Award Lecture 1997) 11/3

Dvorak, R.: On the Dynamics of Bodies in Our Planetary System 11/29

Langer, N., Heger, A., García-Segura, G.: Massive Stars:
the Pre-Supernova Evolution of Internal and Circumstellar Structure 11/57

Ferguson, H.C.: The Hubble Deep Field 11/83

Staveley-Smith, L., Sungeun Kim, Putman, M., Stanimirović, S.:
Neutral Hydrogen in the Magellanic System 11/117

Arnaboldi, M., Capaccioli, M.: Extragalactic Planetary Nebulae
as Mass Tracers in the Outer Halos of Early-type Galaxies 11/129

Dorfi, E.A., Häfner, S.: AGB Stars and Mass Loss 11/147

Kerber, F.: Planetary Nebulae:
the Normal, the Strange, and Sakurai's Object 11/161

Kaufer, A.: Variable Circumstellar Structure of Luminous Hot Stars:
the Impact of Spectroscopic Long-term Campaigns 11/177

Strassmeier, K.G.: Stellar Variability as a Tool in Astrophysics.
A Joint Research Initiative in Austria 11/197

Mauersberger, R., Bronfman, L.: Molecular Gas in the Inner Milky Way 11/209

Zeilinger, W.W.: Elliptical Galaxies 11/229

Falcke, H.: Jets in Active Galaxies: New Results from HST and VLA 11/245

Schuecker, P., Seitter, W.C.: The Deceleration of Cosmic Expansion 11/267

Vrielmann, S.: Eclipse Mapping of Accretion Disks 11/285

Schmid, H.M.: Raman Scattering
and the Geometric Structure of Symbiotic Stars 11/297

Schmidtobreick, L., Schlosser, W., Koczet, P., Wiemann, S., Jütte, M.:
The Milky Way in the UV 11/317

Albrecht, R.: From the Hubble Space Telescope
to the Next Generation Space Telescope 11/331

Heck, A.: Electronic Publishing in its Context
and in a Professional Perspective 11/337

Volume 12 (1999):
Astronomical Instruments and Methods at the Turn of the 21st Century

Strittmatter, P.A.: Steps to the Large Binocular Telescope – and Beyond
(27th Karl Schwarzschild Lecture 1998) 12/1

Neuhäuser, R.: The Spatial Distribution and Origin
of the Widely Dispersed ROSAT T Tauri Stars
(9th Ludwig Biermann Award Lecture 1998) 12/27

Huber, C.E.: Space Research at the Threshold of the 21st Century –
 Aims and Technologies 12/47

Downes, D.: High-Resolution Millimeter and Submillimeter Astronomy:
 Recent Results and Future Directions 12/69

Röser, S.: DIVA – Beyond HIPPARCOS and Towards GAIA 12/97

Krabbe, A., Röser, H.P.:
 SOFIA – Astronomy and Technology in the 21st Century 12/107

Fort, B.P.: Lensing by Large-Scale Structures 12/131

Wambsganss, J.: Gravitational Lensing as a Universal Astrophysical Tool 12/149

Mannheim, K.: Frontiers in High-Energy Astroparticle Physics 12/167

Basri, G.B.: Brown Dwarfs: The First Three Years 12/187

Heithausen, A., Stutzki, J., Bensch, F., Falgarone, E., Panis, J.-F.:
 Results from the IRAM Key Project:
 "Small Scale Structure of Pre-Star-forming Regions" 12/201

Duschl, W.J.: The Galactic Center 12/221

Wisotzki, L.: The Evolution of the QSO Luminosity Function
 between $z = 0$ and $z = 3$ 12/231

Dreizler, S.: Spectroscopy of Hot Hydrogen Deficient White Dwarfs 12/255

Moehler, S.: Hot Stars in Globular Clusters 12/281

Theis, Ch.: Modeling Encounters of Galaxies: The Case of NGC 4449 .... 12/309

Volume 13 (2000): New Astrophysical Horizons

Ostriker, J.P.: Historical Reflections
 on the Role of Numerical Modeling in Astrophysics
 (28th Karl Schwarzschild Lecture 1999) 13/1

Kissler-Patig, M.: Extragalactic Globular Cluster Systems:
 A new Perspective on Galaxy Formation and Evolution
 (10th Ludwig Biermann Award Lecture 1999) 13/13

Sigwarth, M.: Dynamics of Solar Magnetic Fields –
 A Spectroscopic Investigation 13/45

Tilgner, A.: Models of Experimental Fluid Dynamos 13/71

Eislöffel, J.: Morphology and Kinematics of Jets from Young Stars 13/81

Englmaier, P.: Gas Streams and Spiral Structure in the Milky Way 13/97

Schmitt, J.H.M.M.:
 Stellar X-Ray Astronomy: Perspectives for the New Millenium 13/115

Klose, S.: Gamma Ray Bursts in the 1990's –
 a Multi-wavelength Scientific Adventure 13/129

Gänsicke, B.T.: Evolution of White Dwarfs in Cataclysmic Variables 13/151

Koo, D.: Exploring Distant Galaxy Evolution: Highlights with Keck 13/173

Fritze-von Alvensleben, U.:
 The Evolution of Galaxies on Cosmological Timescales 13/189

Ziegler, B.L.: Evolution of Early-type Galaxies in Clusters 13/211

Menten, K., Bertoldi, F.:
   Extragalactic (Sub)millimeter Astronomy – Today and Tomorrow 13/229

Davies, J.I.: In Search of the Low Surface Brightness Universe 13/245

Chini, R.: The Hexapod Telescope – A Never-ending Story 13/257

Volume 14 (2001): Dynamic Stability and Instabilities in the Universe

Penrose, R.: The Schwarzschild Singularity:
   One Clue to Resolving the Quantum Measurement Paradox
   (29th Karl Schwarzschild Lecture 2000) 14/1

Falcke, H.: The Silent Majority –
   Jets and Radio Cores from Low-Luminosity Black Holes
   (11th Ludwig Biermann Award Lecture 2000) 14/15

Richter, P. H.: Chaos in Cosmos 14/53

Duncan, M.J., Levison, H., Dones, L., Thommes, E.:
   Chaos, Comets, and the Kuiper Belt 14/93

Kokubo, E.: Planetary Accretion: From Planitesimals to Protoplanets 14/117

Priest, E. R.: Surprises from Our Sun 14/133

Liebscher, D.-E.: Large-scale Structure – Witness of Evolution 14/161

Woitke, P.: Dust Induced Structure Formation 14/185

Heidt, J., Appenzeller, I., Bender, R., Böhm, A., Drory, N., Fricke, K. J.,
   Gabasch, A., Hopp, U., Jäger, K., Kümmel, M., Mehlert, D.,
   Möllenhoff, C., Moorwood, A., Nicklas, H., Noll, S., Saglia, R.,
   Seifert, W., Seitz, S., Stahl, O., Sutorius, E., Szeifert, Th.,
   Wagner, S. J., and Ziegler, B.: The FORS Deep Field 14/209

Grebel, E. K.: A Map of the Northern Sky:
   The Sloan Digital Sky Survey in Its First Year 14/223

Glatzel, W.:
   Mechanism and Result of Dynamical Instabilities in Hot Stars 14/245

Weis, K.: LBV Nebulae: The Mass Lost from the Most Massive Stars 14/261

Baumgardt, H.: Dynamical Evolution of Star Clusters 14/283

Bomans, D. J.: Warm and Hot Diffuse Gas in Dwarf Galaxies 14/297

Volume 15 (2002): JENAM 2001 – Five Days of Creation: Astronomy with Large Telescopes from Ground and Space

Kodaira, K.: Macro- and Microscopic Views of Nearby Galaxies
   (30th Karl Schwarzschild Lecture 2001) 15/1

Komossa, S.: X-ray Evidence for Supermassive Black Holes
   at the Centers of Nearby, Non-Active Galaxies
   (13th Ludwig Biermann Award Lecture 2001) 15/27

Richstone, D. O.: Supermassive Black Holes 15/57

Hasinger, G.: The Distant Universe Seen with Chandra and XMM-Newton 15/71

Danzmann, K. and Rüdiger, A.:
   Seeing the Universe in the Light of Gravitational Waves 15/93

Gandorfer, A.: Observations of Weak Polarisation Signals from the Sun 15/113

Mazeh, T. and Zucker, S.: A Statistical Analysis of the Extrasolar Planets
and the Low-Mass Secondaries 15/133

Hegmann, M.: Radiative Transfer in Turbulent Molecular Clouds 15/151

Alves, J. F.: Seeing the Light through the Dark:
the Initial Conditions to Star Formation 15/165

Maiolino, R.: Obscured Active Galactic Nuclei 15/179

Britzen, S.: Cosmological Evolution of AGN – A Radioastronomer's View 15/199

Thomas, D., Maraston, C., and Bender, R.: The Epoch(s)
of Early-Type Galaxy Formation in Clusters and in the Field 15/219

Popescu, C. C. and Tuffs, R. J.: Modelling the Spectral Energy Distribution
of Galaxies from the Ultraviolet to Submillimeter 15/239

Elbaz, D.: Nature of the Cosmic Infrared Background
and Cosmic Star Formation History: Are Galaxies Shy? 15/259

Volume 16 (2003): The Cosmic Circuit of Matter

Townes, C. H.: The Behavior of Stars Observed by Infrared Interferometry
(31th Karl Schwarzschild Lecture 2002) 16/1

Klessen, R. S.: Star Formation in Turbulent Interstellar Gas
(14th Ludwig Biermann Award Lecture 2002) 16/23

Hanslmeier, A.: Dynamics of Small Scale Motions in the Solar Photosphere 16/55

Franco, J., Kurtz, S., García-Segura, G.:
The Interstellar Medium and Star Formation: The Impact of Massive Stars .. 16/85

Helling, Ch.: Circuit of Dust in Substellar Objects 16/115

Pauldrach, A. W. A.: Hot Stars: Old-Fashioned or Trendy? 16/133

Kerschbaum, F., Olofsson, H., Posch, Th., González Delgado, D., Bergman, P.,
Mutschke, H., Jäger, C., Dorschner, J., Schöier, F.:
Gas and Dust Mass Loss of O-rich AGB-stars 16/171

Christlieb, N.: Finding the Most Metal-poor Stars of the Galactic Halo
with the Hamburg/ESO Objecrive-prism Survey 16/191

Hüttemeister, S.: A Tale of Bars and Starbursts:
Dense Gas in the Central Regions of Galaxies 16/207

Schröder, K.-P.: Tip-AGB Mass-Loss on the Galactic Scale 16/227

Klaas, U.: The Dusty Sight of Galaxies:
ISOPHOT Surveys of Normal Galaxies, ULIRGS, and Quasars 16/243

Truran, J. W.: Abundance Evolution with Cosmic Time 16/261

Böhringer, H.: Matter and Energy in Clusters of Galaxies as Probes
for Galaxy and Large-Scale Structure Formation in the Universe 16/275

General Index of Contributors

Alberdi, A.	2/177	Deiss, B.M.	3/350
Albrecht, R.	11/331	Dettmar, R.-J.	6/33
Alves, J. F.	15/165	Dietrich, M.	8/235
Andreae, J.	5/58	Dominik, C.	3/199
Appenzeller, I.	14/209	Dones, L.	14/93
Arnaboldi, M.	11/129	Dorfi, E.A.	3/361, 11/147
Arnould, M.	1/155	Dorschner, J.	6/117, 16/171
Aschenbach, B.	4/173	Downes, D.	12/69
Baade, D.	5/125, 9/95	Dreizler, S.	10/219, 12/255
Baade, R.	2/324	Drory, N.	14/209
Baars, J.W.M.	9/111	Duemmler, R.	2/109, 7/207
Bartelmann, M.L.	5/259, 10/61	Duennebeil, G.	3/174
Basri, G.B.	12/187	Duncan, M.J.	14/93
Bastian, U.	9/87	Duschl, W.J.	2/333, 12/221
Baum, E.	3/174	Dvorak, R.	11/29
Baumgardt, H.	14/283	Dziembowski, W.	5/143
Beckers, J.M.	2/90	Edmunds, M.G.	1/139
Beisser, K.	2/221	Eggleton, P.P.	9/221
Bender, R.	14/209, 15/219	Ehlers, J.	10/91
Bensch, F.	12/201	Eislöffel, J.	13/81
Bergman, P.	16/171	Elbaz, D.	15/259
Bertoldi, F.	13/229	Englmaier, P.	13/97
Bodenheimer, P.	6/233	Fahr, H.J.	4/126
Böhm, A.	14/209	Falcke, H.	11/245, 14/15
Böhringer, H.	8/259, 16/275	Falgarone, E.	12/201
Bomans, D.J.	14/297	Ferguson, H.C.	11/83
Boschan, P.	7/207	Fichtner, H.	9/191
Bottinelli, L.	9/127	Fischer, O.	8/103
Britzen, S.	15/199	Fleck, B.	4/90, 10/273
Bronfman, L.	11/209	Fockenbrock, R.	10/297
Buchert, T.	2/267	Fort, B.P.	12/131
Burkert, A.	7/191	Franco, J.	16/85
Camenzind, M.	3/234, 8/201	Freytag, B.	4/43
Capaccioli, M.	11/129	Fricke, K.J.	14/209
Chini, R.	2/180, 13/257	Fritze-von Alvensleben, U.	13/189
Christensen-Dalsgaard, J.	3/313	Gabasch, A.	14/209
Christlieb, N.	16/191	Gandorfer, A.	15/113
Cohen, M.H.	2/177	Gänsicke, B.T.	13/151
Conway, R.G.	10/253	Gail, H.-P.	1/231, 3/156
Cunow, B.	2/109, 7/207	García-Segura, G.	11/57, 16/85
Danzmann, K.	15/93	Gautschy, A.	5/16
Davies, J.I.	13/245	Gehren, T.	1/52
Davis, R.J.	10/253	Geiss, J.	1/1
Degenhardt, U.	8/61	Geller, M.J.	10/159

Gerhard, O.E.	5/174	Kappelmann, N.	10/219
Gilmore, G.	9/263	Kaufer, A.	11/177
Glatzel, W.	4/104, 14/245	Keller, Ch.U.	8/27
Goeres, A.	6/165	Kerber, F.	11/161
González Delgado, D.	16/171	Kerp, J.	7/151
Gouguenheim, L.	9/127	Kerschbaum, F.	16/171
Grebel, E.K.	10/29, 14/223	Kessler, M.	2/53
Grosbol, P.	2/242	Kiessling, M.K.-H.	8/349
Grossmann-Doerth, U.	8/81	King, A.R.	3/14
Gruen, E.	4/157	Kissler-Patig, M.	13/13
Häfner, S.	11/147	Klaas, U.	16/243
Hamann, W.-R.	3/174	Klessen, R. S.	16/23
Hanslmeier, A.	16/55	Kley, W.	3/21
Hanuschik, R.W.	2/148, 4/233	Klose, S.	13/129
Hasinger, G.	3/60, 7/129, 15/71	Knoelker, M.	8/81
Heber, U.	10/219	Koczet, P.	11/317
Heck, A.	11/337	Kodaira, K.	15/1
Heger, A.	11/57	Koeppen, J.	6/179
Hegmann, M.	15/151	Koesterke, L.	3/174
Heidt, J.	14/209	Kokubo, E.	14/117
Heiles, C.	6/19	Kollatschny, W.	5/229
Heithausen, A.	12/201	Komossa, S.	15/27
Helling, Ch.	16/115	Koo, D.	13/173
Henkel, C.	6/69	Kraan-Korteweg, R.C.	2/119
Henkel, R.	1/231	Krabbe, A.	6/103, 12/107
Hensler, G.	9/277	Kratz, K.-L.	1/184
Herbst, E.	1/114	Kraus, U.	3/74
Herbstmeier, U.	7/151	Krause, F.	4/260
Herold, H.	3/74	Krichbaum, T.P.	9/221
Hesse, M.	8/323	Kroll, R.	2/194
Hessman, F.V.	3/32	Kruk, J.	10/219
Hippelein, H.	10/297	Kudritzki, R.-P.	6/271
Hoffmann, M.	4/165	Kuhn, B.	8/277
Hopp, U.	8/277, 14/209	Kümmel, M.	14/209
Horstmann, H.	2/109	Kunze, R.	6/257
Hoyle, F.	6/1	Kurtz, S.	16/85
Huber, C.E.	12/47	Lamb, D.Q.	10/101
Hüttemeister, S.	16/207	Lamers, H.J.G.L.M.	2/24
Ip, W.H.	2/86	Landstreet, J.D.	5/105
Jäger, C.	16/171	Langer, N.	2/306, 11/57
Jäger, K.	11/317, 14/209	Leitherer, C.	7/73
Jahreiß H.	2/72	Lemke, D.	2/53, 10/263
Janka, H.-T.	7/103, 10/201	Leuenhagen, U.	3/174
Jenkner, H.	3/297	Levison, H.	14/93
Jütte, M.	11/317	Liebscher, D.-E.	14/161
Kallrath, J.	9/307	Lobanov, P.A.	9/221

Lucy, L.B.	7/31	Palme, H.	1/28
Luks, T.	7/171	Panis, J.-F.	12/201
Luthardt, R.	5/38	Parker, E.N.	4/1
Maiolino, R.	15/179	Patermann, C.	2/13
Maitzen, H.M.	2/205	Paturel, G.	9/127
Mann, I.	9/173	Pauldrach, A.W.A.	
Mannheim, K.	9/17, 12/167		3/124, 3/140, 6/271, 16/133
Maraston, C.	15/219	Pauliny-Toth, I.I.K.	2/177
Markiewicz, W.J.	6/149	Penrose, R.	14/1
Marsch, E.	4/145	Perley, R.A.	10/253
Martin, R.N.	9/111	Popescu, C.C.	15/239
Mauersberger, R.	6/69, 11/209	Posch, Th.	16/171
Mazeh, T.	15/133	Priest, E.R.	14/133
McCaughrean, M.J.	6/191	Pringle, J.E.	5/97
Meijer, J.	2/109	Puls, J.	3/124, 3/140, 6/271
Meisenheimer, K.		Putman, M.	11/117
	2/129, 10/253, 10/297	Quirrenbach, A.	5/214
Mehlert, D.	14/209	Röser, H.P.	12/107
Menten, K.	13/229	Röser, S.	12/97
Meszaros, P.	10/127	Raedler, K.-H.	8/295
Meyer, F.	3/1	Rauch, T.	10/219
Moehler, S.	12/281	Rayner, J.T.	6/191
Möllenhoff, C.	14/209	Rebetzky, A.	3/74
Moffat, A.F.J.	7/51	Rees, M.J.	2/1, 10/179
Moneti, A.	6/191	Richstone, D.O.	15/57
Montmerle, T.	3/209	Richter, G.A.	5/26
Moorwood, A.	14/209	Richter, P.H.	14/53
Moritz, P.	7/151	Richtler, T.	8/163
Mueller, E.	7/103	Rimmele, Th.	2/105
Mueller, J.	10/191	Roeser, H.-J.	10/253, 10/297
Mutschke, H.	16/171	Ruder, H.	3/74
Najarro, F.	6/271	Rüdiger, A.	15/93
Napiwotzki, R.	11/3	Ruffert, M.	10/201
Naumann, M.	7/207	Saglia, R.	14/209
Neckel, Th.	3/266	Samland, M.	9/277
Neuhäuser, R.	10/323, 12/27	Schilbach, E.	9/87
Neumann, M.	10/253	Schlosser, W.	11/317
Nicklas, H.	14/209	Schmid, H.M.	11/297
Noll, S.	14/209	Schmidt, G.D.	8/147
Nollert, H.-P.	3/74	Schmidt, W.	4/117
Notni, P.	5/200	Schmidtobreick, L.	11/317
Olofsson, H.	16/171	Schmitt, D.	3/86, 8/61
Omont, A.	1/102	Schmitt, J.H.M.M.	5/188, 13/115
Ostriker, J.P.	13/1	Schmutz, W.	3/174
Ott, H.-A.	2/109, 7/207	Schoenfelder, V.	2/47, 9/49
Owocki, S.P.	3/98, 6/271	Schöier, F.	16/171

Schrijver, C.J.	4/18	Thommes, E.	10/297
Schroeder, K.-P.	3/187, 9/2210, 16/227	Thorne, K.S.	10/1
Schuecker, P.	2/109, 7/207, 11/267	Tilgner, A.	13/71
Schuessler, M.	8/11, 8/81	Townes, Ch. H.	16/1
Schulte-Ladbeck, R.	10/135	Trümper, J.E.	8/1
Schwarz, E.	3/174	Truran, J. W.	16/261
Schwenn, R.	1/179	Tucholke, H.-J.	2/109
Schwope, A.D.	3/44, 8/125	Tuffs, R. J.	15/239
Sedlmayr, E.	1/231	Turon, C.	9/69
Seifert, W.	14/209	Ulrich, M.-H.	5/247
Seitter, W.C.	2/109, 7/207, 11/267	Ungerechts, H.	1/210
Seitz, S.	14/209	Ungruhe, R.	7/207
Sigwarth, M.	13/45	van de Hulst, H.	9/1
Soffel, M.	10/191	von der Luehe, O.	2/105
Solanki, S.K.	4/208	Vrielmann, S.	11/285
Spiekermann, G.	7/207	Wagner, S.J.	14/209
Spruit, H.C.	4/197	Wambsganss, J.	12/149
Spurzem, R.	5/161	Wegner, G.	7/235
Stahl, O.	3/286, 14/209	Weinberger, R.	2/167
Stanimirović, S.	11/117	Weis, K.	14/261
Starrfield, S.	5/73	Werner, K.	10/219
Staubert, R.	2/141	Wessolowski, U.	3/174
Staude, H.J.	3/266	White, S.D.M.	7/255
Staude, J.	4/69	Wicenec, A.	4/188
Staveley-Smith, L.	11/117	Wielebinski, R.	4/260, 8/185
Steffen, M.	4/43	Wiemann, S.	11/317
Steiner, O.	8/81	Wilking, B.A.	6/191
Stix, M.	2/248	Williams, D.A.	6/49
Strassmeier, K.G.	11/197	Wilson, R.N.	7/1
Strittmatter, P.A.	12/1	Wisotzki, L.	12/231
Stutzki, J.	1/221, 6/209, 12/201	Wittmann, A.D.	4/61
Sungeun K.	11/117	Woitke, P.	14/185
Sutorius, E.	14/209	Wolf, B.	5/1
Szeifert, T.	14/209	Wolff, B.	10/219
Tammann, G.A.	9/139	Yorke, H.W.	2/283
Teerikorpi, P.	9/127	Zeilinger, W.W.	11/229
Teuber, D.	2/109, 2/229	Zensus, J A.	2/177, 9/221
Theis, Ch.	12/309	Ziegler, B.L.	13/211, 14/209
Theureau, G.	9/127	Zinnecker, H.	6/191
Thomas, D.	15/219	Zucker, S.	15/133